The Handbook of Plant Genome Mapping

Edited by
Khalid Meksem and Günter Kahl

Further Titles of Interest

Günter Kahl

The Dictionary of Gene Technology

Genomics, Transcriptomics, Proteomics

2004
ISBN 3-527-30765-6

Christoph W. Sensen

Essentials of Genomics and Bioinformatics

2002
ISBN 3-527-30541-6

Michael R. Barnes and Ian C. Gray (Editors)

Bioinformatics for Geneticists

2003
ISBN 0470-84393-4

The Handbook of Plant Genome Mapping

Genetic and Physical Mapping

Edited by
Khalid Meksem and Günter Kahl

WILEY-VCH

WILEY-VCH Verlag GmbH & Co. KGaA

Edited by

Prof. Dr. Khalid Meksem
Dept. Plant, Soil, Agriculture
Southern Illinois University
Carbondale, IL 629
USA

Prof. Dr. Günter Kahl
Plant Molecular Biology
Biocentre and GenXPro
Johann Wolfgang Goethe University
Marie-Curie-Strasse 9
60439 Frankfurt am Main
Germany

Cover Illustration:

The Title Page shows (from left to right) a linkage group (arrow-headed) from a genetic map, a blow-up of a specific region (in blue), the positions of molecular markers (RFLPs) along this region, (which is scaled in centiMorgans, cMs), and the two neighboring RFLP markers S10620 (red) and S14162 (green) at position 1.0 cM. These two RFLPs are radioactively labelled and hybridized to an array of BAC clones immobilized on a nylon membrane and (1) used to isolate genomic DNA containing sequences of interest (e. g. genes flanked by both markers) and (2) localized on chromosomal fibres by fibre fluorescent *in situ* hybridization (right). This technique is a variant of the conventional fluorescent *in situ* hybridization (FISH) method for the detection of target sequences, in which the two BAC clones are labelled with different fluorochromes, whose emission light is either in the red (S10620) or yellowish-green range, respectively (S14162). The chromosomal fibres are generated by molecular combing. For details, Chapter 6: "Physical mapping of Plant Chromosomes" by Barbara Hass-Jacobus and Scott A. Jackson

Library of Congress Card No.: applied for

British Library Cataloguing-in-Publication Data:
A catalogue record for this book is available from the British Library.

Bibliographic information published by Die Deutsche Bibliothek
Die Deutsche Bibliothek lists this publication in the Deutsche Nationalbibliografie; detailed bibliographic data is available in the Internet at <http://dnb.ddb.de>.

Printed in the Federal Republic of Germany.
Printed on acid-free paper.

Typesetting hagedorn kommunikation, Viernheim
Printing Strauss Offsetdruck GmbH, Mörlenbach
Bookbinding Litges & Dopf Buchbinderei GmbH, Heppenheim

ISBN-13: 978-3-527-31116-3
ISBN-10: 3-527-31116-5

We dedicate this book to the memory of late

Jozef (Jeff) Stefaan Schell,

who strongly inspired our way to think and to do research

Preface

At a time, when more than 150 bacterial and archaebacterial genomes, two plant genomes and a series of avertebrate and vertebrate genomes including the human genome have been deciphered base by base (some gaps notwithstanding), and more than 400 other genomes are in the mill, a Handbook of Plant Genome Mapping (Genetic and Physical Mapping) might seem a bit out of time. In fact, for many plants and animals genetic maps with vastly different densities are available and being improved continuously. Some of the scientific journals already begin to discourage authors to publish such increasingly dense maps and ask for more detailed informations such as genes isolated by map-based cloning.

In essence, genetic maps are by no means orphanized anymore. Also, if not yet available, genetic maps can be generated with speed and relative ease, provided a good selection of polymorphic parents, a wonderful and numerous segregating progeny, a highly resolving molecular marker system, powerful computer packages, a lot of people and enough money are at hand. So, in a not-too-far future, genetic maps will be commonplace given the relative ease of their generation.

The situation is quite different for physical maps of genomes. First of all, a physical map traditionally depends on the availability of a genetic map. Despite other approaches, the most practical method to establish a physical map still requires a large-insert clone bank on one hand, and a preferably highly dense genetic map on the other. And each and every marker, or marker bundle, that allows easy identification of the underlying DNA clones of whatever make-up (BAC clones seem to have won the race by now, and YAC clones lost because of chimerism and redundancies) is welcome. A highly resolving physical map, however, still requires a lot more input in labour, time, knowledge and funds than a genetic map. It is for this reason, that physical maps are available only for relatively few higher organisms, though common for prokaryota, whose chromosomes are sequenced and directly aligned into an ultimate physical map. Such complete genomic sequences (i. e. complete physical maps) are still an exception for eukaryotes. And, with one single exception, techniques used for the assembly of whole genome sequences still makes use of genetic and physical maps. The exception to this rule is the HAPPY mapping procedure, an ingenious tool with which physical maps can directly and happily be generated. This exception aside, the treadmill for many postdocs is and will be for the foreseeable future, the establishment of ge-

The Handbook of Plant Genome Mapping.
Genetic and Physical Mapping. Edited by Khalid Meksem, Günter Kahl

ISBN 3-527-31116-5

netic maps of the target organism as a prerequisite, including the production of BAC libraries and the physical alignment of the thousands of clones into contigs, at least for a region of interest. And genetically mapped markers will still serve to guide the way.

In appreciation of these facts, we set out to invite internationally renowned and highly competent plant researchers with an undisputed scientific reputation to portray their contributions to the genetic and physical mapping of plant genomes. The present "Handbook of Plant Genome Mapping: Genetic Mapping, Physical Mapping" is the most complete, up-to-date and competently written and compiled treatise of this complex topic. All the authors have striven to report the latest achievements and developments in their fields and did not spare any pains to introduce their areas of research, to detail methodological aspects and to present the state-of-the-art and future perspectives as well. This Handbook reflects the quality of worldwide research on plant genetic and physical mapping.

The editors most cordially appreciate the various contributions to make this book a standard for plant genome mapping for the foreseeable future.

November 2004

Khalid Meksem
Carbondale (IL, USA)

Günter Kahl
Frankfurt am Main (Germany)

Contents

The Handbook of Plant Genome Mapping.
Genetic and Physical Mapping. Edited by Khalid Meksem, Günter Kahl

ISBN 3-527-31116-5

List of Contributors

Bartoš, Jan
Laboratory of Molecular Cytogenetics and Cytometry
Institute of Experimental Botany
Sokolovska 6
77200 Olomouc
Czech Republic

Doležel, Jaroslav
Laboratory of Molecular Cytogenetics and Cytometry
Institute of Experimental Botany
Sokolovska 6
77200 Olomouc
Czech Republic

Gresshoff, Peter M.
ARC Centre of Excellence for Integrative Legume Research and School of Life Sciences
The University of Queensland, St. Lucia
Brisbane Qld 4072
Australia

Hass-Jacobus, Barbara
Department of Agronomy
Agricultural Genomics
Lilly Hall of Life Sciences
915 W. State Street
Purdue University
West Lafayette, IN 47907-2054
USA

Ishihara, Hirofumi
Department of Plant, Soil and General Agriculture
Agriculture Bldg. Room 176
Southern Illinois University at Carbondale
Carbondale, IL 62901-4415
USA

Jackson, Scott A.
Department of Agronomy
Agricultural Genomics
Lilly Hall of Life Sciences
915 W. State Street
Purdue University
West Lafayette, IN 47907-2054
USA

Jesse, Tacco
Keygene N.V., Agro Business Park 90
P.O. Box 216
6700 AE Wageningen
The Netherlands

Kahl, Günter
Plant Molecular Biology, Biocenter and GenXPro
University of Frankfurt am Main
Marie-Curie-Strasse 9
60439 Frankfurt am Main
Germany

The Handbook of Plant Genome Mapping.
Genetic and Physical Mapping. Edited by Khalid Meksem, Günter Kahl

ISBN 3-527-31116-5

Kubaláková, Marie
Laboratory of Molecular Cytogenetics and Cytometry
Institute of Experimental Botany
Sokolovska 6
77200 Olomouc
Czech Republic

Lee, Mi-Kyung
Laboratory for Plant Genomics & GENEfinder Genomic Resources
Department of Soil & Crop Sciences and Institute for Plant Genomics & Biotechnology
Texas A&M University
College Station, TX 77843-2123
USA

Macas, Jiří
Laboratory of Molecular Cytogenetics
Institute of Plant Molecular Biology
Branisovska 31
37005 České Budějovice
Czech Republic

Mast, Andrea
Third Wave Technologies
502 South Rosa Road
Madison, WI 53719-1256
USA

Meksem, Khalid
Department of Plant, Soil and General Agriculture
Agriculture Bldg. Room 176
Southern Illinois University at Carbondale
Carbondale, IL 62901-4415
USA

Nelson, James C.
Department of Plant Pathology
Kansas State University
Manhattan, KS 66506-5502
USA

Nelson, William
Arizona Genome Computational Laboratory
Institute for Biomedical Science and Biotechnology
303 Forbes Building,
University of Arizona
Tucson, AZ 85721
USA

Nguyen, Henry T.
Department of Agronomy,
Plant Sciences Unit
1-87 Agriculture Building
University of Missouri-Columbia
Columbia, MO 65211
USA

Peleman, Johan D.
KEYGENE GENETICS
P. O. Box 216
6700 AE Wageningen
The Netherlands

Peterson, Daniel
Mississippi Genome Exploration Laboratory (MGEL)
Dept. of Plant and Soil Sciences
117 Dorman Hall, Box 9555
Mississippi State University
Mississippi State, MS 39762
USA

Ren, Chengwei
Laboratory for Plant Genomics & GENEfinder Genomic Resources
Department of Soil & Crop Sciences and Institute for Plant Genomics & Biotechnology
Texas A&M University
College Station, TX 77843-2123
USA

Rouppe van der Voort, Jeroen N. A.M.
KEYGENE GENETICS
P. O. Box 216
6700 AE Wageningen
The Netherlands

Santos, Teofila S.
Laboratory for Plant Genomics &
GENEfinder Genomic Resources
Department of Soil & Crop Sciences
and Institute for Plant Genomics &
Biotechnology
Texas A&M University
College Station, TX 77843-2123
USA

Schneider, Katharina
Max-Planck-Institut für Züchtungs-
forschung,
Carl-von-Linne-Weg 10
50829 Cologne
Germany

Shen, Richard
Illumina, Inc.
9885 Towne Centre Drive
San Diego, CA 92121-1975
USA

Scheuring, Chantel
Department of Soil and Crop Sciences
and
Institute for Plant Genomics and Bio-
technology
2123 TAMU
Texas A&M University
College Station, TX 77843-2123
USA

Soderlund, Carol
Arizona Genome Computational
Laboratory
Institute for Biomedical Science and
Biotechnology
303 Forbes Building
University of Arizona
Tucson, AZ 85721
USA

Sørensen, Anker P.
KEYGENE GENETICS
P. O. Box 216
6700 AE Wageningen
The Netherlands

Sun, Shuku
Laboratory for Plant Genomics &
GENEfinder Genomic Resources
Department of Soil & Crop Sciences
and Institute for Plant Genomics &
Biotechnology
Texas A&M University
College Station, TX 77843-2123
USA

Tooke, Nigel
Biotage AB
Kungsgatan 76
753 18 Uppsala
Sweden

Town, Christopher D.
The Institute for Genomic Research
9712 Medical Center Drive
Rockville, MD 20850
USA

van den Boom, Dirk
SEQUENOM, Inc.
3595 John Hopkins Court
San Diego, CA 92121
USA

Wu, Chengcang
Laboratory for Plant Genomics &
GENEfinder Genomic Resources
Department of Soil & Crop Sciences
Crop Biotechnology Center
Texas A&M University
College Station, TX 77843-2123
USA

Wu, Xiaolei
Department of Agronomy,
Plant Sciences Unit
1-87 Agriculture Building
University of Missouri-Columbia
Columbia, MO 65211
USA

Xu, Zhanyou
Laboratory for Plant Genomics &
GENEfinder Genomic Resources
Department of Soil & Crop Sciences and
Institute for Plant Genomics &
Biotechnology
Texas A&M University
College Station, TX 77843-2123
USA

Zhang, Hong-Bin
Laboratory for Plant Genomics &
GENEfinder Genomic Resources
Department of Soil & Crop Sciences and
Institute for Plant Genomics & Biotechnology
Texas A&M University
College Station, TX 77843-2123
USA

Part I
Genetic Mapping

The Handbook of Plant Genome Mapping.
Genetic and Physical Mapping. Edited by Khalid Meksem, Günter Kahl

ISBN 3-527-31116-5

1

Mapping Populations and Principles of Genetic Mapping

Katharina Schneider

Overview

Mapping populations consist of individuals of one species, or in some cases they derive from crosses among related species where the parents differ in the traits to be studied. These genetic tools are used to identify genetic factors or loci that influence phenotypic traits and to determine the recombination distance between loci. In different organisms of the same species, the genes, represented by alternate allelic forms, are arranged in a fixed linear order on the chromosomes. Linkage values among genetic factors are estimated based on recombination events between alleles of different loci, and linkage relationships along all chromosomes provide a genetic map of the organism. The type of mapping population to be used depends on the reproductive mode of the plant to be analyzed. In this respect, the plants fall into the main classes of self-fertilizers and self-incompatibles. This chapter illustrates the molecular basis of recombination, summarizes the different types of mapping populations, and discusses their advantages and disadvantages for different applications.

Abstract

In genetics and breeding, mapping populations are the tools used to identify the genetic loci controlling measurable phenotypic traits. For self-pollinating species, F_2 populations and recombinant inbred lines (RILs) are used; for self-incompatible, highly heterozygous species, F_1 populations are mostly the tools of choice. Backcross populations and doubled haploid lines are a possibility for both types of plants. The inheritance of specific regions of DNA is followed by molecular markers that detect DNA sequence polymorphisms. Recombination frequencies between traits and markers reveal their genetic distance, and trait-linked markers can be anchored, when necessary, to a more complete genetic map of the species. For map-based cloning of a gene, populations of a large size provide the resolution required.

The Handbook of Plant Genome Mapping.
Genetic and Physical Mapping. Edited by Khalid Meksem, Günter Kahl

ISBN 3-527-31116-5

Due to intensive breeding and pedigree selection, genetic variability within the gene pools of relevant crops is at risk. Interspecific crosses help to increase the size of the gene pool, and the contribution of wild species to this germ plasm in the form of introgression lines is of high value, particularly with respect to traits like disease resistance. The concept of exotic libraries with near-isogenic lines, each harboring a DNA fragment from a wild species, implements a systematic scan of the gene pool of a wild species.

To describe the complexity of genome organization, genetic maps are not sufficient because they are based on recombination, which is largely different along all genomes. However, genetic maps, together with cytogenetic data, are the basis for the construction of physical maps. An integrated map then provides a detailed view on genome structure and enforces positional cloning of genes and ultimately the sequencing of complete genomes.

1.1 Introduction

Since Mendel formulated his laws of inheritance in 1865, it is a core component of biology to relate genetic factors to functions visible as phenotypes. At Mendel's time, genetic analysis was restricted to visual inspection of the plants. Pea (*Pisum sativum* [Fabaceae]) was already a model plant at the time, and Mendel studied visible traits such as seed and pod color, surface structure of seeds and pods (smooth versus wrinkled), and plant height. These traits are, in fact, the first genetic markers used in biology. In 1912 Vilmorin and Bateson described the first work on linkages in *Pisum*. However, the concept of linkage groups representing chromosomes was not clear in *Pisum* until 1948, when Lamprecht described the first genetic map with 37 markers distributed on 7 linkage groups (summarized in Swiecicki et al. 2000). Large collections of visible markers are today available for several crop species and for *Arabidopsis thaliana* (Koornneef et al. 1987; Neuffer et al. 1997).

In the process of finding more and more genetic markers, the first class of characters scored at the molecular level was isoenzymes. These are isoforms of proteins that vary in amino acid composition and charge and that can be distinguished by electrophoresis. The technique is applied to the characterization of plant populations and breeding lines and in plant systematics, but it is also used for genetic mapping of variants, as shown particularly in maize (Frei et al. 1986, Stuber et al. 1972). However, due to the small number of proteins for which isoforms exist and that can be separated by electrophoresis, the number of isoenzyme markers is limited.

The advance of molecular biology provided a broad spectrum of technologies to assess the genetic situation at the DNA level. The first DNA polymorphisms described were restriction fragment length polymorphism (RFLP) markers (Botstein et al. 1980). This technique requires the hybridization of a specific probe to restricted genomic DNA of different genotypes. The whole genome can be covered by RFLP and, depending on the probe, coding or non-coding sequences can be analyzed. The next generation of markers was based on PCR: rapid amplified polymorphic DNA (RAPD) (Williams et al. 1990; Welsh and McClelland 1990) and amplified fragment length polymorphism (AFLP) (Vos et al. 1995). Recently, methods have been developed to detect single nucleotide polymorphisms (summarized in Rafalski 2002). Because these methods have the potential for automatization and multiplexing, they allow the establishment of high-density genetic maps.

Whereas RAPD and AFLP analyses are based on anonymous fragments, RFLP and SNP analyses allow the choice of expressed genes as markers. Genes of a known sequence and that putatively influence the trait of interest can be selected and mapped. In this way function maps can be constructed (Chen et al. 2001; Schneider et al. 2002). Phenotypic data of the segregating population, correlated to marker data, prove or disprove potential candidate genes supporting mono- and polygenic traits.

The basis for genetic mapping is recombination among polymorphic loci, which involves the reaction between homologous DNA sequences in the meiotic prophase. Currently, the double-strand-break repair model (Szostak et al. 1983) is acknowledged to best explain meiotic reciprocal recombination (Figure 1.1). In this model, two sister chromatids break at the same point and their ends are resected at the 5′ ends. In the next step the single strands invade the intact homologue and pair with their complements. The single-strand gaps are filled in using the intact strand as template. The resulting molecule forms two Holliday junctions. Upon resolution of the junction, 50 % of gametes with recombinant lateral markers and 50 % non-recombinants are produced. In the non-recombinants, genetic markers located within the region of strand exchange may undergo gene conversion, which can result in nonreciprocal recombination, a problem interfering in genetic mapping. In plants, gene conversion events were identified by Büschges et al. (1997) when cloning the *Mlo* resistance gene from barley.

The likelihood that recombination events occur between two points of a chromosome depends in general on their physical distance: the nearer they are located to each other, the more they will tend to stay together after meiosis. With the increase of the distance between them, the probability for recombination increases and genetic linkage tends to disappear. This is why genetic linkage can be interpreted as a measure of physical distance. However, taking the genome as a whole, the fre-

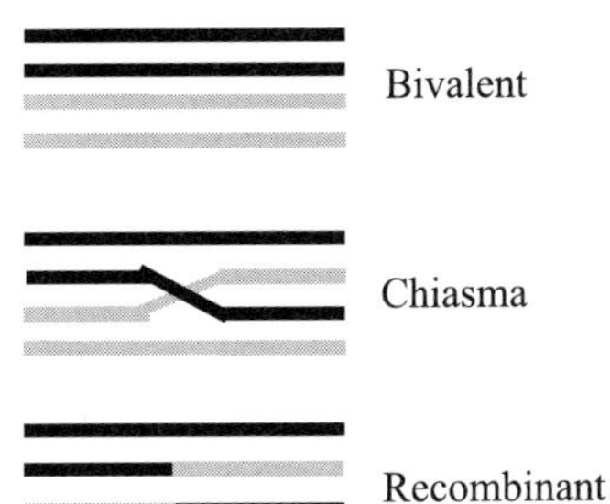

Figure 1.1. Generation of recombinants by chiasma formation. In the meiotic prophase, two sister chromatids of each parent (labeled in red and green, respectively) align to form a bivalent. A chiasma is formed by a physical strand exchange between two non-sister chromatids. Breakage and reunion of reciprocal strands leads to the generation of recombinants.

quency of recombination is not constant because it is influenced by chromosome structure. An example is the observation that recombination is suppressed in the vicinity of heterochromatin: here, the recombination events along the same chromatid appear to be reduced, an observation called positive interference. It reduces the number of double recombinants when, for example, three linked loci are considered.

Linkage analysis based on recombination frequency and the order of linked loci is evaluated statistically using maximum likelihood equations (Fisher 1921; Haldane and Smith 1947; Morton 1955). Large amounts of segregation data are routinely processed by computer programs to calculate a genetic map; among the most popular are JoinMap (Stam 1993) and MAPMAKER (Lander et al. 1987).

1.2 Mapping Populations

The trait to be studied in a mapping population needs to be polymorphic between the parental lines. Additionally, a significant trait heritability is essential. It is always advisable to screen a panel of genotypes for their phenotype and to identify the extremes of the phenotypic distribution before choosing the parents of a mapping population. It is expected that the more the parental lines differ, the more genetic factors will be described for the trait in the segregating population and the easier their identification will be. This applies to monogenic as well as to polygenic traits.

A second important feature to be considered when constructing a mapping population is the reproductive mode of the plant. There are two basic types. On the one hand are plants that self naturally, such as *Arabidopsis thaliana*, tomato, and soybean, or that can be manually selfed, such as sugar beet and maize; on the other hand are the self-incompatible, inbreeding-sensitive plants such as potato. Self-incompatible plants show high genetic heterozygosity, and for these species it is frequently not possible to produce pure lines due to inbreeding depression. Usually only self-compatible plants allow the generation of lines displaying a maximum degree of homozygosity. In conclusion, the available plant material determines the choice of a mapping population. Other factors are the time available for the construction of the population and the mapping resolution required. Based on these concepts, this section will be divided into seven parts:

1. mapping populations suitable for self-fertilizing plants,
2. mapping populations for cross-pollinating species,
3. two-step strategies for mapping mutants and DNA fragments,
4. chromosome-specific tools for mapping,
5. mapping in natural populations/breeding pools,
6. mapping genes and mutants to physically aligned DNA, and
7. specific mapping problems.

1.2.1 Mapping Populations Suitable for Self-fertilizing Plants

If pure lines are available or can be generated with only a slight change of plant vigor, the mapping populations that can be used consist of F_2 plants, recombinant inbred lines (RIL), backcross (BC) populations, introgression lines assembled in exotic libraries, and doubled haploid lines (DH).

1.2.1.1 F_2 Populations

The simplest form of a mapping population is a collection of F_2 plants (Figure 1.2). This type of population was the basis for the Mendelian laws (1865) in which the foundations of classic genetics were laid. Two pure lines that result from natural or artificial inbreeding are selected as parents, parent 1 (P1) and parent 2 (P2). Alternatively, doubled haploid lines can be used to avoid any residual heterozygosity (see Section 1.2.1.5). If possible, the parental lines should be different in all traits to be studied. The degree of polymorphism can be assessed at the phenotypic level (e. g., morphology, disease resistance) or by molecular markers at the nucleic acid level. For inbreeding species such as soybean and the Brassicaceae, wide crosses between genetically distant parents help to increase polymorphism. However, it is required that the cross lead to fertile progeny. The progeny of such a cross is called the F_1 generation. If the parental lines are true homozygotes, all individuals of the F_1 generation will have the same genotype and have a similar phenotype. This is the content of Mendel's law of uniformity. An individual F_1 plant is then selfed to produce an F_2 population that segregates for the traits different between the parents. F_2 populations are the outcome of one meiosis, during which the genetic material is recombined. The expected segregation ratio for each codominant marker is 1:2:1 (homozygous like P1:heterozygous:homozygous like P2). It is a disadvantage that F_2 populations cannot be easily preserved, because F_2 plants are frequently not immortal, and F_3 plants that result from their selfing are genetically not iden-

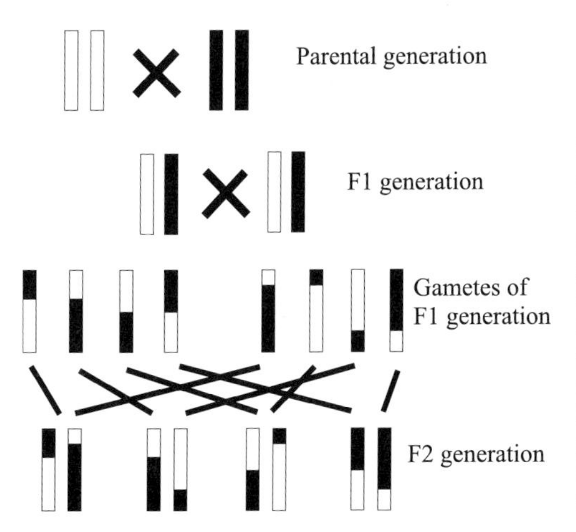

Figure 1.2. The generation of an F_2 population. Two chromosomes are shown as representatives of the diploid parental genome. In the parental generation, the genotypes are homozygous (represented by equal colors) and in the F_1 generation are heterozygous. For gamete formation the genetic material undergoes meiosis, leading to recombination events in F_1 gametes. Correspondingly, F_2 plants vary largely in their genetic constitution.

tical. For species like sugar beet, there is a possibility of maintaining F_2 plants as clones in tissue culture and of multiplying and re-growing them when needed. A particular strategy is to maintain the F_2 population in pools of F_3 plants. Traits that can be evaluated only in hybrid plants, such as quality and yield parameters in sugar beet or maize, require the construction of testcross plants by crossing each F_2 individual with a common tester genotype (example given in Schneider et al. 2002). Ideally, different common testers should produce corresponding results to exclude the specific effects of one particular tester genotype.

To produce a genome-wide map as an overview, a population of around 100 F_2 individuals is recommended as a compromise between resolution of linked loci and cost/feasibility.

For mapping quantitative trait loci (QTLs), Monte Carlo simulations have shown that at least 200 individuals are required (Bevis 1994). For higher resolution, as required for positional cloning of selected genes, progenies of several thousand plants are developed. For example, more than 3400 individuals were analyzed to obtain a detailed map around a fruit weight locus in tomato (Alpert and Tanksley 1996).

1.2.1.2 **Recombinant Inbred Lines**

Recombinant inbred lines (RILs) are the homozygous selfed or sib-mated progeny of the individuals of an F_2 population (Figures 1.2, 1.3). The RIL concept for mapping genes was originally developed for mouse genetics. In animals, approximately 20 generations of sib mating are required to reach useful levels of homozygosity. In plants, RI lines are produced by selfing, unless the species is completely self-incompatible. Because in the selfing process one seed of each line is the source for the next generation, RILs are also called single-seed descent lines. Self-pollination allows the production of RILs in a relatively short number of generations. In fact, within six generations, almost complete homozygosity can be reached. Along each chromosome, blocks of alleles derived from either parent alternate. Because recombination can no longer change the genetic constitution of RILs, further segregation in the progeny of such lines is absent. It is thus one major advantage that these lines constitute a permanent resource that can be replicated indefinitely and

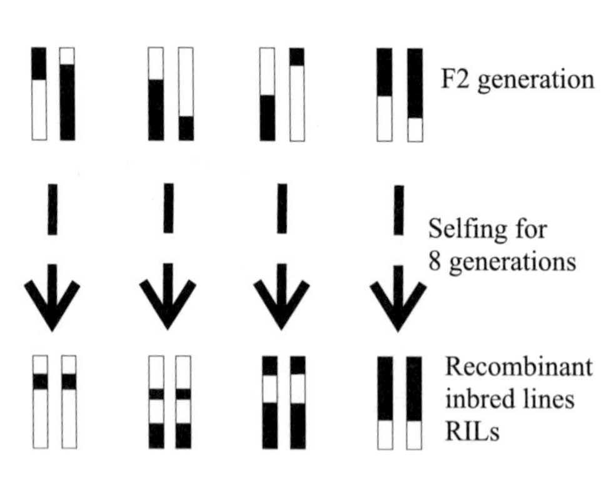

Figure 1.3. The generation of RILs. For the generation of RILs, plants of an F_2 population are continuously selfed. In each generation meiotic events lead to further recombination and reduced heterozygosity until completely homozygous RILs with fragments of either parental genome are achieved.

be shared by many groups in the research community. A second advantage of RILs is that because they undergo several rounds of meiosis before homozygosity is reached, the degree of recombination is higher compared to F_2 populations. Consequently, RIL populations show a higher resolution than maps generated from F_2 populations (Burr and Burr 1991), and the map positions of even tightly linked markers can be determined. In plants, RILs are available for many species, including rice and oat (Wang et al 1994; O'Donoughue et al. 1995).

In *Arabidopsis thaliana*, 300 RILs have become a public mapping tool (Lister and Dean 1993). *Arabidopsis* RILs were constructed by an initial cross between the ecotypes Landsberg *erecta* and Columbia, and a dense marker framework was established. Every genomic fragment that displays a polymorphism between Landsberg *erecta* and Columbia can be mapped by molecular techniques.

1.2.1.3 Backcross Populations

To analyze specific DNA fragments derived from parent A in the background of parent B, a hybrid F_1 plant is backcrossed to parent B. In this situation, parent A is the donor of DNA fragments and parent B is the recipient. The latter is also called the recurrent parent. During this process two goals are achieved: unlinked donor fragments are separated by segregation and linked donor fragments are minimized due to recombination with the recurrent parent. To reduce the number and size of donor fragments, backcrossing is repeated and, as a result, so-called advanced backcross lines are generated. With each round of backcrossing, the proportion of the donor genome is reduced by 50 % (see Figure 1.4). Molecular markers help to monitor this process and to speed it up. In an analysis of the chro-

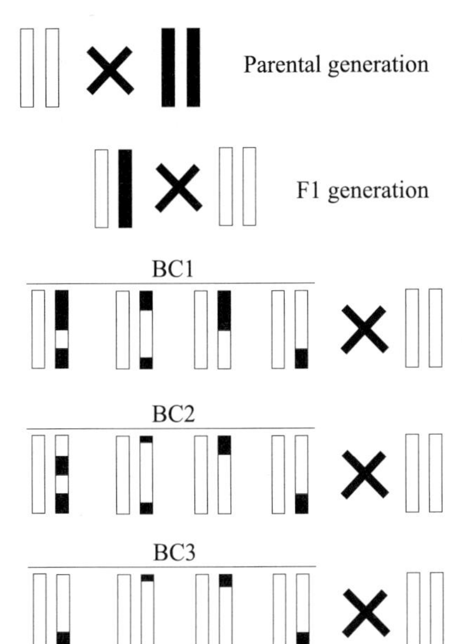

Figure 1.4. The generation of advanced backcross lines (BC). BC lines originate from an F_1 plant that is repeatedly backcrossed to the same recurrent parent. With each round of backcrossing, the number and size of genomic fragments of the donor parent are reduced until a single DNA fragment distinguishes the BC line from the recurrent parent (white: recurrent parent; black: donor parent).

mosomal segments retained around the *Tm-2* locus of tomato, it was estimated that marker-assisted selection reduced the number of required backcrosses from 100 – in the case of no marker selection – to two (Young and Tanksley 1989). The progeny of each backcross is later screened for the trait introduced by the donor. In the case of dominant traits, the progeny can be screened directly; in the case of recessive traits, the selfed progeny of each backcross plant has to be assessed.

Lines that are identical, with the exception of a single fragment comprising one to a few loci, are called nearly isogenic lines (NILs). The generation of NILs involves several generations of backcrossing assisted by marker selection. To fix the donor segments and to visualize traits that are caused by recessive genes, two additional rounds of self-fertilization are required at the end of the backcrossing process. If two NILs differ in phenotypic performance, this is seen as the effect of the alleles carried by the introgressed DNA fragment. The procedure is quite helpful in the functional analysis of the underlying genes. The strategy is particularly valuable for those species for which no transformation protocol is established to produce transgenics for the alleles of interest. A further advantage is that in NILs genomic rearrangements, which may happen during transformation, are avoided.

Backcross breeding is an important strategy if a single trait, such as resistance, has to be introduced into a cultivar that already contains other desirable traits. The only requirement is that the two lines be crossable and produce fertile progeny. Lines incorporating a fragment of genomic DNA from a very distantly related species are called introgression lines, whereas lines incorporating genetic material from a different variety are indicated as intervarietal substitution lines.

1.2.1.4 **Introgression Lines: Exotic Libraries**

The breeding of superior plants consists of combining positive alleles for desirable traits on the elite cultivar. One source for such alleles conferring traits such as disease resistance or quality parameters is distantly related or even wild species. If the trait to be introduced is already known, the introgression can be performed in a direct way supported by marker-assisted selection. However, the potential of wild species to influence quantitative traits often is not yet assessed. In this case, backcross breeding is a method to identify single genetic components contributing to the phenotype. NILs are constructed by an advanced backcross program, and their phenotypic effects are assayed. For example, in the work of Tanksley et al. (1996), loci from the wild tomato species *Lycopersicon pimpinellifolium* were shown to have positive effects on tomato fruit size and shape.

To assess the effects of small chromosomal introgressions at a genome-wide level, a collection of introgression lines, each harboring a different fragment of genomic DNA, can be generated. Such a collection is called an exotic library, which is achieved by advanced backcrossing. This corresponds to a process of recurrent backcrossing (ADB) and marker-assisted selection for six generations and to the self-fertilization of two more generations to generate plants homozygous to the introgressed DNA fragments (summarized in Zamir 2001). An example is the introgression lines derived from a cross between the wild green-fruited species *L. pennellii* and the tomato variety M82 (Eshed and Zamir 1995). The lines, after the ADB program, will resemble the cultivated parent, but introgressed fragments with even subtle phenotypic effects can be easily identified. In other words, phenotypic assessment for all traits of interest will reveal genomic fragments with positive effects on measurable traits. The introgressed fragments are obviously defined by the use of molecular markers.

In this context, it should be noted that recombination is reduced in interspecific hybrids with respect to intraspecific ones because differences in DNA sequence lead to reduced pairing of the chromosomes during meiosis. This, in turn, causes a phenomenon called linkage drag, which describes the situation when larger-than-expected fragments are retained during backcross breeding (Young and Tanksley 1989). The following example illustrates this concept. For the *Tm2a* resistance gene introgressed into tomato from the distantly related *Lycopersicum peruvianum* species, the ratio of physical to genetic distance is more than 4000 kb cM^{-1}, whereas the average ratio in the cultivated species is about 700 kb cM^{-1} (Ganal et al. 1989).

1.2.1.5 **Doubled Haploid Lines**

Doubled haploid lines contain two identical sets of chromosomes in each cell. They are completely homozygous, as only one allele is available for all genes. Doubled haploids can be produced from haploid lines. Haploid lines either occur spontaneously, as in the case of rape and maize, or are artificially induced. Haploid plants are smaller and less vital than diploids and are nearly sterile. It is possible to induce haploids by culturing immature anthers on special media. Haploid plants can later be regenerated from the haploid cells of the gametophyte. A second option is microspore culture. In cultivated barley it is possible to induce the generation of haploid embryos by using pollen from the wild species *Hordeum bulbosum.* During the first cell divisions of the embryo, the chromosomes of *H. bulbosum* are eliminated, leaving the haploid chromosomal set derived from the egg cell. Occasionally in haploid plants the chromosome number doubles spontaneously, leading to doubled haploid (DH) plants. Such lines can also be obtained by colchicine treatment of haploids or of their parts. Colchicines prevent the formation of the spindle apparatus during mitosis, thus inhibiting the separation of chromosomes and leading to doubled haploid cells. If callus is induced in haploid plants, a doubling of chromosomes often occurs spontaneously during endomitosis and doubled haploid lines can be regenerated via somatic embryogenesis. However, in vitro culture conditions may reduce the genetic variability of regenerated materials to be used for genetic mapping.

Doubled haploid lines constitute a permanent resource for mapping purposes and are ideal crossing partners in the production of mapping populations because they have no residual heterozygosity. Examples of their use in wheat, barley, and rice are found in Chao et al. (1989), Heun et al. (1991), and McCouch et al. (1988).

1.2.2
Mapping Populations for Cross-pollinating Species

If pure lines cannot be generated from a species due to self-incompatibility or inbreeding depression, heterozygous parental plants are used to derive mapping populations such as F_1 and backcross lines (BC). This is the case for several tree species such as apple, pear, and grape and for potato. For the tree species crosses between different cultivars are used to produce F_1 progenies to be genotyped (Maliegaard et al. 1998; Yamamoto et al. 2002; Grando et al. 2003). In potato, the heterozygosity of parental lines used for one cross was evaluated to correspond to 57–59 % (Gebhardt et al. 1989). In the foundation cross population, different alleles are contributed from either parent to individual F_1 plants. The linkage among markers is assessed by the production of a genetic map for either parent. In potato, Gebhardt et al. (1989, 1991) reported the construction of a backcross population in which an individual F_1 plant was pollinated with one parent. To maintain the identity of the F_1 genotypes of the mapping population, parental lines and each of their F_1 progenies were propagated clonally.

1.2.3
Two-step Strategies for Mapping Mutants and DNA Fragments

Mapping mutants always requires the construction of a segregating population. In a first step, tightly linked or co-segregating markers are selected. In a second step, the map position of these markers is determined. Whereas the first step requires a population segregating for the trait of interest, the linked marker can be anchored in a reference population for which a dense marker framework is available. Such a method is particularly applicable when a large set of mutants needs to be mapped in a limited time.

This concept was implemented by Castiglioni et al. (1998), who mapped mutations in barley by an AFLP-based procedure. The procedure takes advantage of the very high diversity index of AFLP markers that allows the screening of a whole genome with a limited number of PCR primer combinations. In the cited work, a genetic map comprising 511 AFLP markers was derived from the cross between the lines Proctor × Nudinka. To map morphological mutations, mutants were crossed to Proctor and to Nudinka, respectively, and 30–50 mutant plants were selected for AFLP analysis. The specific presence or absence of AFLP bands in the mutant population was correlated to identical fragments mapped in the Proctor × Nudinka cross, thus inferring their map position and, consequently, anchoring the mutant locus to the same map.

A different approach to enrich for linked markers is based on the concept of bulked segregant analysis (BSA), introduced by Michelmore et al. (1991). BSA requires only a population segregating for the trait of interest. Two bulks or pools of segregating genotypes are selected in which either the mutant or the wild-type phenotype is present and homozygous. This implies that within each pool the individuals are identical with respect to the genomic region in which the responsible gene maps, but the genetic constitution of the rest of the genome is random, while between the pools the selected region is molecularly dissimilar. The bulks are screened for polymorphisms by molecular techniques such as RFLP, RAPD, or AFLP. In their first BSA analysis, Michelmore et al. (1991) identified markers linked to a gene conferring resistance to downy mildew in lettuce. They generated bulks of 17 F_2 individuals homozygous for alternate alleles of the resistance locus DM5/8 and analyzed them with 100 arbitrary RAPD primers to detect around 900 loci. Three RAPD markers linked to the resistance locus were identified. This work shows that markers can be reliably identified in a 0- to 25-cM window to either side of the locus of interest. The method can be applied iteratively, in the sense that new bulks are constructed based on each new marker linked more closely to the gene. The linkage of each marker with the tagged locus is verified by analyzing single plants of the segregating population. In the work cited, the BSA pools were from the mapping population. However, the BSA markers could also be anchored in a different population, according to the two-step procedure outlined before.

1.2.4
Chromosome-specific Tools for Mapping

Chromosome-specific tools allow a segregating population to be genotyped in a way that each chromosome is directly scanned for linkage. The first tools of this kind were mutant lines with one or more visible mapped mutations. For *Arabidopsis thaliana* such multiple marker lines are available: the line W100, for example, contains mutations identifying each arm of the five chromosomes (Koornneef et al. 1987). The marker lines with the genotypes *aa*, *bb*, to *zz* are crossed to the mutant line with the genotype *mm*. The progeny is selfed to generate an F_2 population. In this population the frequency of double mutants *aa/mm*, *bb/mm*, to *zz/mm* is counted. It is expected that in the case of recessive inheritance of both mutations, 1/16 of all progeny are double mutants. If there are less than the expected number of double mutants, this is taken as evidence of reduced recombination due to the linkage between the locus *m* and one of the markers tested. The more the two loci are in physical proximity, the fewer the recombination events are.

In the past decade, numerous molecular marker sets have been developed. For a defined mutation in a given genetic background, a cross to a contrasting genotype needs to be performed, and an F_2 is generated from which plants displaying the phenotype are selected for molecular analysis. In *Arabidopsis*, different ecotypes such as Columbia and Landsberg *erecta* are well established, and marker sets for cleaved amplified polymorphic sequence (CAPS) and simple sequence repeat (SSR) analysis have been developed (Konieczny and Ausubel 1993; Bell and Ecker 1994). Marker loci covering the entire genome are tested in the mutant F_2 plants, and, again, reduced recombination between the mutant allele and a marker allele indicates linkage to the marker locus.

1.2.5
Mapping in Natural Populations/Breeding Pools

The natural variation between individuals of one species can be exploited for mapping. In the case of crop plants, sets of different breeding lines can fulfill this purpose. This approach is suited to map complex traits that are influenced by the action of many genes in a quantitative way. Such loci are defined as quantitative trait loci (QTL). It is important that such a collection of different accessions contain a whole spectrum of phenotypes for a given trait; in particular, the availability of extreme phenotypes is advantageous. The underlying idea is that genomic fragments naturally present in a particular genotype are transmitted as non-recombining blocks and that markers, like single-nucleotide polymorphisms (SNPs) and insertions/deletions, can easily follow the inheritance of such blocks. These units are also called haplotypes, and their existence reveals a state of linkage disequilibrium (LD) among allelic variants of tightly linked genes. The existence of haplotypes has clearly been shown for maize and sugar beet (Ching et al. 2002, Schneider et al. 2001).

Two strategies exist for searching for haplotypes specifically associated with extreme phenotypes. On the one hand, the whole genome can be scanned for phenotype-marker associations, if sufficient sequence information is available for marker development. However, as the linkage disequilibrium often does not extend for more than 2000 bases (reviewed in Buckler and Thornsberry 2002), this approach is generally too time-consuming and costly. Alternatively, the focus is on candidate genes that, upon their predicted physiological function, are likely to influence the trait under investigation. The latter approach has been successfully applied in maize. An association between a marker and a trait exists if one marker allele or haplotype is significantly associated with a particular phenotype when studied in unrelated genotypes. In maize this has been shown for *dwarf 8*, a locus influencing plant height (Thornsberry et al. 2001), and for six genes of the starch metabolism influencing kernel quality (Whitt et al. 2002). The advantage of this approach is that it does not require the construction of experimental populations. Particularly for self-pollinating species, inbred individuals of natural ecotypes are practically immortal, and phenotyping needs to be performed only once. Natural populations are particularly informative because usually more than two alleles exist for each marker locus. For crop species, different breeding lines representing extreme phenotypes can be used for the same purpose.

As unrelated lines of natural populations are genetically separated by many generations, the corresponding large number of meiotic events leads to a high rate of recombinations. Therefore, and with the limit that LD blocks still exist, trait-supporting loci can be mapped with high precision, largely exceeding the resolution of F_2 populations. Association mapping can thus greatly accelerate QTL positional cloning approaches. However, it requires thorough statistical assessment to investigate the relatedness of the lines and the overall population structure. Only if the population structure is homogenous can an association between a haplotype and a phenotype be considered realistic.

1.2.6
Mapping Genes and Mutants to Physically Aligned DNA

The distances in genetic maps are based on recombination frequencies. However, recombination frequencies are not equally distributed over the genome. In heterochromatic regions such as the centromeres, recombination frequency is indeed quite reduced. In these cases, cytogenetic maps can provide complementary information because they are based on the fine physical structure of chromosomes. Chromosomes can be visualized under the microscope and be characterized by specific staining patterns, e. g., with Giemsa C, or can be based on morphological structures such as the centromeres, the nucleolus-organizing region (NOR), the telomeres, and the so-called knobs, heritable heterochromatic regions of particular shape. Cytogenetic maps allow association of linkage groups with chromosomes and determination of the orientation of the linkage groups with respect to chromosome morphology. In species such as maize, wheat, barley, and *Arabidopsis thaliana*, lines carrying chromosome deletions, translocation breakpoints, or trisomics can be generated as valuable tools for the cytogenetic approach (Helentjaris et al. 1986; Weber and Helentjaris 1989; Sandhu et al. 2001; Künzel et al. 2000; Koornneef and Vanderveen 1983). Numerical aberrations in chromosome number, together with marker data generated, e. g., by RFLP analysis, can clearly identify chromosomes. Defined translocation breakpoints can also localize probes to specific regions on the arms of chromosomes.

More recently, techniques have been developed to localize nucleic acids *in situ* on the chromosomes. During the pachytene, a stage during the meiotic prophase, the chromosomes are 20 times longer than at mitotic metaphase. They display a differentiated pattern of brightly fluorescing heterochromatin segments. It is possible to identify all chromosomes based on chromosome length, centromere position, heterochromatin patterns, and the positions of repetitive sequences such as 5S rDNA and 45S rDNA visualized by fluorescence *in situ* hybridization (FISH) (Hanson et al. 1995), as shown in *Medicago trunculata* by Kulikova et al. (2001). In tomato this approach has been successful in mapping two genes near the junction of euchromatin and pericentromeric heterochromatin (Zhong et al. 1999). Refined multicolor FISH even allows the mapping of single-copy sequences (Desel et al. 2001). In this context, cytogenetic maps based on FISH provide complementary information for the construction of physical maps to position BAC clones (Islam-Faridi et al. 2002) and other DNA sequences along the chromosomes (see Figure 1.5).

1.2.7
Specific Mapping Problems

A loss in genetic diversity inevitably causes problems for the breeding of new varieties. The genus *Lycopersicon*, which comprises modern tomato cultivars, is an example of this development (Miller and Tanksley 1990). When tomato was introduced from Latin America to Europe by Spanish explorers, presumably only lim-

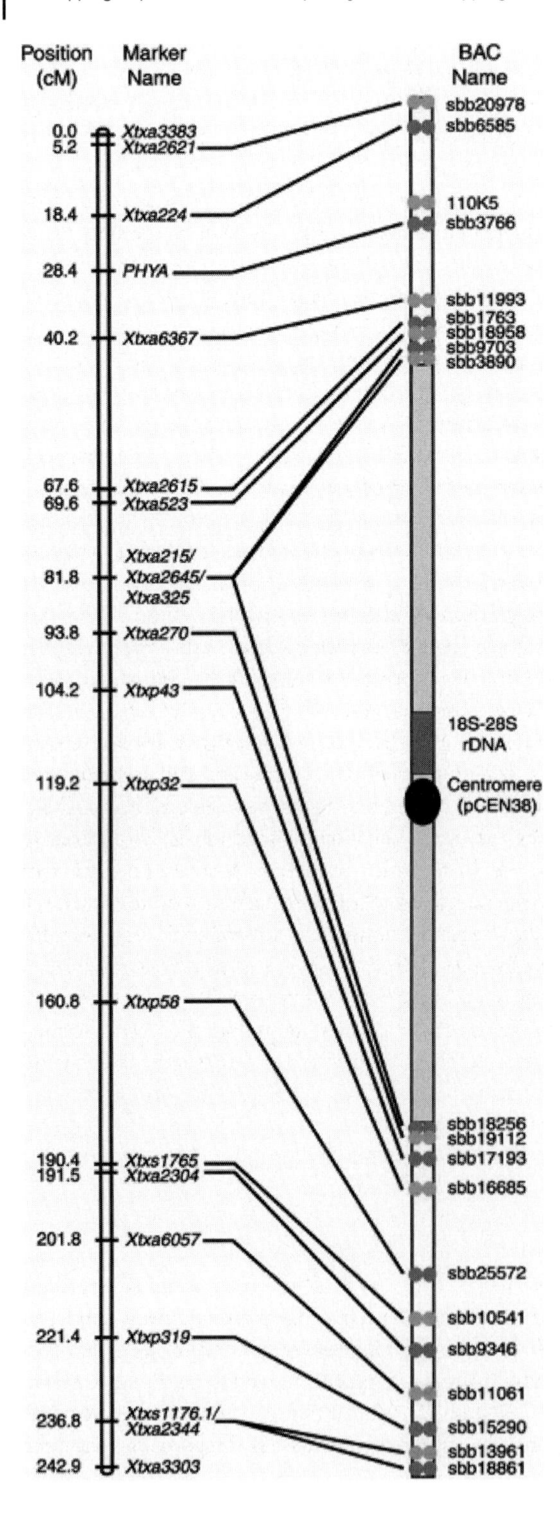

Figure 1.5. Alignment of the genetic and physical map of sorghum chromosome 1 (figure taken from Islam-Faridi et al. 2002, courtesy of *Genetics*). A diagrammatic representation of the cytogenetic locations of 20 sorghum BACs, 18S-28S rDNA, and CEN38 (a centromere-associated clone; on right).

ited numbers of seeds (and accessions) were transferred and became the basis of today's modern cultivars. This created a bottleneck. Breeding methods such as single-seed descent and pedigree selection also promote genetic uniformity. As the tomato cultivars are generally self-compatible, this contributes even further to a decrease DNA polymorphism. For RFLP analysis, the degree of polymorphic probes has been found to be exceptionally low in tomato crosses between modern cultivars (Miller and Tanksley 1990). A further indication of the low genetic diversity of tomato genotypes is the small number of microsatellite alleles in a set of tomato varieties (Areshchenkova and Ganal 2002). Self-incompatible species of the *Lycopersicon* genus show a much higher genetic distance within and between accessions, indicating the role of the mode of reproduction in the maintenance of genetic variability. The use of landraces that are not genetically uniform is one option to increase genetic polymorphism. Given that almost all species of the *Lycopersicon* genus can be crossed with cultivated tomato, the construction of inter- rather than intraspecific crosses and populations is essential for introducing new genetic factors into the breeding pool of this crop.

A second problem that is often encountered in genetic mapping is distorted segregation. This term describes a deviation from the expected Mendelian proportion of individuals in a given genotypic class within a segregating population (Lyttle 1991). That one allelic class can be underrepresented due to a dysfunction of the concerned gametes is well known for plants (Xu et al. 1997). This can occur in pollen, in megaspores, or in both organs and can be explained either by the selective abortion of male and female gametes or by the selective fertilization of particular gametic genotypes. A selection process during seed development, seed germination, and plant growth can also be active. Gametophyte loci leading to a distorted segregation have been identified in rice (summarized in Xu et al. 1997). They are supposed to be responsible for the partial or total elimination of gametes carrying one of the parental alleles. A marker locus linked to a gametophyte locus, also referred to as a gamete eliminator or pollen killer, can also show distorted segregation. Self-incompatibility loci preventing self-pollination are also a direct cause of distorted segregation, as is discussed for potato (Gebhardt et al. 1991). Breeding programs that aim at the generation of specific recombinants are directly affected if one locus is close to a region affected by segregation distortion.

1.3 Discussion

In the field of plant breeding, genetic mapping is still the most valuable approach to identifying the genetic factors that underlie particularly quantitatively inherited traits. Genetically linked markers can be used in marker-assisted breeding to identify individuals with the desirable level of relevant characters at an early stage. High-throughput technologies based on SNP detection (Rafalski 2002) allow the scoring of thousands of data points in a short time. This has the potential to reduce greatly the size of field trials. Therefore, it is important to have mapping tools at hand as described in this chapter.

Important factors to be balanced in the experimental design are the mapping resolution, the required time, and human resources. RILs, NILs, and introgression lines are very laborious and time-consuming to construct, although they allow applied assessments with high precision. Alternatively, local maps of the regions harboring mutations are produced in F_2 populations and cross-linked to populations with high mapping resolution (Castiglioni et al. 1998). Working with natural populations is an alternative that circumvents the construction of experimental mapping populations and still maintains a high resolution because it takes advantage of the multiple meiotic events that occurred during plant evolution (Rafalski 2002). Any marker in the region of linkage disequilibrium that surrounds a genetic factor responsible for a trait may be indicative of the level of the expression of a trait.

The next step is the identification of the genetic factor itself. Unless a gene is tagged by the candidate gene approach, positional cloning involving the partial construction of a physical map is the method of choice. This requires the construction of contigs based on large-insert clones such as BACs, which are assembled according to fingerprinting data. Local, but also genome-wide, physical maps are in progress or have already been obtained for plants with model character such as *Arabidopsis thaliana*, as well as for crops such as rice and sorghum (Mozo et al. 1999; Tao et al. 2001; Klein et al. 2003). In plants for which the whole genome has been sequenced, such as *Arabidopsis thaliana* (The Arabidopsis Genome Initiative 2000) and rice (Yu et al. 2002), this is becoming increasingly easier. Alternatively, new genetic markers located in underrepresented regions of the genome can be developed from the complete DNA sequence and applied in mapping. Once cloned, a gene becomes subject to transgenic approaches. Complementation of a phenotype missing a specific trait is the ultimate proof of gene function, and superior varieties can be created accordingly. Alternatively, association studies for the gene identified can be performed to detect superior alleles of the locus linked to specific markers.

In summary, mapping populations are the basic tools for understanding the effect of selected genetic factors and the organization of the genome of a species as a whole. They are the backbone of genomics research that aims to decipher large, complex genomes at the physical or even sequence level.

Acknowledgments

The author would like to thank Francesco Salamini for fruitful discussions in writing this chapter.

References

Alpert KB, Tanksley SD (1996) High-resolution mapping andisolation of a yeast artificial chromosome contig containing fw2.2: A major fruit weight quantitative trait locus in tomato. PNAS USA 93, 15503–15507.

The Arabidopsis Genome Initiative (2000) Nature 408, 796–815.

Areshchenkova T, Ganal MW (2002) Comparative analysis of polymorphism and chromosomal location of tomato microsatellite markers isolated from different sources. Theor Appl Genet 104, 229–235.

Beavis WD (1994) The power and deceit of QTL experiments: lessons from comparative QTL studies. Proc. 49th Ann. Corn and Sorghum Ind. Res. Conf., Washington, DC American Seed Trade Association.

Bell C, Ecker J (1994) Assignment of 30 microsatellite loci to the linkage map of Arabidopsis. Genomics 19, 137–144.

Botstein D, White RL, Skolnick M, Davis RW (1980) Construction of a genetic linkage map in man using restriction fragment length polymorphisms. Am. J. Hum. Genet. 32:314–331.

Buckler ES, Thornsberry JM (2002) Plant molecular diversity and applications to genomics. Curr. Opin. Plant Biol. 5:107–111.

Burr B, Burr FA (1991) Recombinant inbreds for molecular mapping in maize: theoretical and practical considerations. Trends in Genet. 7, 55–60.

Büschges R, Hollrichter K, Panstruga R et al. (1997) The barley Mlo gene: a novel control element of plant pathogen resistance. Cell 88, 695–705.

Castiglioni P, Pozzi C, Heun M, Terzi V, Müller KJ, Rohde W, Salamini F (1998) AFLP-based procedure for the efficient mapping of mutations and DNA probes in barley. Genetics 149: 2039–2056.

Chao S, Sharp PJ, Worland AJ, Warham EJ, Koebner RMD, Gale MD (1989) RFLP-based genetic maps of wheat homoeologous group 7 chromosomes. Theor. Appl. Genet. 78, 495–504.

Chen X, Salamini F, Gebhardt C (2001) A potato molecular-function map for carbohydrate metabolism and transport. Theor. Appl. Genet. 102, 284–295.

Ching A, Caldwell KS, Jung M, Dolan M, Smith OS, Tingey S, Morgante M, Rafalski AJ (2002) SNP frequency, haplotype structure and linkage disequilibrium in elite maize inbred lines. BMC Genetics 3, 19.

Desel C, Jung C, Cai D, Kleine M, Schmidt T (2001) High-resolution mapping of YACs and the single-copy gene Hs1(pro-1) on Beta vulgaris chromosomes by multi-colour fluorescence in situ hybridization. Plant Mol. Biol. 45:113–122.

Eshed Y and Zamir D (1995) An introgression line population of Lycopersicon pennellii in the cultivated tomato enables the identification and fine mapping of yield-associated QTL. Genetics 141, 1147–1162.

Fisher, RA (1921) On the mathematical foundations of theoretical statistics. Philos. Trans. R. Soc. London, Ser. A 122, 309–368.

Frei OM, Stuber CW, Goodman MM (1986) Use of allozymes as genetic markers for predicting performance in maize single cross hybrids. Crop Sci. 26, 37–42.

Ganal MW, Young ND, Tanksley SD (1989) Pulsed field gel electrophoresis and physical mapping of large DNA fragments in the Tm-2a region of chromosome 9 in tomato. Mol. Gen. Genet. 215: 395–400.

Gebhardt C, Ritter E, Debener T, Schachtschabel U, Walkemeier B, Uhrig H, Salamini F (1989) RFLP mapping and linkage analysis in Solanum tuberosum. Theor. Appl. Genet. 78, 65–75.

Gebhardt C, Ritter E, Barone A, Debener T, Walkemeier B, Schachtschabel U, Kaufmann H, Thompson RD, Bonierbale MW, Ganal MW, Tanksley SD, Salamini F (1991) RFLP maps of potato and their alignment with the homoeologous tomato genome. Theor. Appl. Genet. 83, 49–57.

Grando MS, Bellin D, Edwards KJ, Pozzi C, Stefanini M, Velasco R (2003) Molecular linkage maps of Vitis vinifera L. and Vitis riparia Mchx. Theor. Appl. Genet. 106, 1213–1224.

Haldane JBS and Smith CAB (1947) A new estimate of the linkage between the genes for colour-blindness and haemophilia in man. Ann. Eugen. 14, 10–31.

Hanson RE, Zwick M, Choi S, Islam-Faridi MN, McKnight TD (1995) Fluorescent in situ hybridization of a bacterial artificial chromosome. Genome 38, 646–657.

Helentjaris T, Weber DF, Wright S (1986) Use of monosomics to map cloned DNA fragments in maize. Proc. Natl. Acad. Sci. USA 83, 6035–6039.

Heun M, Kennedy AE, Anderson JA, Lapitan NLV, Sorrells ME, Tanksley (1991) Construction of a restriction fragment length polymorphism map for barley (Hordeum vulgare). Genome 34, 437–447.

Islam-Faridi MN, Childs KL, Klein PE, Hodnett G, Menz MA, Klein RR, Rooney WL, Mullet JE, Stelly DM, Price HJ (2002) Molecular cytogenetic map of sorghum chromosome 1: fluorescence in situ hybridization analysis with mapped bacterial artificial chromosomes. Genetics 161, 345–353.

Klein PE, Klein RR, Vrebalov J, Mullet JE (2003) Sequence-based alignments of sorghum chromosome 3 and rice chromosome reveals extensive conservation of gene order and one major chromosomal rearrangement. The Plant J. 34, 605–521.

Konieczny A, Ausubel FM (1993) A procedure for mapping Arabidopsis mutations using co-dominant; ecotype-specific PCR-based markers. The Plant J. 4, 403–410.

Koornneef M, Vanderveen HJ (1983) A marker line, that allows the detection of linkage on all Arabidopsis chromosomes. Genetica 61, 41–46.

Koornneef M, Hanhart CJ, von Loenen Martinet EP, Peeters AJM, van der Veen JH (1987) Trisomics in Arabidopsis thaliana and the location of linkage groups. Arabidopsis Inf. Serv. Frankfurt 23, 46–50.

Kulikova O, Gualtieri G, Geurts R, Kim DJ. Cook D, Huguet T, de Jong HJ, Fransz PF, Bisseling T (2001) Integration of the FISH pachytene and genetic maps of Medicago truncatula. The Plant J. 27, 49–58.

Künzel G, Korzun L, Meister A (2000) Cytologically integrated physical restriction fragment length polymorphism maps for the barley genome based on translocation breakpoints. Genetics 154, 397–412.

Lander ES, Green P, Abrahamson J, Barlow A, Daly MJ, Lincoln SE, Newburg L (1987) MAPMAKER: an interactive computer package for constructing primary linkage maps of experimental and natural populations. Genomics 1, 174–181.

Lister C, Dean C (1993) Recombinant inbred lines for mapping RFLP and phenotypic markers in Arabidopsis thaliana. Plant J. 4, 745–750.

Lyttle TW (1991) Segregation distorters. Annu. Rev. Genet. 25, 511–557.

Maliepaard C et al. (1998) Aligning male and female linkage maps of apple (Malus pumila Mill.) using multi-allelic markers. Theor. Appl. Genet. 97, 60–73.

McCouch SR, Kochert G, Yu ZH, Wang ZY, Khush GS, Coffman WR, Tanksley SD (1988) Molecular mapping of rice chromosomes. Theor. Appl. Genet. 76, 815–829.

Michelmore RW, Paran I, Kesseli RV (1991) Identification of markers linked to disease-resistance genes by bulked segregant analysis: A rapid method to detect markers in specific genomic regions by using segregating populations. Proc. Natl. Acad. Sci. USA 88, 9828–9832.

Miller JC, Tanksley SD (1990) RFLP analysis of phylogenetic relationships and genetic variation in the genus Lycopersicon. Theor Appl Genet 80, 437–448.

Morton N (1955) Sequential tests for the detection of linkage. Am. J. Hum. Genet. 7, 277–318.

Mozo T, Dewar K, Dunn P, Ecker JR, Fischer S, Kloska S, Lehrach H, Marra M, Martinssen R, Meier-Ewert S, Altmann T (1999) A complete BAC-based physical map of the Arabidopsis thaliana genome. Nat. Genet. 22, 271–275.

Neuffer MG, Coe E, Wessler S (1997) Mutants of Maize. Cold Spring Harbor Laboratory, New York.

O'Donoughue LS, Kianian SF Rayapati PJ Penner GA Sorrells ME et al. (1995) Molecular linkage map of cultivated oat (Avena byzantina X A. sativa cv. Ogle). Genome 38:368–380.

Rafalski A (2002) Applications of single nucleotide polymorphisms in crop genetics. Curr. Opinion in Plant Biol. 5:94–100.

Sandhu D, Champoux JA, Bondareva SN, Gill KS (2001) Identification and physical localization of useful genes and markers to a major gene-rich region on wheat group 1S chromosomes. Genetics 157, 1735–1747.

Schneider K, Weisshaar B, Borchardt DC, Salamini F (2001) SNP frequency and allelic haplotype structure of Beta vulgaris expressed genes. Mol. Breeding 8, 63–74.

Schneider K, Schäfer-Pregl R, Borchardt DC, Salamini F (2002) Mapping QTLs for sucrose content, yield and quality in a sugar beet population fingerprinted by EST-related markers. Theor. and Appl. Genet. 104, 1107–1113.

Stam, P (1993) Construction of integrated linkage maps by means of a new computer package: JOINMAP. The Plant Journal 5, 739–744.

Stuber CW, Goodman MM, Moll RH (1972) Improvement of yield and ear number resulting from selection of allozyme loci in a maize population. Crop Sci. 22, 737–740.

Swiecicki WK, Wolko B, Weeden NF (2000) Mendel's genetics, the Pisum genome and pea breeding. Vortr. Pflanzenzüchtg 48, 65–76.

Szostak JW, Orr-Weaver TL, Rothstein RJ, Stahl FW (1983) The Double-Strand-Break Repair Model for Recombination. Cell 33, 25–35.

Tanksley SD, Grandillo S, Fulton TM, Zamir D, Eshed Y, Petiard V, Lopez J, Beckbunn T, (1996) Advanced backcross QTL analysis in a cross between an elite processing line of tomato and its wild relative L. pimpinellifolium. Theor. Appl. Genet. 92:213–224.

Tao QZ, Chang YL. Wang JZ, Chen HM, Islam-Faridi MN, Scheuring C. Wang B, Stelly DM, Zhang HB (2001) Bacterial artificial chromosome-based physical map of the rice genome constructed by restriction fingerprint analysis. Genetics 158, 1711–1724.

Thornsberry JM, Goodman MM, Doebley J, Kresovich S, Nielsen D, Buckler ES (2001) Dwarfs polymorphisms associate with variation in flowering time. Nat. Genet. 28, 286–289.

Vos P, Hogers R, Bleeker M, Reijans M, van der Lee T, Fornes M, Frijters A, Pot J, Peleman J, Kuiper M, Zabeau M (1995) AFLP: a new technique for DNA fingerprinting. Nucleic Acids Res. 23:4407–4414

Wang GL, Mackill DJ, Bonman JM, McCouch SR, Champoux MC, Nelson RJ (1994) RFLP mapping of genes conferring complete and partial resistance to blast in a durably resistant rice cultivar. Genetics 136:1421–1434.

Weber D, Helentjaris T (1989) Mapping RFLP loci in maize using B-A translocations. Genetics 121, 583–590.

Welsh J, McClelland M (1990) Fingerprinting genomes using PCR with arbitrary primers. Nucleic Acids Res. 18, 7213–7218.

Whitt SR, Wilson LM, Tenaillon MI, Gaut BS, Buckler ES (2002) Genetic diversity and selection in the maize starch pathway. PNAS 99, 12959–12962.

Williams JGK, Kubelik AR, Livak, KJ, Rafalski JA, Tingey SV (1990) DNA polymorphisms amplified by arbitrary primers are useful as genetic markers. Nucleic Acids Res. 18:6531–6535.

Xu Y, Zhu L, Xiao J, Huang N, McCouch SR (1997) Chromosomal regions associated with segregation distortion of molecular markers in F2 backcross, doubled haploid and recombinant inbred populations in rice (Oryza sativa L). Mol Gen Genet 253, 535–545.

Yamamoto T, Kimura T, Shoda M, Imai T, Saito T, Sawamura Y, Kotobuki K, Hayashi K, Matsuta N (2002) Genetic linkage maps constructed by using an interspecific cross between Japanese and European pears. Theor. Appl. Genet. 106, 9–18.

Young ND, Tanksley SD (1989) RFLP analysis of the size of chromosomal segments retained around the Tm-2 locus of tomato during backross breeding. Theor. Appl. Genet. 77, 353–359.

Yu J et al. (2002) A draft sequence of the rice genome (Oryza sativa L. ssp. indica). Science 296, 79–92.

Zamir D (2001) Improving plant breeding with exotic genetic libraries. Nature Rev. 2, 983–989.

Zhong XB, Bodeau J, Fransz PF, Williamson VM, van Kammen A, de Jong JH, Zabel P (1999) FISH to meiotic pachytene chromosomes of tomato locates the root-knot nematode resistance gene Mi-1 and the acid phosphatase gene Aps-1 near the junction of euchromatin and pericentromeric heterochromatin of chromosome arms 6S and 6L, respectively. Theor. Appl. Genet. 98, 365–370.

2
Molecular Marker Systems for Genetic Mapping

Henry T. Nguyen and Xiaolei Wu

Abstract

Genetic variation is a force of genome evolution. Naturally occurring polymorphism due to varied DNA sequences throughout plant and animal genomes between and within species is the basis of genetic variation. The manipulation of molecular markers is based on these naturally occurring polymorphisms. Since the discovery of the primary structure of DNA, it has been characterized in a number of species. Many molecular marker detection systems that have the ability to distinguish variation present in genomic DNA sequences have been developed for genetic analysis. Among these techniques, RFLP, RAPD, SSR, and AFLP are most commonly used in genetic linkage analysis. The implications of these powerful and novel methods have led to a new plant breeding revolution: marker-assisted selection (MAS). Numerous genes have been targeted and cloned in all major crop species. Genetic linkage analysis has become a routine tool to identify and characterize genes that regulate agronomically important multigenic traits. This chapter introduces the main marker detection systems that are used in genetic linkage analysis, and the focus will be on the following aspects for each marker system: (1) developmental history, (2) advantages and disadvantages, (3) detection methods, and (4) applications in plant genome mapping.

2.1
Introduction

Successful investigation of many value-added traits in crops has attributed to the advances in our understanding of the organization of plant genomes and genetics through the aid of molecular markers. One of the most productive approaches to retrieving important genomic information is the use of genetic mapping techniques to locate genes responsible for a particular trait. Genetic maps for a wide range of plant species have been established using various molecular marker systems, such as restriction fragment length polymorphism (RFLP) (Botstein et al.

The Handbook of Plant Genome Mapping.
Genetic and Physical Mapping. Edited by Khalid Meksem, Günter Kahl

ISBN 3-527-31116-5

1980), random amplified polymorphic DNA (RAPD) (Williams et al. 1990; Caetano-Anolles et al. 1991), simple sequence repeat or microsatellite (SSR) (Litt and Luty 1989), sequence-tagged sites (STS) (Palazzolo et al. 1991), amplified fragment length polymerphism (AFLP) (Vos et al. 1995), single-nucleotide polymorphism (SNP) (Lai et al. 1998), sequence-characterized amplified region (SCAR) (Williams et al. 1991), and cleaved amplified polymorphic sequences (CAPS) (Lyamichev et al. 1993), each of which has its own unique advantages and disadvantages that fit different purposes (Table 2.1). However, RFLP, RAPD, SSR, and AFLP markers are most commonly used in plant species for genetic mapping.

Polymorphism of nucleotide sequences at the same locus is the basis for developing molecular markers used in genome mapping and DNA fingerprinting. Polymorphism is revealed by molecular detection techniques, which are grouped in hybridization-based RFLP techniques, and PCR-based techniques, such as RAPD, SSR, and AFLP. When using molecular markers for genetic mapping, the simplicity of the genotyping procedure, the cost-effectiveness, and the information content that molecular markers provide must be considered for large-scale genotyping. Based on these comprehensive considerations, only four categories (RFLP, RAPD, SSR, and AFLP) of molecular markers commonly used in genetic mapping will be described in this chapter. SNP markers, known as "the third generation of molecular markers," are abundant across whole genomes and have the potential to dissect genes or quantitative trait loci (QTLs) in QTL mapping and association genetics. SNPs will be described in another chapter.

Table 2.1. Comparison of the most broadly used marker systems.

Marker/ technique	*PCR-based*	*Polymorphism (abundance)*	*Dominance*	*Repro-ducibility*	*Auto-mation*	*Running cost*
RFLP	No	Low/medium	Codominant	High	Low	High
RAPD	Yes	Medium/high	Dominant	Low	Medium	Low
SCARS/CAPS	Yes	High	Codominant	High	Medium	Medium
AFLP	Yes	High	Dominant	High	Medium/ high	Medium
SSR	Yes	High	Codominant	High	Medium/ high	Low
ISSR	Yes	High	Dominant	High	Medium/ high	Low
STS	Yes	High	Codominant/ dominant	High	Medium/ high	Low
SRAP/EST	Yes	Medium	Codominant	High	Medium	Low
IRAP/REMAP	Yes	High	Codominant	High	Medium/ high	Low
SNP	Yes	Extremely high	Codominant/ dominant	High	High	Low

2.2
DNA-based Markers Popularly Used in Genetic Mapping

2.2.1
RFLP

Among the various molecular markers developed, RFLPs were the earliest molecular markers (known as the first generation of molecular markers) to be used in animal and plant genome mapping projects. RFLPs are caused by DNA sequence changes such as single-nucleotide mutations, insertions or deletions of DNA fragments ranging from one to several hundred base pairs, or DNA rearrangements of large chromosomal fragments. These changes cause the gain, loss, or movement of some restriction sites, which is the basis for generating RFLPs. RFLPs have proven to be robust and accurate for genotyping. The technique relies on digesting genomic DNA with specific restriction enzymes and hybridizing with probes. The probes used in RFLP could be fragments of genomic DNA, cDNA, or expressed tag sequences (ESTs) that have been developing through EST projects in crop species. Since RFLP markers are bi-allelic codominant, a unique locus and chromosomal position can be identified with a specific probe. However, RFLP analysis is labor-intensive and time-consuming. Improvements to simplify RFLP for high-throughput procedures have been achieved in current approaches. To avoid using the time-consuming Southern blotting technique in RFLP assays, RFLP detection is now replaced by PCR-RFLP assays, or RFLP markers are converted into SCAR markers by sequencing the two ends of genomic DNA clones and designing oligonucleotide primers based on the end sequences. These primers are used directly on genomic DNA in a PCR reaction to amplify the polymorphic region. If no amplified fragment length polymorphism is noticed, then PCR fragments are subjected to restriction digestion with restriction enzymes to detect RFLPs within the amplified fragment.

SCARs have the advantage of being inherited in a codominant fashion, in contrast to RAPDs, which are inherited in a dominant manner. Amplified fragment length polymorphism (AFLP) is based on PCR amplification of restriction fragments generated by specific restriction enzymes and oligonucleotide adapters of a few nucleotide bases. This method generates a large number of restriction fragment bands facilitating the detection of polymorphisms but is also inherited in a dominant manner. The number of DNA fragments can be controlled by choosing different base numbers and the composition of nucleotides in adapters. This approach is very useful in saturation mapping and for discrimination between varieties. Comparisons of the DNA mapping techniques RFLP, RAPD, and AFLP show that AFLP is the most efficient technique in detecting polymorphism with high reproducibility. This makes AFLP DNA analysis an attractive technique for identifying polymorphisms and fine mapping. However, most AFLPs are dominant markers, and the procedure is too complicated to be suitable for genotyping or for MAS.

2.2.1.1 **Conventional RFLP Analysis**

1. DNA preparation: Adjust the concentration of all DNA samples to a uniform scale (500 ng μL^{-1}) after estimating the DNA concentration using 0.8 % agarose gel electrophoresis. The high purity of the DNA sample is a key factor in the success of DNA digestion by any restriction enzyme.
2. Restriction enzyme digestion: Transfer an equal volume aliquot of DNA prepared in step 1 into separate microcentrifuge tubes or wells of a 96-well PCR plate. Prepare a bulk mixture for the digestion reaction. Add the restriction enzyme to the mixture last. The recipe for restriction digestion of genomic DNA is as follows:

DNA	10 μg
10× buffer	1×
Restriction enzyme	3.5 U μg^{-1} DNA
dd H_2O to	30 μL

   Incubate at the optimum temperature of each restriction enzyme for at least 6 h.
3. Agarose gel electrophoresis and Southern blotting: Following the incubation period, add 5 μL of 6× gel loading buffer to each tube and mix well. Load all of the reaction mixture into a 0.8 % agarose gel and start running under low voltage (0.5 volt cm^{-1}) with size standards for 16~18 h. Following separation of DNA fragments by agarose gel electrophoresis, the DNA is then transferred to Hybond N+ membranes by Southern blotting.
4. Probe labeling: To label probes with ^{32}P, the random primer labeling method (Feinberg and Vogelstein 1984) is usually used.
5. Hybridization and autoradiography.

2.2.1.2 **PCR-RFLP**

Although RFLP mapping is a powerful technique that has been extensively used in a number of species, it is inconvenient, requiring large amounts of DNA. With the advent of the PCR technique, PCR-based RFLP markers have been developed that require far less DNA and allow for rapid analysis and easy high-throughput genotyping. PCR-RFLP is a well-established method and is widely used as a means to rapidly and reliably detect DNA sequence variations. These variations result in unique restriction sites that discriminate a pair of alleles in a three-step process: PCR amplification of the DNA sequence of interest, restriction endonuclease digestion, and electrophoresis-based analysis of the resulting fragments. The markers generated by PCR-RFLP are known as CAPS (cleaved amplified polymorphic sequences) and are proven to be powerful for the mapping of mutant loci and SNP genotyping with medium-throughput and low-cost features (Neff et al. 1998).

1. PCR amplification: The recipe (in final concentration) is as follows:

PCR buffer	1×
$MgCl_2$	2.5
dNTPs	200 μM (each)
Forward primer	1 μM
Reverse primer	1 μM
Taq DNA polymerase	1 U
DNA template	1–10 ng

   Thermocycler programmed for the PCR run: Initial denaturation: 95 °C for 5 min, then 40 cycles: 95 °C for 30 s (denaturation); 50~60 °C for 30 s (annealing, the temperature is dependent on T_m of the primers); 72 °C for 5 min (elongation); and finally hold at 4 °C.
2. Endonuclease digestion.
   a. Transfer 10 μL of PCR products into each microtube.
   b. Prepare a digestion master mix. For each sample use 2 μL of appropriate restriction buffer, 5~10 U of enzyme required, and dd H_2O to give a total volume of 20 μL per sample.
   c. Dispense 10 μL of the digestion mix per sample.
   d. Incubate on a thermocycler at appropriate temperature for 3 h.
3. Electrophoresis.
   a. Load 10 μL of digested products plus 2 μL of 6× PCR-loading buffer in 2~3% agarose gels with ethidium bromide (depending on the size discrimination needed) covered with TBE buffer, and carry out electrophoresis at a constant voltage.
   b. Visualize gels on a UV transilluminator and capture image by a camera.

2.2.1.3 Mismatch PCR-RFLP

In the CAPS technique, a limitation exists in that only mutations creating or disrupting a restriction enzyme recognition site can be detected. However, the majority of single-nucleotide changes do not generate restriction site differences. A similar technique of CAPS, called dCAPS (derived cleaved amplified polymorphic sequence), has been developed using PCR reactions with primers in which one or more mismatched nucleotides are introduced to discriminate between alleles that differ by single-nucleotide changes, which do not generate restriction-site differences (Neff et al. 1998). The amplified PCR products obtained from mismatched PCR primers are used to create restriction sites digested by an appropriate restriction enzyme and resolved by gel electrophoresis. Heterozygous plants can also be rapidly discriminated using this method because of the codominant nature of dCAPS. The dCAPS marker is thus also applicable for fine genetic mapping. The protocol of dCAPS is similar to CAPS.

2.2.2 RAPD

Random amplification polymorphic DNA (RAPD) was the pioneer marker of PCR-based molecular markers developed for genetic-mapping and DNA-fingerprinting purposes. Since then, many new modifications of the PCR-based molecular marker techniques have been developed for different purposes. PCR-based markers are the second generation of molecular markers. They have been extensively developed and utilized in genetic mapping, genomic fingerprinting, and diversity studies because they are relatively simple and sequence information is not necessary for primer design.

When a single random oligonucleotide primer (approximately 10 bases) is used in a PCR reaction, it anneals to homologous sequences in the genome. Amplicons throughout the genome are targeted and amplified if the primer also anneals to sequences on complementary strands not far away from the 3′ end of the other primers (Welsh and McClelland 1990). Since PCR-amplified fragments are generated primers hybridizing to genomic sites at which they fortuitously match or almost match, multiple products are produced under conditions of low stringency with a single random primer, which are separated on an agarose gel stained with ethidium bromide. However, a single primer can generate a relatively complex pattern that varies among plant materials; in most cases, a single primer provides one to three intense bands that may differ between parents of a mapping population. Therefore, only reproducible, intense bands in each pattern can be used as molecular markers. RAPD has been less popular in genetic mapping and genomic fingerprinting because of several disadvantages in the procedure. There is the problem of reproducibility not only among laboratories but also within a laboratory over time, making the marker information difficult to share and repeat. When RAPD PCR amplification is under less stringent conditions, virtually every aspect of PCR can affect reproducibility. Changes in PCR parameters affect most notably the presence of low-intensity bands, but can also affect the position and intensity of high-intensity bands.

1. Put 25~50 ng template DNA into each microtube.
2. Prepare a master mix for PCR reactions:

PCR buffer	1×
$MgCl_2$	1.5 mM
dNTPs	200 μM (each)
10-mer oligo primer	1 μM
Taq DNA polymerase	1 U
DNA template	50 ng

3. Place the tubes into a PCR thermocycler and run the following program: 95 °C for 3 min, then 40 cycles; 95 °C for 1 min, 36 °C for 1.5 min, and 72 °C for 1.5 min, followed by a final extension at 72 °C for 10 min.
4. Run gel electrophoresis of 10 μL of amplified products plus 2 μL of PCR-loading buffer in 1~2 % agarose gel for 2 h at 50 V cm^{-1} gel.

2.2.3 SSR Markers

Simple sequence repeat (SSR) or microsatellite is one of the most important categories of molecular markers. It comprises the core marker system of the PCR-based molecular markers and is widely used for DNA fingerprinting, genetic mapping, MAS, and studies of genetic diversity and population genetics (Hearne et al. 1992; Zietkiewicz et al. 1994). Microsatellite markers are abundant, highly polymorphic, and codominant and distinguish multiple alleles in a plant species due to variation in the number of repeat units (motif), which are composed of 1~6-bp short DNA sequences, such as dinucleotide repeats [(AT)n and (CT)n] and trinucleotide repeats (ATT), and disperse mainly in the regions between genes and un-coding regions throughout genomes (Li et al. 2002).

The analysis of microsatellite polymorphism is PCR-based. Two conserved and locus-specific PCR primers flanking each microsatellite repeat are used for PCR amplification. Several detection methods have been developed based on primer labeling, gel utility, staining, and detection equipment. The PCR amplification can be done with either two normal primers or one normal primer and another primer modified at the 5′ end labeled by radioactive [γ-^{32}P] ATP using T4 polynucleotide kinase, which is simple and cheap but produces a lower resolution than using a fluorescent dye (fluorophore). PCR products can be detected using agarose gel, non-denaturing polyacrylamide gel, and denaturing polyacrylamide gel. The agarose gel detection system using more than 3 % concentrated agarose and ethidium bromide staining is commonly used in MAS. In most cases, a resolution of at least 10 bp can be reached. However, if the difference in size of SSR PCR products is less than 10 bp, a polyacrylamide gel-based system for electrophoresis using isotope-labeled primer is preferred. Another way to detect SSR products is carried out on a DNA sequencer system that detects fluorescent dye labels by laser excitation. The dyes fluoresce at certain wavelengths, and this information is collected and analyzed by software programs that score SSR alleles as peaks on a graphical display. Internal size standards are run within each lane for accurate allele sizing, down to a separation of only 1 bp. Three different dyes can be used as labels to analyze at least three markers in a single gel lane on an ABI DNA sequencer. This number can be increased if the allele size range of SSR loci is well characterized. Using this information, alleles with the same dye that are run within the same gel lane are correctly scored because they fit in a size range of their corresponding SSR locus. Based on these features, multiplexing can be conducted by including several primer pairs in a single PCR amplification (multiplex PCR), by pooling single PCR amplifications, or by combining these approaches. Therefore, a fluorescence-based genotyping system using a sequence analyzer has been widely adopted. It offers both precise estimation of fragment size and analysis of data by a semiautomated, high-throughput system that includes software for accurate and timely allele assignment. This genotyping system can also be used for AFLP analysis.

Fluorescence provides a number of advantages over other labeling methods: (1) fluorescent compounds and substrates have a longer shelf life than radioisotopes (which must be used immediately to avoid radiolytic decay); (2) safety, handling, and disposal issues are significantly less demanding and expensive than those for radioisotopes and silver staining; (3) fluorescent detection methods are much faster than autoradiography and silver stain development; (4) the fluorescent signal is linear over a wider concentration range, allowing more flexibility in quantitative analysis; (5) variations in SSR sizes and fluorophores allow multiplexed markers to be detected simultaneously; (6) background can be minimized by specific selection of the fluorescent label; and (7) it uses electronic data collection, including spectral software for fragment size determination and calling of allele and genotype.

2.2.3.1 **Conventional SSR Analysis**

A: Agarose Gel Detection System

1. Run PCR for each sample using the following recipe:

10 × PCR buffer	2.0 μL
$MgCl_2$ (50 mM)	1.0 μL
dNTPs (2mM)	2.0 μL
PVP (20%)	2.0 μL
BSA (10 mg mL^{-1})	0.2 μL
Unlabeled forward primer (20 nM)	0.4 μL
Unlabeled reverse primer (20 nM)	0.4 μL
Taq DNA polymerase (5 units μL^{-1})	0.2 μL
H_2O	9.8 μL
DNA (50~100ng)	2.0 μL

2. Make a 3% agarose gel with 5% ethidium bromide.
3. Add 5 μL of 6 × loading dye to each well with PCR product. Mix well.
4. Load 10 μL of each sample into separate wells of the agarose gel.
5. Let the gel run for approximately 3 h at 60–75 volts.
6. Put the gel into an Alpha Imager and then photograph the gel or save the image into a file.

B: Denaturing Gel Detection System (Radioactive Method)

Based on the difference of radioactive labeling methods, two methods have been used for SSR polymorphism detection. One method is to label [γ-^{32}P] ATP to the 5′ end of one of the two PCR primers using T4 polynucleotide kinase. The other commonly used method is to directly add [α-^{32}P]dATP (Bui and She 1996) or [α-^{33}P]dATP into the PCR mixture (Cregan et al. 1994; Reineke and Karlovsky 2000).

Labeled PCR primer using T_4 polynucleotide kinase

1. Label the 5′ end of one of the two PCR primers using T4 polynucleotide kinase. Mix the following

Oligonucleotide primer	50 pmoles
10× kinase buffer	2.0 μL
[γ-^{32}P] ATP (7000 Ci mmoles^{-1})	2.0 μL
T4 polynucleotide kinase (1 unit μL^{-1})	1.0 μL
dd H_2O to	20 μL

and incubate at 37 °C for 1 h.
2. Separate the labeled oligonucleotides from unincorporated [γ-^{32}P] ATP by centrifugation through small columns of Sephadex G-75.
3. Perform PCR amplification for each sample. The recipe of preparing PCR for each sample is as follows:

10× PCR buffer	1.0 μL
$MgCl_2$ (25 mM)	1.0 μL
dNTPs (2mM)	1.0 μL
Unlabeled 5′ primer (20 μM)	0.1 μL
Unlabeled 3′ primer (20 μM)	0.1 μL
Labeled primer	1/100th
Taq DNA polymerase (5 units μL^{-1})	0.1 μL
dd H_2O to	8 μL
DNA (50~100ng)	2.0 μL

4. Aliquot 8 μL of mix (except 2.0 μL of template DNA) to each PCR tube.
5. Prepare a standard sequencing gel containing 6 % polyacrylamide, 7.7 M urea.
6. Add 2 μL of 6 × loading dye to each PCR reaction and load 2~4 μL of PCR products on the gel.
7. Electrophorese the gel at 60~70 W for 2~3 h depending on the sizes of the amplified products.
8. Visualize the SSR bands with autoradiography.

Direct isotope incorporation into PCR products

1. PCR amplification:

Tris-HCl pH 9.0	10 mM
KCl	50 mM
$MgCl_2$	1.5 mM
dNTPs	0.2 mM
5′ primer	0.15 μM
3′ primer	0.15 μM
Triton X-100	0.1 %
[α-^{32}P]dATP (10 μCi μL^{-1})	0.1 μL
Taq DNA polymerase (5 units μL^{-1})	0.2 μL
DNA (50–100 ng)	2.0 μL
H_2O	to 10 μL

The PCR reaction can be performed with a PCR thermocycler using the following program: Denaturation at 94 °C for 3 min, 35 cycles at 94 °C for 1 min; 47~60 °C (depending on the primers) for 1 min; 68 °C for 1 min; and a final extension at 72 °C for 10 min.
2. Prepare a standard sequencing gel containing 6% polyacrylamide, 6.0 M urea.
3. Load PCR products on the gel, electrophorese, and take an autoradiograph.

C: SSR Detection System Using a Sequencer
This protocol can be a valuable tool for efficient fragment analysis for high-throughput laboratories. The detection of DNA fragments including SSRs can be performed on a gel-based DNA sequencer (such as ABI 377 or LI-COR IR^2) or a capillary-based DNA sequencer (such as ABI 3100 with 16 capillaries, CEQ 8000 with eight capillaries, ABI 3700 with 96 capillaries, or Aurora with a 24-, 48-, 96-, or 192-capillary array for higher throughput). Three methods of fluorescence labeling will be described: fluorophore-modified primer, PCR products with internal fluorophore-modified dNTP, and the post-PCR fluorescence-labeling method.

Fluorophore-modified primer

1. Forward primers can be labeled with 6-carboxyfluorescein (6-FAM; blue), hexachloro-6-carboxyfluorescein (HEX; yellow green), NED (yellow), or tetrachloro-6-carboxyfluorescein (TET; green).
2. The PCR reaction mixture contains:

Genomic DNA	5–30 ng
Tris-HCl, pH 8.3	10 mM
KCl	50 mM
$MgCl_2$	1.5 mM
dNTPs	0.2 mM
Labeled forward primer	2.5–10 pmol
Unlabeled reverse primer	2.5–10 pmol
Taq DNA polymerase	0.5–1 unit
dd H_2O to	10 μL

3. PCR reaction can be performed with a PCR thermocycler using the following program: denaturation at 94 °C for 2 min, 30 cycles at 94 °C for 1 min; 47~60 °C (depending on the primers) for 1 min; 72 °C for 1 min; and a final extension at 72 °C for 10 min.

Note: Following amplification, an aliquot of PCR (0.5~1.5 μL) is mixed with 10 μL of formamide and a ROX (red)-labeled internal size standard (GENESCAN-500 or GENESCAN-400), then denatured at 94 °C for 4 min, and finally chilled on ice. For fluorescent multiplex PCR, the quantity of template DNA needs to be minimized, and primer concentration for each SSR must be optimized to get a good resolution of peaks and uniform signals for each color. For pooling of PCR products that are performed separately in different PCR amplification with each labeled primer, the aliquot of PCR products of each primer needs to be optimized before combining samples based the SSR sizes and fluorescent dye colors.

PCR products with internal fluorophore-modified dNTP

This method is an alternative to fluorescent-labeling PCR products. It involves the use of fluorescent-labeled dNTPs, which are internally incorporated during the PCR amplification reaction. It offers a cost-effective way to survey polymorphism between parents for genetic mapping using all SSR makers because it is not necessary to synthesize or purchase fluorophore-modified primers before knowing the polymorphic information.

1. The PCR reaction mixture contains:

Genomic DNA	30–50 ng
Tris-HCl, pH 8.3	10 mM
KCl	50 mM
$MgCl_2$	2.5 mM
dNTPs	0.2 mM
Forward primer	8 pmol
Reverse primer	8 pmol
Taq DNA polymerase	1 unit
f-dCTP	0.5–1 pmol
dd H_2O to	10 μL

2. The PCR reaction can be performed with a PCR thermocycler using the following program: denature at 94 °C for 2 min followed by 30 cycles at 94 °C for 1 min; 47–60 °C (depending on the primers) for 1 min; 72 °C for 1 min; and a final extension at 72 °C for 10 min.
3. Following amplification, an aliquot of PCR (0.5-1.5 μL) is mixed with 10 μL of formamide and a ROX (red)-labeled internal size standard (GENESCAN-500 or GENESCAN-400) and then denatured at 94 °C for 4 min and chilled on ice.

Note: The fluorescent-labeled dNTP (deoxy-nucleotide triphosphate) is usually dCTPs ([R110]dCTP, [R6G]dCTP, [TAMARA]dCTP, or [Cy5]dCTP). Multiplex loading can be performed on a capillary sequencer based on different colors. Although cost-effective, this method is not suitable for precise estimation of the length of PCR products because the amplified fragments are chemically heterogeneous, which is caused by variability in the number of incorporated fluorescent nucleotides.

Post-PCR Fluorescence-labeling Method

This method is cost-effective because it does not use expensive fluorescent-labeled primers. It also has the advantage of avoiding ambiguity in the analysis, which may arise from the addition of non-template nucleotides by *Taq* DNA polymerase.

1. The PCR reaction mixture contains:

Genomic DNA	30–50 ng
Tris-HCl, pH 8.3	10 mM
KCl	50 mM
$MgCl_2$	2.5 mM
dNTPs	0.2 mM
Forward primer	1 μM
Reverse primer	1 μM
Taq DNA polymerase	1 unit
dd H_2O to	10 μL

2. The PCR reaction can be performed with a PCR thermocycler using the following cycling profile: denaturation at 94 °C for 2 min; 30 cycles at 94 °C for 30 s; 47–60 °C (depending on the primers) for 30 s; 72 °C for 1 min; and a final extension at 72 °C for 5 min.
3. Label PCR products.

PCR products or pooled samples	4 μL
Labeling reaction mix	4 μL
Including: Tris-HCl, pH 8.7	5 mM
$MgCl_2$	10 mM
Klenow fragment DNA polymerase I	0.1 U μL^{-1}
f-dCTP or f-dUTP	2 μM

4. Incubate the mixture for 15 min at 37 °C.
5. Add 0.8 μL 0.2 M EDTA to stop the reaction.
6. Add 0.8 μL calf intestinal alkaline phosphatase (2 U) and incubate the mixture for an additional 30 min at 37 °C to degrade the labeled nucleotide.
7. Take 4 μL of each PCR product of samples that should be independently labeled for multiplex loading.
8. Mix 1 μL of the pooled sample with 0.2 μL of size standard and 10 μL formamide.
9. Denature at 95 °C for 4 min on a thermocycler.
10. Electrophorese using ABI 3100 DNA sequencer.

Note: Based on agarose gel electrophoresis results, combinations of microsatellite markers for multiplex electrophoresis can be performed using fluorescent dyes to label the PCR products, provided that there are no overlapping peaks for the same color in the image profile. This method is more flexible than other fluorescent labeling methods, but the SSR profile and size must be known in advance.

2.2.3.2 **ISSR**

Inter-simple sequence repeats (ISSR) is a RAPDs-like marker system that does not require any prior knowledge of genome sequence (Godwin et al. 1997). ISSR can access variation in the numerous microsatellite regions by using primers that are anchored at the 5′ or 3′ end of a repeat region and extend into the flanking region. This technique then allows amplification of the genomic segments between inversely oriented repeats (ISSRs). Generally, a series of single primers are used to generate a series of fragments that are size-separated on either an agarose gel or a polyacrylamide gel (Nagaraju et al. 2002). There are three steps involved in ISSR marker analysis: designing oligonucleotide primers, amplifying ISSR segments by PCR, and separating the PCR products. Oligonucleotide primers of ISSR-PCR are usually designed based on the repeat sequence motifs of microsatellites, such as (GA)n or (CA)n, following the rule of the adjunction of either the 5′ or 3′ flanking anchor, such as $(GA)_8RGY$ or $(CA)_8RG$. These anchors ensure a higher resolution and better reproducibility of the bands in ISSR profiles. The multi-locus marker system is useful for fingerprinting, genetic diversity analysis, and genetic mapping. This technique is rapid and can differentiate between closely related individuals. The advantages of this technique include multiple polymorphic loci, high throughput, and low cost. Disadvantages are that ISSR markers are dominant, less reproductive, and poorly productive for some primer combinations. Adding an additional PCR re-amplification step into the standard ISSR-PCR protocol can prominently increase reproducibility and productivity (Wiesner and Wiesnerová 2003).

A: ISSR-PCR Protocol

1. The PCR reaction mixture contains:

Genomic DNA	20 ng
Tris-HCl, pH 8.3	10 mM
KCl	50 mM
$MgCl_2$	4.0 mM
dNTPs	0.2 mM
ISSR primer	0.2 μM
Taq DNA polymerase (TaKaRa)	1 unit
dd H_2O to	25 μL

2. The PCR reaction can be performed with a PCR thermocycler using the following cycling profile: denaturation at 94 °C for 3 min, 35 cycles at 94 °C for 1 min; 55 °C for 30 s; 72 °C for 1 min; and a final extension at 72 °C for 3 min.
3. Electrophorese.

B: FISSR-PCR Protocol

1. The PCR reaction mixture contains:

Genomic DNA	5 ng
Tris-HCl, pH 8.3	10 mM
KCl	50 mM
$MgCl_2$	2.5 mM
dNTPs	1 mM
Gelatin	0.01 %
Triton X-100	0.01 %
ISSR primer	4 μM

Taq DNA polymerase (Ampli *Taq* Gold, Perkin-Elmer)	0.25 unit
Fluorescent dUTP (TAMARA, R110 or R6G)	0.4 μM
Add H_2O to	5 μL

2. The PCR reaction can be performed with a PCR thermocycler using the following cycling profile: denature at 94 °C for 10 min, followed by 35 cycles at 94 °C for 30 s; 50 °C for 30 s; 72 °C for 1 min; and a final extension at 72 °C for 10 min.

2.2.3.3 STMP

SSR markers have extensive utility in many areas, but a major limitation of SSRs is present in discovering and characterizing of individual loci when the DNA sequence is not available. The procedure involves constructing a genomic library, screening positive clones with SSR-specific probes, and characterizing copy number, polymorphism, and chromosomal position of each SSR, which is too complicated. STMP (sequence-tagged microsatellite profiling) is a technique for rapidly generating large numbers of simple sequence repeat (SSR) markers from genomic DNA or cDNA (Hayden and Sharp 2001; Hayden et al. 2002). The technique is based on the principle of AFLP to produce a pool of amplified *Pst* I-*Mse* I restriction fragments and then to enrich DNA fragments containing SSRs by PCR using the primer combination of a single adapter primer with a biotinylated oligonucleotide primer that anchors to the SSR 5′ ends. After pre-amplification, only fragments containing target SSRs are captured using streptavidin-coated magnetic beads. These fragments can be cloned for sequencing or directly sequenced after fractioning and cleaning up of PCR products using specific adapter primers. Longer fragments in the pool of SSR-contained amplicons allow STMP markers to be converted into SSR markers. This technique eliminates the need for library screening to identify SSR-containing clones and provides a more efficient way to develop SSR markers compared to traditional methods.

2.2.4 AFLP

2.2.4.1 Conventional AFLP Analysis

Typical AFLP is a PCR-based DNA fingerprinting technique, usually comprised of the following steps:

1. Restriction of the DNA with two restriction endonucleases, preferably a hexa-cutter and a tetra-cutter, both generating 5′ protruding ends (overhang).
2. Ligation of two double-stranded adapters complimentary to the ends of the restriction fragments.
3. Amplification of a subset of the restriction fragments using two primers complementary to the adapter and restriction site sequences and extended at their 3′

ends by selective nucleotides. The forward primer is labeled with radioactive ^{33}P.

4. Gel electrophoresis of the amplified restriction fragments on denaturing polyacrylamide gels.
5. Visualization of the DNA fingerprints by means of autoradiography or phosphoimaging.

The availability of many different restriction enzymes and corresponding primer combinations provides a great deal of flexibility, enabling AFLPs to be used in a large number of applications, such as the assessment of genetic diversity, population structure, phylogenetic relationships, genetic mapping, QTL mapping, fine physical mapping, and transcript profiling for gene expression (Mueller and Wolfenbarger 1999).

AFLP combines the advantages of RFLP and PCR to provide greatly enhanced performance in terms of reproducibility, resolution, time efficiency, and polymorphism detection at the whole-genome level (Scottdl et al. 1998), and no prior DNA sequence information is required, rendering it a particularly useful discovery tool.

AFLP Protocol

1. Adjust the concentration of genomic DNA samples to a uniform concentration.
2. Restriction digestion of DNA:

Genomic DNA	200~500 ng
Tris-acetate, pH 7.5 at 37 °C	10 mM
K-acetate	50 mM
Mg-acetate	10 mM
BSA	0.1 mg mL^{-1}
EcoR I	5 U
Mse I	5 U
Add dd H_2O to	12.5 μL

3. Incubate at 37 °C for 5~8 h.
4. Ligation of adapters:
   a. Adapter preparation: *EcoR* I-adapter: 5′-CTCGTAGACTGCGTACC (*EcoR* I-adapterF) 3′-CATCTGACGCATGGTTAA-5′ (*EcoR* I-adapterR) *Mse* I-adapter: 5′-GACGATGAGTCCTGAG (*Mse* I-adapterF) 3′-TACTCAGGACTCAT-5′ (*Mse* I-adapterR). Dissolve ssDNA adapter oligonucleotides (non-phosphorylated) in 10 mM Tris, pH 8.5, and adjust the concentration of each oligonucleotide to a 200 pmoles μL^{-1} stock concentration. Mix equal amounts of each complementary oligonucleotide together to a working concentration of 20 pmoles μL^{-1} for each.
   b. Ligation cocktail:

Tris-acetate, pH 7.5 at 37 °C	10 mM
K-acetate	50 mM

Mg-acetate	10 mM
ATP	0.4 mM
Adaptor 1	50 μM
Adaptor 2	50 μM
T4 DNA ligase	1 U
dd H_2O to	12.5 μL

Add 12.5 μL of ligation mixture to each restriction mixture. Incubate at 16 °C overnight. Add 100 μL of 10 mM Tris (pH 8.5) to dilute the template DNA for pre-amplification.

5. Pre-amplification
   a. Pre-amplification recipe:

1:5 diluted template DNA	2.5 μL
Tris-HCl, pH 8.4	10 mM
KCl	50 mM
$MgCl_2$	1.5 mM
dNTP	0.2 mM
Pre-selective primer 1	0.75 μM
Pre-selective primer 2	0.75 μM
Taq DNA polymerase	1 U
Deionized formamide	2%
dd H_2O to	20 μL

   b. PCR thermocycle profile: 94 °C, 2 min; 94 °C, 30 s; 50 °C, 30 s; 72 °C, 1 min; repeat steps 2–4 20 times; and hold at 4 °C.
6. Selective amplification:
   a. Primer labeling by γ-^{33}P-ATP: In order to detect the selective amplified PCR products, one of the two selective amplification primers should be labeled with γ-^{32}P-ATP or γ-^{33}P-ATP at the P end of the primer (oligonucleotides), which is catalyzed by T4 polynucleotide kinase. The recipe for a 50-μL labeling solution is as follows:

EcoR I selective primer	10 μL
γ-^{33}P-ATP	10 μL
Tris-HCl, pH 7.6	70 mM
$MgCl_2$	10 mM
KCl	100 mM
2-mercaptoethanol	1 mM
T4 polynucleotide kinase	10 U
Add dd H_2O to	50 μL

   Incubate at 37 °C for 1 h and then keep at 70 °C for 10 min.

   b. AFLP PCR recipe:

1:5 diluted pre-amplified template DNA	2.5 μL
Tris-HCl, pH 8.4	10 mM
KCl	50 mM
$MgCl_2$	1.5 mM

dNTP	0.2 mM
EcoR I primer (labeled)	4 pmol
Mse I primer	20 pmol
Taq DNA polymerase	1 U
Add dd H_2O to	20 μL

c. PCR thermocycle profile: 94 °C, 2 min; 94 °C, 30 s; 65 °C, 30 s, –0.7 °C per cycle starting next cycle; 72 °C, 1 min; repeat steps 2–4 12 times; 94 °C, 30 s; 56 °C, 30 s; 72 °C, 1 min; repeat steps 6–8 25 times; 72 °C, 5 min; hold at 4 °C

7. Electrophoresis of PCR products:
   a) Mix each reaction with an equal volume of 2× formamide buffer (98 % formamide, 10 mM EDTA, pH 8.0, plus bromophenol blue and xylene cyanol for tracking dye).
   b) Incubate samples at 93 °C for 4~ 5 min and immediately place on ice.
   c) Load 3 μL sample on gel (6.0 % acrylamide:bisacrylamide [19:1], 7.5 M urea, 1× TBE buffer).
   d) Run at 40–50 W with 1× TBE as the running buffer until the xylene cyanol is about 2–3 cm from the bottom of the gel and stop.
8. Autoradiography: Dry gel on filter paper and then generate autoradiograph using Kodak BioMax film (or similar product from another supplier) at –80 °C for three days.

Note: This pre-amplification step can decrease the background noise that is observed on autoradiography films.

2.2.4.2 f-AFLP

The fluorescence-based AFLP (f-AFLP) approach utilizes an automated DNA sequencer to simultaneously collect image data of multiplexed AFLP PCR products labeled with different fluorescent dyes (Myburg et al. 2001). This approach has the advantage of being more amenable to semiautomated and high-throughput genotyping than is typical AFLP. The procedures of restriction, ligation, and pre-amplification are almost the same as with typical AFLPs (see typical AFLP section), except that lower amounts of DNA (100~300 ng) should be used, and the dilution ratio of pre-amplification template DNA and selective amplification DNA is preferably 1:10 because fluorescence is more sensitive than radioactivity.

1. Selective amplification:

1:10 diluted pre-amplified template DNA	2.0 μL
Tris-HCl, pH 8.4	10 mM
KCl	50 mM
$MgCl_2$	1.5 mM
dNTP	0.2 mM
EcoR I primer (labeled)	0.4 pmol
Mse I primer	6 pmol
Taq DNA polymerase	1 U
Add dd H_2O to	10 μL

2. Prepare samples for electrophoresis on a capillary sequencer.
   a. Aliquot 1.0~1.5 μL of PCR products for each sample. Multiplex loading is allowed if the PCR reactions are performed using the same set of samples and different combinations of AFLP primers labeled with different colors (ABI detection system can promise four-color multiplex loading).
   b. Mix 0.2 μL of 400-bp or 500-bp size standard and 10 μL of Hi-Di Formamide for each sample of the loading plate, and then dispense 10.2 μL of mixture into each sample.
   c. Denature at 95 °C for 5 min and put the plate on ice.
3. Electrophorese on a DNA sequencer following the instructions of the manufacturer.

2.2.4.3 **cDNA-AFLP and HiCEP**

The cDNA-AFLP method has been commonly used to display the expression profile of the transcriptomes of a specific tissue, stress response, or developmental stage, and to construct a transcriptome map (Brugmans et al. 2002). It has also been used to develop the transcriptome-derived cDNA-AFLP markers specifically for targeting the coding regions of genes across the whole genome and, like AFLP, it does not require prior sequence information. The cDNA-AFLP protocol was derived from AFLP using a combination of two restriction enzymes, one with frequent cut sites and another one with rare cut sites. Unlike AFLP, there are some important aspects to consider when deciding on which enzymes to use for cDNA, especially the expected and observed frequency of cut sites, coverage (percentage of expressed genes that are observable), and false positives. Extremely rare or extremely frequent cutting restriction enzymes will likely not produce fingerprints suitable for interpretation. Rare cutting restriction enzymes, such as *EcoR* I, *Pst* I, and *Ase* I, which are usually used in AFLP protocol, would probably not generate many cuts in cDNA, and the power of cDNA-AFLP would be low. However, if only frequent 4-bp cutters are used and redundant bands are not eliminated as more than one band is produced, the method would suffer from a high rate of false positives. Based on applications of cDNA-AFLP, there are two kinds of cDNA-AFLP methods that have been developed. One is simply an application of the AFLP approach with minor modification to the cDNA; another is a specific approach for cDNA, which is known as "high-coverage expression profiling" (HiCEP) (Fukumura et al. 2003). The major advantages of cDNA AFLP technology are that a large fraction of the expressed genes is targeted, the technique is very sensitive, and it is consequently able to detect low-abundance transcripts.

Conventional cDNA-AFLP Analysis

1. mRNA isolation:
   a. Harvest young tissue (1–5 g) and grind to a fine powder under liquid nitrogen and store at –80 °C until RNA isolation.
   b. Isolate total RNA.

2. cDNA synthesis:
   a. Extract the poly(A)+RNA from 10 μg total RNA using poly(d)[T] 25 V oligonucleotides coupled to paramagnetic beads.
   b. Synthesize first- and second-strand cDNA according to standard protocols.
   c. Adjust the volumes for mRNA isolation and cDNA synthesis to facilitate handling in a 96-well plate format such that the whole sample can be processed simultaneously.
   d. The final volume after cDNA synthesis is 50 μL. Then, 5 μL of the reaction mix is analyzed on a 1% agarose gel to estimate the concentration, which is equalized to 10 ng μL^{-1}.
3. Template preparation: With the exception of what enzymes are used, the procedures for template preparation are similar to AFLP.

High-coverage Expression Profiling (HiCEP)

1. Digest a total of 1.5 μg of mRNA with DNase I (1.5 U mL^{-1} at 25 °C for 15 min).
2. Synthesize the first strand of cDNA using a Superscript First-Strand Synthesis System for RT-PCR (Invitrogen) with 100 pmol of 5′ biotinylated oligo(dT) primer.
3. Synthesize the second-strand cDNA.
4. Digest the double-stranded cDNA (dscDNA) with 50 U *Msp*I.
5. Ligate to 5.0 μg of *Msp*I adapter (5′-AATGGCTACACGAACTCGGTTCATGACA-3′ and 5′-CGTGTCATGAACCGAGTTCGTGTAGCCATT-3′) with 400 U T4 DNA ligase.
6. Use magnetic separation of the ligated products bearing biotin at the 5′ terminus, which will be bound to magnetic beads coated with streptavidin (Dynabeads M-280 Streptavidin).
7. Wash twice with 1.0 mL of washing buffer (5 mM Tris-HCl pH 7.5, 0.5 mM EDTA, 1.0 M NaCl).
8. Digest the cDNA fragments on the magnetic beads with 20 U *Mse* I.
9. Collect the supernatant, including the digested fragments.
10. Follow with ligation with 10.2 pmol *Mse* I adapter (5′-AAGTATCGTCACGAGGCGTCCTACTGCG-3′ and 5′-TACGCAGTAGGACGCCTCGTGACGATACTT-3′) using 400 U T4 DNA ligase in the presence of 2 U *Mse* I in 15 μL of reaction mixture.
11. Add 1485 μL of 0.1× TE (1.0 mM Tris-HCl pH 8.0, 0.1 mM EDTA), and use 1.0 μL of the resulting solution as a template for selective PCR.
12. Selective PCR and electrophoresis. The mixture in 20 mL of solution includes:
    a. pmol of *Msp* I-NN primer (5′-label- ACTCGGTTCATGACACGGNN-3′)
    b. pmol of *Mse* I-NN primer (5′-AGGCGTCCTACTGCGTAANN-3′)
    c. 40 nmol of dNTPs
    d. 1× *Taq* DNA polymerase buffer
    e. 1 U *Taq* DNA polymerase

PCR conditions: 95 °C for 1 min, 28 cycles of 95 °C for 20 s, 71.5 °C for 30 s, and 72 °C for 1 min, followed by 60 °C for 30 min.

2.2.4.4 TE-AFLP

TE-AFLP, called three-endonuclease AFLP (van Der Wurff et al. 2000), is derived from the typical AFLP method. Three endonucleases and two sets of adapters are used in TE-AFLP rather than two endonucleases as in AFLP. Adding the additional endonuclease increases the discriminatory power by reducing the number of restriction fragments, and provides discrimination at restriction sites of the third endonuclease. This method can simplify the two-step amplification to one-step amplification in fingerprinting complex genomes.

1. Restriction-ligation:

Genomic DNA	20 ng
Tris-HCl, pH 7.6	10 mM
NaCl	50 mM
$MgCl_2$	10 mM
DTT	1 mM
ATP	0.5 mM
Adapters	4 pmol
T4 DNA ligase	0.5 Weiss units
Xba I	6 U
*Bam*H I	1.25 U
*Rs a*I	1 U
Add H_2O to	20 μL

   The reaction tube or plate is incubated for 1.5 h at 30 °C.

2. PCR amplification:

Digestion-ligation solution as template	0.5 μL
Tris-HCl, pH 8.5	15 mM
KCl	50 mM
$MgCl_2$	1.5 mM
dNTPs	0.2 mM
labeled forward primer	10 pmol
Unlabeled reverse primer	10 pmol
Taq DNA polymerase	2.5 unit
Add H_2O to	50 μL

   PCR program: denaturation at 95 °C for 2.5 min; 10 cycles at 95 °C for 30 s, 70 °C for 30 s, and 72 °C for 1 min; 40 cycles at 95 °C for 30 s, 60 °C for 30 s, and 72 °C for 1 min; and a final extension at 72 °C for 20 min.

3. Electrophoresis: Same as for f-AFLP.

2.2.4.5 MEGA-AFLP

MEGA-AFLP (multiplex-endonuclease genotyping approach; Agbo et al. 2003) is also a derivation of the conventional AFLP technique. In this approach, four or more endonucleases in combination with one pair of adapters/primers are used for accessing more robust loci of restriction sites than can the two endonucleases used in conventional AFLP. The principle of this method is that some endonucleases create cohesive ends that are compatible with the overhang sites created by other endonucleases. Only one pair of adapters is used for ligation to allow amplification of the fragments by a pair of cognate primers in PCR, permitting robust and stringent reaction conditions.

1. Double digestion.
   a. The first digestion uses *Bglα* and *EcoR* I.

Genomic DNA	100 ng
Tris-HCl, pH 7.5 at 37 °C	50 mM
NaCl	100 mM
$MgCl_2$	10 mM
BSA	0.1 mg mL^{-1}
Bglα	10 U
EcoR I	10 U
H_2O to	20 μL

   The reaction is incubated for 4 h at 37 °C.
   b. Add 30 mL 2-propanol to precipitate digested fragments.
   c. The digests are dissolved in 10 μL distilled water for the second digestion.
   d. Second digestion:

Digested fragments	10 μL
Tris-HCl, pH 7.5 at 37 °C	10 mM
NaCl	50 mM
$MgCl_2$	10 mM
BSA	0.1 mg mL^{-1}
BclI	10 U
MunI	10 U
Add H_2O to	20 μL

   The reaction is incubated for 4 h at 37 °C.
   e. The digests are precipitated using 2-propanol and again dissolved in 10 μL distilled water.
2. Ligation:

Digested DNA	10 μL
Tris-HCl, pH 7.5	330 mM
$MgCl_2$	25 mM
DTT	5 mM
ATP	5 mM
Bglα adapters	20 pmol
MunI adapters	20 pmol
T4 DNA ligase	400 U
Add H_2O to	20 μL

   The mixture is incubated for 2 h at 25 °C.

3. Pre-selective amplification:

1:1 diluted ligation product	4 μL
Tris-HCl, pH 9.0	10 mM
KCl	5 mM
Triton X-100	0.1%
Gelatin	0.01% w/v
$MgCl_2$	2.5 mM
dNTPs (for each)	0.2 mM
*Bgl*II pre-amp primer	5 pmol
*Mun*I pre-amp primer	5 pmol
Taq DNA polymerase	1 U
Add H_2O to	20 μL

PCR program: denature at 95 °C for 2 min; 20 cycles of PCR at 95 °C for 30 s, 56 °C for 30 s, and 72 °C for 2 min; and 40 cycles at 95 °C for 30 s, 60 °C for 30 s, and 72 °C for 1 min.

4. Selective amplification:

1:20 diluted pre-amplification	4 μL
Tris-HCl, pH 9.0	10 mM
KCl	5 mM
Triton X-100	0.1%
Gelatin	0.01% w/v
$MgCl_2$	2.5 mM
dNTPs (for each)	0.2 mM
*Bgl*II primer	5 pmol
Labeled *Mun*I primer	5 pmol
Taq DNA polymerase	1 U
Add H_2O to	20 μL

PCR program: denature at 95 °C for 2 min; 20 cycles of PCR at 95 °C for 30 s, 56 °C for 30 s, and 72 °C for 2 min; 40 cycles at 95 °C for 30 s, 60 °C for 30 s, and 72 °C for 1 min; and a final extension at 60 °C for 30 min.

5. Electrophoresis: Same as for f-AFLP.

2.2.4.6 MITE-AFLPs

Plant genomes have numerous families of transposable elements (TEs) that frequently comprise the majority of genomic DNA and spread throughout the genome. Miniature inverted-repeat transposable elements (MITEs) are one of the major transposable element families recently discovered in a number of plant genomes (Casa et al. 2000). Each family of TEs has its own structural features that can be utilized to distinguish one from the other, and to identify molecular markers preferentially anchored in TEs. The MITE-AFLP technique has been developed to assess genetic variation and phylogenetic analysis in species such as rice, wheat, and maize because of the presence of MITE sequences in high copies across whole genomes. It has been demonstrated that MITEs tend to be located in the 3′

end of genes and are relatively polymorphic across inbred lines. The MITE-AFLP technique is amenable for studying genetic variations and relationships within or among species (Park et al. 2003).

1. Genomic DNA digestion: Digest 100 ng genomic DNA with *Mse* I.
2. Ligation with an adapter of *Mse* I: 5′-GACGATGAGTCCTGAG-3′ and 5′-TACT-CAGGACTCAT-3′.
3. Pre-amplification: Use *Mse* I pre-amp primer 5′-GACGATGAGTCCTGAGTAA-3′and one consensus sequence domain of MITEs. For example, for MITE element *Pangrangja*, the primer is 5′-AARCAGTTTGACTTTGATC-3′ (R=A,G). For MITE element Hbr, the primer is 5′-GATTCTCCCCACAGCCAGATTC-3′.

Digested ligation product	5 μL
10× buffer	5 μL
$MgCl_2$	1.5 mM
dNTPs (for each)	0.2 mM
Mse I pre-amp primer	0.5 μM
MITE primer	0.5 μM
Taq DNA polymerase	1.5 U
Add H_2O to	50 μL

PCR program: 72 °C for 2 min; denaturation at 95 °C for 3 min; 25 cycles of PCR at 94 °C for 30 s, 53 °C for 30 s, and 72 °C for 1 min; and a final extension at 72 °C for 5 min.

4. Selective amplification:

1:50 diluted pre-amplified products	3 μL
10× buffer	5 μL
$MgCl_2$	1.5 mM
dNTPs (for each)	0.2 mM
Mse I selective primer	0.5 μM
MITE primer	0.5 μM
Taq DNA polymerase	1.0 U
Add H_2O to	30 μL

PCR program: denature at 94 °C for 5 min; 10 cycles of touchdown PCR at 94 °C for 30 s, 62 °C for 30 s, and 72 °C for 1 min with a decrease of 1 °C of the annealing temperature in each cycle; 26 cycles at 94 °C for 30 s, 53 °C for 30 s, and 72 °C for 1 min; and a final extension at 72 °C for 5 min.

2.2.4.7 **AFLP Conversion**

Although AFLP markers can be used for many applications, many AFLP markers are redundant and dominant and cannot perform as well in high-throughput genotyping in an automated format as SSR and STS markers allow in MAS. There is a strong need to convert specific AFLP markers into other easy-to-use, codominant, and sequence-specific markers (Meksem et al. 2001; Nicod and Largiader 2003; Weerasena et al. 2003), such as CAPS markers or SCAR or SNP markers, because the AFLP assay is more complex and AFLP primers cannot be directly used for amplifying the polymorphic markers on genomic DNA or cDNA. Therefore, it is necessary to have a reliable and efficient protocol for conversion of AFLP markers into high-throughput, single-locus PCR markers. However, the design of new PCR primers for a locus-specific marker requires information on the DNA sequence of the AFLP band, and a locus-specific marker should have the capability to distinguish between diverse alleles. In order to isolate an AFLP fragment

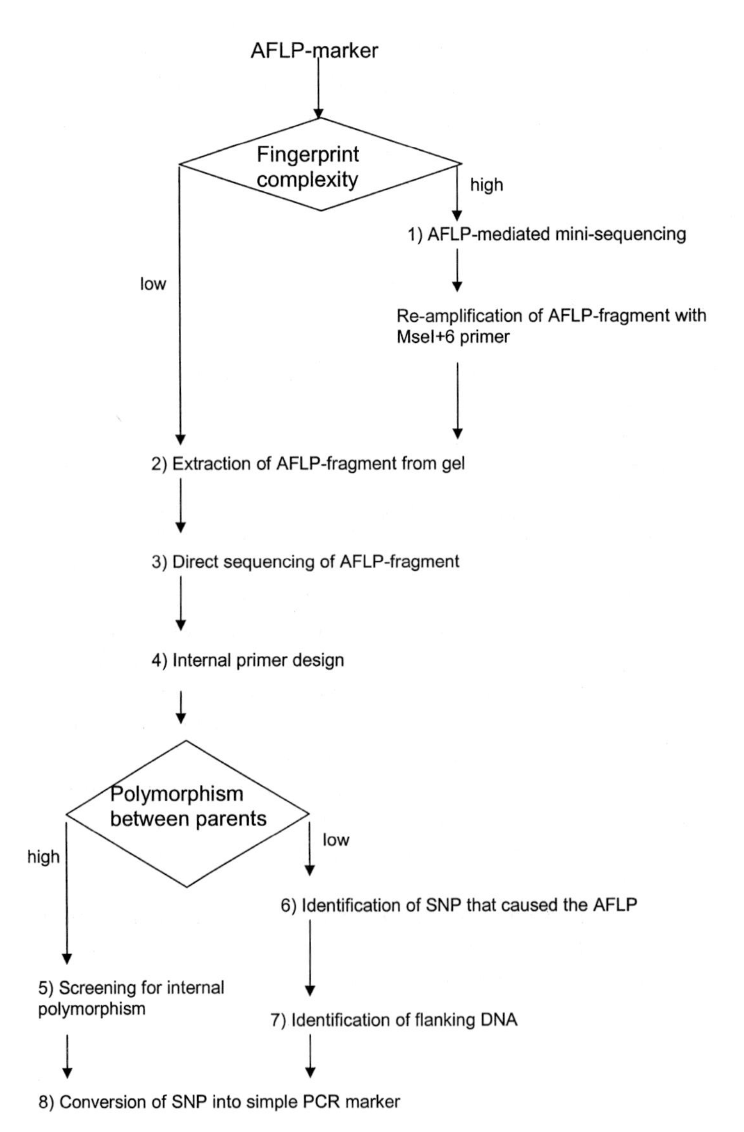

Figure 2.1. Flowchart of the steps of the protocol to convert any AFLP marker into a single-locus PCR-based marker assay (Brugmans et al. 2003).

from multiple fragments with sizes almost the same as the AFLP fragment of interest, more complicated procedures have to be taken. Polymorphic AFLP bands are excised from polyacrylamide gels by cutting out the band. The DNA in the gel fragment is extracted and then subjected to re-amplification using the same primer combinations as in the selective amplification. Re-amplified AFLP bands are either sequenced directly or cloned. Based on the internal sequence variations of AFLP bands, PCR primers can be designed for genomic amplification of the internal sequence, which can be used in generating sequence-specific CAPS, dCAPS, or STS markers. When searching for internal polymorphisms with a lower success rate, an additional step will be performed for identification of the SNP in the restriction sites or selective nucleotides that caused the original AFLP. As long as the AFLP marker is a single-locus marker, there is no obstacle to exploit the unique AFLP causing SNP to allow for the conversion of any AFLP marker into a simple PCR-based marker. The flowchart of the protocol steps is overviewed in Figure 2.1 (Brugmans et al. 2003).

2.2.5 REMAP and IRAP

The retrotransposon-microsatellite amplified polymorphism (REMAP) and inter-retrotransposon amplified polymorphism (IRAP) techniques are retrotransposon-based DNA marker systems (Kalendar et al. 1999). Retrotransposons play an important role in genome components of many plant species, particularly in grasses (Vicient et al. 2001). Their abundance, dispersion across the nuclear genome, and insertional activity indicate that they are related to plant genome evolution, and also provide an excellent basis for the development of DNA-based marker systems that rely on PCR to generate fingerprints. The fingerprints are multi-locus profiles revealing polymorphisms in the insertion of members of given families of retrotransposons. The IRAP technique generates PCR products from retrotransposons inserted near enough to each other so as to allow efficient amplification, and generally uses primers matching the outer segments of the long terminal repeats (LTRs). The REMAP technique uses one LTR primer, together with a primer designed for annealing to the 3′ end of a stretch of simple sequence repeats (SSR), and detects retrotransposons inserted near SSRs. Therefore, REMAP can amplify a pool of three different sequences: fragments situated between an LTR and a microsatellite locus, sequences situated between two microsatellite loci, and sequences situated between retroelements. REMAP and IRAP are efficient at detecting insertion events. The following protocol is an example developed from the *BARE-1* retrotransposon family (Kalendar et al. 1999).

2.2.5.1 IRAP

1. Design primers: The primers should be located to conserve stretches of the *BARE*-1 LTR, particularly at the primers' 3′ ends.
2. PCR:

Genomic DNA	20 ng
Tris-HCl pH 8.8	20 mM
KCl	50 mM
Tween-20	0.01 %
Glycerol	5 %
$MgCl_2$	2.0 mM
dNTPs (for each)	0.2 mM
Forward primer	5 pmol
Reverse primer	5 pmol
Taq DNA polymerase	1.0 U
Add H_2O to	20 μL

   PCR program: 94 °C for 2 min; 1 cycle at 94 °C for 30 s, 60 °C for 30 s, ramp +0.5 °C s^{-1} to 72 °C; 30 cycles of 72 °C for 2 min +3 s; 72 °C for 10 min; and 4 °C.
3. Electrophoresis: Load PCR products on 2 % NuSieve 3:1 agarose gel and detect by ethidium bromide staining.

2.2.5.2 REMAP

1. Primer design: One of the primers should match the retrotransposon sequence, like *BARE*-1a, the reverse primer 5′-*GGAATTC*ATAGCATGGATAATAAAC-GATTATC-3′ corresponding to nt 369–393 of the *BARE*-1a LTR with the addition of a tail containing an *Eco*R I site at the 5′ end. Forward primers can be designed for microsatellite sequences based on two dinucleotide repeats, (GA)n and (CT)n, and three trinucleotide repeats, (CAC)n and (GTG)n.
2. PCR reaction and program and product resolution are the same as in IRAP.

2.2.6 SRAP

Sequence-related amplified polymorphism (SRAP) is a new PCR-based marker technique that has the features of simplicity, reliability, moderate throughput ratio, and high percentage of fragments from coding regions. It preferentially targets open reading frames (ORFs) in the genome and results in a moderate number of codominant markers. SRAP markers have been successfully used to construct a genetic map and a transcriptome map in *Brassica oleracea* (Li and Quiros 2001) and to evaluate genetic diversity of germ plasms (Budak et al. 2003; Ferriol et al. 2003). Sequencing of the markers generated by the cDNA-SRAP approach allows it to be a potentially powerful tool for aligning genes of a well-characterized model species to other crops for gene discovery.

1. Primer design: The design of SRAP primers is based on the fact that there is a high percentage of "GC" content within coding regions in a genome and a high percentage of "AT" content at the 3′ untranslated regions (3′ UTR). Each PCR reaction needs two primers. The forward primer consists of 17 bases that contain a core sequence of 14 bases rich in G and C and three selective nucleotides at the 3′ end. This primer preferentially anneals onto coding regions that tend to be rich in C and G. The reverse primer consists of 18 bases that contain a core sequence of 15 bases rich in A and T and three selective nucleotides at the 3′ end. This primer preferentially anneals onto un-coding regions that tend to be rich in T and A. Examples of forward primers include me1, 5′-TGAGTCCAAACCGGATA-3′ and me2, 5′-TGAGTCCAAACCGGAGC-3′, and examples of reverse primers include em1, 5′-GACTGCGTACGAATTAAT-3′ and em2, 5′-GACTGCGTACGAATTTGC-3′.
2. The PCR reaction mixture contains:

Genomic DNA or cDNA	20 ng
Tris-HCl, pH 8.3	10 mM
KCl	50 mM
$MgCL_2$	1.5 mM
dNTPs	0.2 mM
Forward primer	0.3 μM
Reverse primer	0.3 μM
Taq DNA polymerase	1 unit
Add H_2O to	25 μL

   PCR program: 94 °C for 5 min; 5 cycles at 94 °C for 1 min, 35 °C for 1 min, and 72 °C for 2 min; 30 cycles at 94 °C for 1 min, 50 °C for 1 min, and 72 °C for 2 min; 72 °C for 5 min; and 4 °C.
3. Electrophoresis: Separation of amplification fragments can be accomplished on 12% polyacrylamide gels (acrylamide:bisacrylamide [29:1], 1× TBE) at 500 V for 11 h or run on a sequencer.

2.3 Discussion

The foregoing text reviews the main molecular maker detection techniques – hybridization-based and amplification-based – that are currently in vogue to study genomic diversity and to construct genetic maps in many plant species. In utilizing any DNA marker technique for a particular organism, there are many factors that need to be taken into account. These broadly include polymorphism, dominance, efficiency, simplicity, reliability, cost, reproducibility, prior sequence information, and task. As reviewed, several techniques such as RFLP-PCR, RAPD, SSR, and AFLP have been introduced as amplification-based protocols to further enhance the resolution and detection of polymorphic DNA. However, the RAPD protocols are relatively simple, straightforward, and less time-consuming compared with RFLP and AFLP, but they are not very reliable or reproducible. Each of these techniques is best suited for specific tasks and should be chosen accordingly. SSR markers have the greatest value for poorly polymorphic self-pollinated species and co-dominant features, whereas AFLP has great utility in the construction of genetic maps in complex genomes but is a dominant marker system. Indeed, the wider potential utility of the deviations of AFLP such as cDNA-AFLP and MITE-AFLP should unequivocally target genes of interest and transposable repeat elements that are spread throughout the genome. AFLPs have a great advantage in terms of ease of use in a laboratory. Indeed, standard oligonucleotide primer sets considerably reduce the effort and consumables. Once the technique has been set to work in a laboratory, data can be produced for different species by using exactly the same reagents and conditions. However, the drawback is that AFLP markers are generally dominant and generate anonymous loci. The dominance problem can be partially overcome by the possibility of quickly generating high-density maps, establishing the linkage relationship between markers from a linkage group and a phenotype, and focusing only on that one particular region, leaving the rest of the genome aside. One major problem with RAPDs is their low reproducibility, depending highly on the PCR conditions. In contrast, AFLP markers can be a good choice for QTL mapping, high-density genetic map construction, and fine mapping. With the advance of SNP markers, genetics study will become easier.

References

Agbo EC, Duim B, Majiwa PA, Buscher P, Claassen E, Te Pas MF (2003) Multiplex-endonuclease genotyping approach (MEGA): a tool for the fine-scale detection of unlinked polymorphic DNA markers. Chromosoma. 111(8):518–524.

Botstein B, White RL, Skolnick M, Davis RW (1980) Construction of a genetic linkage map in man using restriction fragment length polymorphisms. Am J Hum Genet 32:314–331.

Brugmans B, Fernandez del Carmen A, Bachem CW, van Os H, van Eck HJ, Visser RG (2002) A novel method for the construction of genome wide transcriptome maps. Plant J. 31:211–222.

Brugmans B, van der Hulst RG, Visser RG, Lindhout P, van Eck HJ (2003) A new and versatile method for the successful conversion of AFLP markers into simple single locus markers. Nucleic Acids Res. 31:e55.

Budak H, Shearman RC, Parmaksiz I, Gaussoin RE, Riordan TP, Dweikat I (2003) Molecular characterization of Buffalograss germplasm using sequence-related amplified polymorphism markers. Theor Appl Genet. 2003 Sep 13 [Epub ahead of print].

Bui MM and She JX (1996) Analysis of microsatellite polymorphism using the polymerase chain reaction. In "Practical protocols in molecular biology" edited by Yongming Li and Yuqi Zhao, 1996, Science Press, Beijing, New York.

Caetano-Anolles G, Bassam BJ, Gresshoff PM (1991) DNA amplification fingerprinting using very short arbitrary oligonucleotide primers. BioTechnology 9:553–557.

Casa AM, Brouwer C, Nagel A, Wang L, Zhang Q, Kresovich S, Wessler SR (2000) Inaugural article: the MITE family heartbreaker (Hbr): molecular markers in maize. Proc Natl Acad Sci U S A. 97:10083–9.

Cregan PB, Bhagwat AA, Akkaya MS, Jiang RW. 1994, Microsatellite fingerprinting and mapping of soybean. Methods Mol. Cell Biol. 5:49–61.

Feinberg AP and Vogelstein B. (1984) A technique for radiolabeling DNA restriction fragments to a high specific activity. Anal. Biochem. 132:6–13.

Ferriol M, Pico B, Nuez F (2003) Genetic diversity of a germplasm collection of Cucurbita pepo using SRAP and AFLP markers. Theor Appl Genet. 107:271–82.

Fukumura R, Takahashi H, Saito T, Tsutsumi Y, Fujimori A, Sato S, Tatsumi K, Araki R, Abe M. (2003) A sensitive transcriptome analysis method that can detect unknown transcripts. Nucleic Acids Res. 31:e94.

Godwin ID, Aitken EA, Smith LW (1997) Application of inter simple sequence repeat (ISSR) markers to plant genetics. Electrophoresis 18:1524–8.

Hayden MJ, Sharp PJ. (2001) Sequence-tagged microsatellite profiling (STMP): a rapid technique for developing SSR markers. Nucleic Acids Res. 29:E43–3

Hayden MJ, Good G, Sharp PJ. (2002) Sequence tagged microsatellite profiling (STMP): improved isolation of DNA sequence flanking target SSRs. Nucleic Acids Res. 30:e129

Hearne CM, Ghosh S, Todd JA (1992) Microsatellites for linkage analysis of genetic traits. Trends Genet 8:288–294.

Kalendar R, Grob T, Regina M, Suoniemi A, Schulman A. (1999) IRAP and REMAP: two new retrotransposon-based DNA fingerprinting techniques. Theor Appl Genet 98:704–711.

Lai E, Riley J, Purvis I, Roses A. (1998) A 4-Mb high-density single nucleotide polymorphism-based map around human APOE. Genomics 54:31–8.

Li G, Quiros CF (2001) Sequence-related amplified polymorphism (SRAP), a new marker system based on a simple PCR reaction: its application to mapping and gene tagging in Brassica. Theor Appl Genet (2001) 103:455–461.

Li YC, Korol AB, Fahima T, Beiles A, Nevo E (2002) Microsatellites: genomic distribution, putative functions and mutational mechanisms: a review. Mol Ecol. 11:2453–65.

Litt M, Luty JA. (1989) A hypervariable microsatellite revealed by in vitro amplification of a dinucleotide repeat within the cardiac muscle actin gene. Am J Hum Genet. 44:397–401.

Lyamichev V, Brow MAD, Dahlberg JE. (1993) Structure-specific endonucleolytic cleavage of nucleic acids by eubacterial DNA polymerases. Science 260:778–783.

Meksem K, Ruben E, Hyten D, Triwitayakorn K, Lightfoot DA. (2001) Conversion of AFLP bands into high-throughput DNA markers. Mol Genet Genomics. 265:207–14.

Mueller UG,Wolfenbarger LL (1999) AFLP genotyping and fingerprinting. Trends in Ecology and Evolution 14:389–394.

Myburg AA, Remington DL, O'Malley DM, Sederoff RR, and Whetten RW (2001) High-throughput AFLP analysis using infrared dye-labeled primers and an automated DNA sequencer. BioTechniques 30:348–357.

Nagaraju J, Kathirvel M, Subbaiah EV, Muthulakshmi M, Kumar LD (2002) FISSR-PCR: a simple and sensitive assay for highthroughput genotyping and genetic mapping. Mol Cell Probes 16:67–72.

Neff MM, Neff JD, Chory J, Pepper AE (1998) dCAPS, a simple technique for the genetic analysis of single nucleotide polymorphisms: experimental applications in Arabidopsis thaliana genetics. Plant J. 14:387–92.

Nicod JC , Largiader CR (2003) SNPs by AFLP (SBA): a rapid SNP isolation strategy for non-model organisms. Nucleic Acids Res. 31: e19

Palazzolo MJ, Sawyer SA, Martin CH, Smoller DA, Hartl DL (1991) Optimized strategies for sequence-tagged-site selection in genome mapping. Proc Natl Acad Sci U S A. 88:8034–8.

Park KC, Kim NH, Cho YS, Kang KH, Lee JK, Kim NS (2003) Genetic variations of AA genome Oryza species measured by MITE-AFLP. Theor Appl Genet. 107:203–9.

Reineke A and Karlovsky P (2000) Simplified AFLP protocol: replacement of primer labeling by the incorporation of alpha-labeled nucleotides during PCR. BioTechniques 28:622–623.

Scottdl, Walker MD, Clarck C, Prakash CS and Deahl KL (1998) Rapid assessment of primer combinations and recovery of AFLP^{TM} products using ethidium bromide staining *Plant Molecular Biology Reporter* 16:41–47.

van Der Wurff AW, Chan YL, van Straalen NM, Schouten J (2000) TE-AFLP: combining rapidity and robustness in DNA fingerprinting. Nucleic Acids Res. 28:E105.

Vicient CM, Jaaskelainen MJ, Kalendar R, Schulman AH. (2001) Active retrotransposons are a common feature of grass genomes. Plant Physiol. 125:1283–92.

Vos P, Hogers R, Bleeker M, Reijans M, Van de Lee T, Hornes M, Frijters A, Pot J, Pelman J,

M. Kuiper M, Zabeau M (1995) AFLP: A new technique for DNA fingerprinting. Nucleic Acids Research. 23:4407–4414.

Weerasena JS, Steffenson BJ, Falk AB (2003) Conversion of an amplified fragment length polymorphism marker into a codominant marker in the mapping of the *Rph15* gene conferring resistance to barley leaf rust, *Puccinia hordei* Otth. Theor Appl Genet. 2003 Oct 2 [Epub ahead of print]

Welsh J, McClelland M (1990) Fingerprinting genomes using PCR with arbitrary primers. Nucl Acids Res 8:7213–7218.

Wiesner I and Wiesnerová D (2003) Insertion of a reamplification round into the ISSR-PCR protocol gives new flax fingerprinting patterns. Cellular & Molecular Biology Letters 8:743–748.

Williams JGK, Kubelik AR, Livak KJ, Rafalski JA, Tingey SV (1990) DNA polymorphisms amplified by arbitrary primers are useful as genetic markers. Nucl Acids Res 18:6531–6535.

Williams MNV, Pande N, Nair S, Mohan M, Bennett J. (1991) Restriction fragment length polymorphism analysis of polymerase chain reaction products amplified from mapped loci of rice (*Oryza sativa* L.) genomic DNA. Theor Appl Genet 82:489–498.

Zietkiewicz E, Rafalski A, Labuda D (1994) Genomic fingerprinting by simple sequence repeat (SSR)-anchored polymerase chain reaction amplification. Genomics 20:176–183.

3
Methods and Software for Genetic Mapping

James C. Nelson

Overview

The aims of this chapter are to explain in general terms how plant genetic marker genotype data are used for constructing linkage maps; to describe the evolution and variety of statistical methods for combining marker and map data with phenotypic data to identify putative genes controlling quantitative trait loci (QTLs); to survey the software resources available for these two kinds of mapping; and finally to offer remarks on current directions in these research areas. The chapter is aimed at readers with basic but not extensive statistical- and quantitative-genetics knowledge, who wish to turn genetic data intelligently into a map and/or QTL model using available software tools or programming the methods for themselves. Mathematical details are omitted, and reference to mammalian genetics where much theory originated is out of our scope.

Abstract

Multilocus genetic mapping may be separated into three problems: grouping, ordering, and distance estimation, of which the second is computationally the most interesting. With the increasing abundance of DNA markers and availability of DNA sequences for several species, the practical importance of these problems declines and interest shifts to comparative mapping and recombination studies. QTL mapping has evolved from the calculation of simple correlation statistics to the identification of elaborate multigene models by methods such as Markov Chain Monte Carlo (MCMC) for exploring high-dimension parameter spaces. Recent trends are integration of QTL data with other genomic data, such as for gene expression and metabolic pathways, and expansion of QTL models to exploit linkage disequilibrium in populations and pedigrees. Computer software applications for linkage and QTL mapping are numerous but are limited in analytical scope, quality of interface, and opportunities for extension. An open-source development platform might allow better exploitation of theoretical advances.

The Handbook of Plant Genome Mapping.
Genetic and Physical Mapping. Edited by Khalid Meksem, Günter Kahl

ISBN 3-527-31116-5

3.1 Introduction

Preceding chapters have described the genetic materials and laboratory operations used to develop a set of DNA-marker genotype data from which a genetic map can be developed. Chapter 1 introduced genetic recombination as the biological event that governs the segmental distribution of grandmaternal and grandpaternal DNA along individual chromosomes and that provides us with a measure of genetic distance between a pair of loci in the form of linkage.

Linkage mapping in self-pollinating crops poses few practical problems, though, like any problem with a kernel of formal complexity, it retains theoretical attraction. The rich catalog of mapped crop species described nearly a decade ago (Paterson 1996) has only expanded, building on itself as genomic relationships between related taxa emerge. For outcrossing and autopolyploid systems, software tools remain very basic or lacking. Beyond our scope is the current interest in interspecies map comparison and in the use, treated elsewhere in this volume, of genetic maps as assembly frameworks for physical maps.

A more fertile area for theoretical advances is QTL mapping. It has long been evident that a more challenging problem than estimating the parameters of a QTL model is identifying the model in the first place, i. e., how many QTLs, where they are on the map, and how they are interacting. Much progress has been made in recent years, but here too, satisfactory software tools for applying these advances lag well behind theory. A perennial challenge has been to assess epistasis between QTLs, because of the potentially limitless number of gene interactions. Much interest now is in combining highly parallel gene expression assays with QTL mapping as a way of grounding gene-interaction models in fine-scale experimental observation. Also of growing interest is the information about genetic control of quantitative traits that is provided by linkage disequilibrium in natural as well as experimental populations and pedigrees.

3.1.1 Methods and Tools for Genetic Linkage Mapping in Plants

3.1.1.1 Statement of the Problem

Linkage between two loci is the probability r of each crossover occurring between them at a single meiosis. We estimate it by tabulating the distinct genotype combinations observed at these loci (each of which has a probability dependent on r and on the crossing design of our experiment) and then finding a value for r that maximizes the likelihood of the observed frequencies of these combinations. Reference to maximum likelihood (ML) methods will recur below; for mathematical details of much of the statistics described, see Liu (1998). An additive measure of distance can be obtained from linkage via an algebraic mapping function.

Let us now suppose that we have estimated the r for each pair of markers in our dataset and, if our chosen method requires it, converted these linkages into distances. Our aims are to group loci into candidate chromosome segments – linkage groups – and arrange the loci within these groups at relative positions most compatible with their linkages. Sometimes we will have multiple experimental datasets with disparate crossing designs but some common markers. For any design, we want some way of evaluating the quality of the resulting maps, i. e., how much better one map is than others, and which loci account for the difference.

We'll ignore that in practice we often have more information than is supplied by the marker dataset alone. We'll also neglect complicating phenomena such as genetic interference, recombination suppression, and special chromosome types and aberrations.

3.1.2
Locus Grouping

Grouping is a matter of setting admission rules and requiring any candidate locus ineligible for any existing group to initiate a new group. Usual admission rules are based on upper linkage thresholds and lower limit of detection (LOD) score thresholds for linkage with some other member of the group. These LODs are measures of informativeness, based on r and the number of observations used to estimate it. The resulting set of groups is uniquely determined by the admission thresholds, which the analyst will set to give a balance between number and size of groups that experience and prior knowledge of the genetic material will suggest. Too-large groups lead to unstable locus orders; too-small groups result in a fragmented map.

3.1.3
Locus Ordering

Ordering is the central problem in linkage mapping, and also the most interesting in the sense that for groups of even modest size there is no sure way to find the best of $N!/2$ possible orders. Here "best" depends on the criterion (objective function) adopted to evaluate candidate orders based on sums or products of the individual linkage or LOD statistics for adjacent locus pairs. What is tricky is proposing candidate orders by exploring the "order space" in some manner expected to lead to progressively better solutions. Preliminary ordering based on two-point linkage distances can be refined by re-estimation of multipoint linkages along with local reversals ("rippling") in search of a globally improved order.

One of the simplest algorithms, seriation (Ellis 1997; Doerge 1996; Crane and Crane 2004a), involves growing an order outward from the most tightly linked locus pairs. It is "greedy," in the sense that each successive addition is made to optimize the current order without consideration of the loci not yet added or removal of any previously added. A more elaborate greedy algorithm is MAPMAKER's (Lander et al. 1987) method of finding all three-locus orders then excluding the most unlikely and proceeding by evaluating permissible multilocus orders built from the remaining ones. Final optimizing is then done by rippling. The method of JoinMap (Stam 1993; Stam and van Ooijen 1995) is also sequential, adding the most informative markers one at a time, accepting only if a goodness-of-fit test shows an improvement, and shuffle-optimizing at each step. A similar strategy is used by Map Manager QTX (Manly and Olson 1999), hereafter abbreviated MMQTX, which unites grouping and ordering operations by identifying the best (if any) insertion point for a new marker as it is added to a group and optimizes by rippling after all markers are distributed.

A few methods of map construction are not "bottom-up" but instead work by updating, with various swap/evaluate operations, some initial order of all loci in a group, using approaches derived from the well-known and closely analogous "traveling salesman problem" (TSP). The TSP requires finding the shortest tour passing once through each of a given set of cities. Simulated annealing (SA), used by GMendel (Liu and Knapp 1990), employs a "temperature" parameter that governs the amount of change in a configuration that may be applied at each step, as well as the probability of acceptance of a configuration with a lower (more unfavorable) score than the current one. As the configuration stabilizes at some temperature, the system is "cooled," changes become less extreme, and unfavorable changes are less readily accepted. CarthaGene (Schiex and Gaspin 1997) offers SA as well as greedy methods, all derived from TSP heuristics, and also implements a genetic algorithm (GA) for "evolving" a population of locus orders. GAs are intended to emulate nature's system for combinatorial optimization of species fitness using sexual recombination with some mutation rate, followed by selection based on a fitness function. Another evolutionary algorithm, but one relying on asexual reproduction and thus only on mutational variation, was reported to be fast and effective (Mester et al. 2003).

Quality assessment in mapping can be incorporated into methods for successive assembly described above. CarthaGene provides a "framework"-building method permitting the controlled exclusion of "bad" (unstably ordered) markers. In MAPMAKER, suspect loci, such as those expanding the map by introducing excessive double crossovers, can be identified by a "drop-one" feature. Where a fast ordering algorithm is used (Mester et al. 2003), this operation can be automated with bootstrapping, or a Monte Carlo algorithm such as SA can be run repeatedly as in GMendel. Most of the above programs will produce a number of "best" maps. Then the distribution of "best" scores (e.g., multipoint likelihood or LOD) itself is a guide to reliability, with similar scores being evidence for at least one unstable local order. Inspection often reveals the responsible markers.

Two novel ordering methods treat locus ordering as the reduction of a high-dimensional ($N-1$, in the case of N loci) space to one dimension, with the constraint of minimum loss of position information. Metric multidimensional scaling (Ukai et al. 1990) is used by MAPL (Ukai 1997) to collapse the genetic distance matrix into a 2D plot of points having the property that distances between all pairs of points are optimally close in a least-squares sense to their distances in the original matrix. The analyst is asked to select an end locus, and the rest of the loci are then added in a presumably greedy way, though there seems to be no built-in way to assess quality. The second algorithm (Newell et al. 1995) uses distance-geometric methods to convert the distance matrix into points in $N-1$-dimensional space and then a numerical gradient-descent method to arrive at least-squared-error 1D coordinates. Uncertainty in a marker position is calculated from the uncertainty in its distances from the other markers and can be displayed graphically. The computer program offered, DGMAP, does not appear to have been used for plants or to be available any longer.

3.1.4 Multilocus Distance Estimation

Once a locus order has been obtained, the problem remains of computing interlocus distances. Naive methods retain the original distances between adjacent markers, an unsatisfactory resort since these were based on ML approximations and partial information to begin with. One improvement described by Jensen and Jorgensen (1975), adapted by JoinMap, and reinvented by Newell et al. (1995) consists of calculating the distances having least-squared errors from the two-point distances, while giving more weight to distance estimates based on more information. MAPMAKER and CarthaGene update the linkage estimate directly, using an EM algorithm (a class of convergent algorithm for estimating mutually dependent unknown parameters by updating each in turn from the others). Both methods increase the likelihood of the final map. GMendel uses a simpler and somewhat less stable method that adjusts the raw distance estimate between two loci to show least absolute deviation from that expected from the unweighted distances between all flanking loci including the two themselves.

3.1.5 Using Variant and Mixed Cross Designs

3.1.5.1 Outbreeding Species

Line crosses in diploid or allopolyploid species tolerant of inbreeding account for common mapping designs in plants. For obligate out-breeders, however, linkage estimation must distinguish between coupling and repulsion phase, for both codominant and dominant markers, and must accommodate as many as four (in a diploid) alleles segregating at a locus. Though methods were elaborated by Ritter et al. (1990), Ritter and Salamini (1996), and Maliepaard et al. (1997), available software does not handle phase-unknown data. The analyst must infer the seven possible marker segregation types from two-locus genotype frequencies. The next step is either to build separate maps in MAPMAKER and join them by hand with "allelic bridges" – markers common to two classes segregating alleles from both parents – or to submit the class-coded markers to software packages such as JoinMap or CarthaGene, which accept multiple segregation types in a single cross. MMQTX will handle mixed segregation and will flip phase, but it cannot accommodate more than two alleles at a locus.

3.1.5.2 **Autopolyploid Species**

For autopolyploids, where allele doses in conventional crossing designs can amount to half the ploidy number, the phase problem is compounded by dose uncertainty. Dose can be inferred from segregation frequencies and linkage estimated for simplex (single-dose) (Wu et al. 1992) and higher-plex markers in tetraploids (Hackett et al. 1998) and octoploids (Ripol et al. 1999). General mapping software will not make these calculations, and one must either build a map solely from simplex markers and add the higher-plex markers by hand or pre-compute linkages and their LODs and supply them to a program such as JoinMap for assembly. Polylink (He et al. 2001) computes single two-point *r* values for even-ploidy designs, but more useful is TetraploidMap (Hackett and Luo 2002; Hackett et al. 2003; Julier et al. 2003), which computes two-point linkages and LODs, identifies linkage groups, and uses SA and rippling for ordering. It can identify double reduction (segregation of sister chromatids to the same gamete, owing to multivalent meiotic pairing) but cannot compute recombination in this case. Wu et al. (2001) have provided a full model.

3.1.5.3 **Combining Datasets**

It is also desirable to compute a linkage map from independent datasets containing common markers. JoinMap computes common two-point linkages based on the assumption of identical underlying linkages in all populations, supplying LOD scores based on the variation in observed values for a given locus pair. CarthaGene permits the assumption of only a common order but not necessarily common interlocus distances, leading to population-specific maps containing the merged markers. GMendel emits heterogeneity tests aimed at alerting the analyst to the danger in pooling but computes joint *r* values based on pooled observed and expected two-locus genotype frequencies, as an extension of the one-population ML methods.

Increasing interest in comparative mapping between species has led to a variety of software for map viewing and comparison, based solely on locus positions in the maps to be compared. These are outside of our focus on construction and evaluation of linkage maps from segregation data, though one application, MapInspect (van Berloo and Kraakman 2002), deserves note as a tool for assessing "tension" indicated by large differences in two-point distances expected to be similar in different maps. This tension is analogous to that between observed and multipoint distances that objective functions seek to minimize within any given dataset.

3.1.6
Linkage-mapping Software Availability, Interfaces, and Features

Of the software packages mentioned above, all but JoinMap, a commercial program, are freely available. With respect to user interfaces (UIs), MAPMAKER (except for the Apple Macintosh version), FLIPPER (Crane and Crane 2004b), GMendel, and TetraploidMap all have command-line UIs, while the others mentioned have simple to elaborate graphical UIs (GUIs). MAPMAKER and CarthaGene are scriptable. MMQTX offers an elaborate set of analyses including QTL mapping (a feature also provided in MAPL), while the others are single-purpose. All handle standard crossing designs (backcross, F_2, doubled haploids, recombinant inbreds), while MMQTX handles a further suite of designs including advanced intercrosses and backcrosses. MMQTX and CarthaGene accept radiation-hybrid data as well. MMQTX, GMendel, TetraploidMap, and JoinMap provide tests for deviation of segregation from expectation, and the first two also provide statistical corrections to linkage estimation in this case.

3.2 Methods and Tools for QTL Mapping in Plants

3.2.1 Statement of the Problem

Like the linkage-mapping section, this section gives only a brief explanation of the mathematical basis of the practical statistical operations involved. Detailed treatment can be found in Liu (1998) and Lynch and Walsh (1998). The focus will be on methods of theoretical interest, whether or not they are much used in practice.

A QTL may be defined as a gene whose genotype cannot be unequivocally determined from phenotype. For QTL searches we start with DNA marker–genotype data and trait data for a set of progeny derived from a planned cross and, ordinarily, a genetic map constructed from either the same genotype data or a different set. Our aim is to locate QTLs influencing our trait and to estimate their allelic effects, i. e., additive and dominance effects at individual QTLs and interaction (epistasis) among these effects at two or more QTLs. As with linkage mapping, we often have more information than is contained in a single experimental dataset. Variables to consider are mating design, type of trait, and the environmental variation inherent in phenotypic data, which must be modeled and can be reduced by suitable replication. As always, we want to evaluate the stability of solutions.

In statistical terms our aim is to identify and solve a model whose predictor variables are either DNA markers or "virtual" QTLs and whose response variable is the trait. A virtual QTL genotype is a function of marker genotypes and can be expressed as an array of genotype probabilities or an expectation. For any specified model, estimates of the coefficients may be obtained by least-squares (LS) or ML methods, but these will be wrong (Knapp et al. 1992) if the model poorly describes the true genetical control of the trait. To specify a model is to write terms for all QTL effects as well as QTL number and location. But since these last two are unknown, we must properly include them in the model search. We must either test all possible models in "model space" or find some way to wander the model space in a generally upwards direction (towards higher likelihood), a process that requires estimating the model coefficients anew at every step. We must also have a way to rank models that balances explanatory power against complexity. The methods for these operations are loosely classified into LS, ML, and Bayesian methods.

A recent review of QTL mapping methods (Doerge 2002) offers a somewhat more general view of the sequence of ideas described below. A tutorial (Jones et al. 1997) directed at biologists introduces the QTL problem more simply. For a detailed comparison of QTL-mapping software features, see Manly and Olson (1999).

3.2.2 Single-marker Association

3.2.2.1 Metric Traits

The notion of correlating genetic marker phenotypes with quantitative-trait phenotypes is usually traced to Sax (1923) and reappears in later work cited by Soller et al. (1979) done before DNA markers were available. The simple principle is that since the genotype of a marker should be correlated with the genotype at a linked QTL, the marker should also show a statistical influence on the trait, though one declining with increasing genetic distance from the QTL. This influence can be tested by a contrast of phenotypic means of the marker-genotype classes, using *t* tests, ANOVA, or regression. For regression, which requires a numerical explanatory variable, the genotype of an individual at a marker locus may be expressed as the quantity of the reference allele (in a diploid, 0, 1, or 2) that it carries. In designs where all three genotypes are present the effect can be partitioned into additive and dominance effects. In designs where more than two alleles may segregate at a locus, regression may be replaced by a general linear model or nonparametric equivalent such as the Kruskal-Wallis test.

The single-marker test does not require a genetic map and cannot distinguish a nearby weak QTL from a more distant strong one. ML methods proposed to do so by exploiting the expected variation within the genotype classes (Zhuchenko et al. 1979; Weller 1986; Luo and Kearsey 1989; Simpson 1989) have given way to more powerful interval methods. Still, in a sufficiently marker-saturated map region (Darvasi et al. 1993), in designs where recombination has been limited, or where no map is available, single-marker testing is as good as any other method for finding QTLs one at a time.

3.2.2.2 Categorical Traits

When plants are scored as belonging to discrete classes, e. g., of disease infection type or growth habit, comparison of means is impossible. Binary-scored traits can be treated as genetic loci and handled with linkage analysis, but where the data fall into more classes, some form of categorical test is needed. Unordered classes can be handled with chi-square tests of independence from cross-tabulated genotype classes, but more commonly phenotype classes have a natural order (the usual case with disease data) and can be handled with nonparametric rank tests. Qualitatively, recoding these ordinal class assignments as numerals 1 through (the number of classes) and handling them by regression seems to work. However, better methods using interval mapping have been developed for such data.

3.2.3
Interval Mapping: Simple (SIM)

3.2.3.1 ML Methods

In a landmark paper, Lander and Botstein (1989) showed how we can estimate the effects of a QTL at any tested location between a given pair of mapped markers. With the flanking markers providing a probability distribution for the QTL genotypes and the assumed normal distribution of the trait within QTL genotype classes providing the probability of the phenotypic observations for each class, the EM algorithm is used iteratively to find values for the trait means and variances in these classes that maximize the likelihood of the phenotype/flanking-marker genotype combinations observed in progeny individuals. A QTL mapping procedure, implemented in MAPMAKER/QTL (Lincoln et al. 1992), "walks" along chromosomes, performing the ML calculation at regularly spaced points one or two centimorgans apart, and the resulting LOD scores are plotted to reveal candidate QTL sites of highest likelihood.

Aware that this algorithm falls short of solving a complete QTL model, the authors provided a way to incorporate a second possible QTL by searching the genome with the effects of a primary QTL fixed (Lincoln et al. 1992). But since even single-QTL searches are computationally intensive and slow, simultaneously modeling multiple QTLs with an ML treatment was impractical. Yet reliance on a simple model can give deceptive estimates of effect. Linked QTLs having positive effects from the same parent appear as a single broad "ghost QTL" peak between them, while those with opposite effects are invisible or insignificant (Martinez and Curnow 1992). Other programs implementing ML-SIM are QTL Cartographer (Wang et al. 2004), MAPL, and MultiQTL (Korol 2004).

A qualitatively different ML approach to SIM was adapted by Xu and Atchley (1995) from a method from human genetics that requires only the estimation of the identity-by-descent (IBD) proportion of alleles shared by pairs of individuals at a map position. For a QTL at this position, high IBD should be accompanied by low phenotypic difference. Such a "random model" algorithm, which models the variance rather than the magnitude of QTL effects, has been implemented for plant designs in the WWW-based software QTL Express (Seaton et al. 2002). Its advantages over "fixed-model" methods are that it requires no knowledge of linkage phase or number of alleles at loci and is readily adapted to complicated pedigree designs.

3.2.3.2 Least-squares (Regression) and Nonparametric Methods

LS methods are much easier and faster to compute than ML methods and allow more straightforward modeling of a large variety of effects, mating designs, and generations with usually negligible loss of estimation accuracy and precision.

The LS analogue of ML-SIM was described by Haley and Knott (1992) and Martínez and Curnow (1992), who showed how, at a QTL position, simple regression of phenotypic values on the expectation of genotype given flanking markers, expressed in terms of additive effect a and dominance d, leads to estimates of these effects. The first authors extended the method to outbreeding designs (Haley et al. 1994). Still, while these ideas are the basis of several modern software packages, the method of building two-QTL models via two-dimensional searches over all QTL positions is unrealistic for larger models. Equally rarely applied in practice is the three-marker test for linked QTLs described by Martínez and Curnow (1992). These authors (Martínez and Curnow 1994) went on to describe how to compensate for a missing flanking-marker genotype in an individual, by basing QTL expectations instead on the nearest available non-missing flanking markers, and this was followed by a more general algorithm (Jiang and Zeng 1997) for producing the genotype probability distribution at any QTL position, that employs all the information available from incompletely informative or dominant flanking markers. Computer programs implementing LS-SIM are numerous and include MQTL (Tinker and Mather 1995), MMQTX, QGene (Nelson 1997), QTL Express, and MCQTL (Jourjon et al. 2004).

A different LS approach (Kearsey and Hyne 1994) involves regressing the difference between genotype-class means at every marker locus upon a function of the recombination probability between that locus and a QTL position. Like the above LS-SIM methods, it plots the resulting statistic (for example, residual sum of squares) against the map at each of the regularly spaced QTL positions tested. The method, which trades away the information from trait variation within marker classes for that from variation between marker-class means, is easy to implement in statistical programming languages and sometimes is used in practice and even extended (Charmet et al. 1998). A less-used bivariate-regression approach described by Whittaker et al. (1996) regresses phenotype on both flanking markers at once and infers from the coefficients both the presence and the location of a QTL, thereby dispensing with a full "walk."

Several other extensions of SIM fall into the "rarely used" class. Among these are a nonparametric method (Kruglyak and Lander 1995) for traits not following the normal distribution expected by LS methods. A logistic-regression approach was developed (Hackett and Weller 1995) for ordinal rather than metric traits, although simulations (Rebaï 1997) have since shown that treating the discrete trait data as continuous and using ordinary regression is usually about as accurate. The multiple-allele problem inherent in linkage mapping in autopolyploids also complicates QTL mapping, and Xie and Xu (2000) and Hackett et al. (2001a) have introduced ways of employing codominant markers.

Regression on a conditional QTL expectation, as in Haley and Knott (1992), inflates the residual-error estimate by the error in approximating the QTL genotype. Of several remedies, one is an iteratively reweighted LS method of Xu (1998a), while others are inherent in the composite methods discussed next.

3.2.4
Interval Mapping: Composite (CIM)

The failures of SIM in the presence of multiple, especially linked, QTLs are the result of its testing the wrong hypothesis at each map position, i. e., that of "QTL at the test position" vs. "no QTL anywhere." The correct test (Jansen 1993) is that of a multiple-QTL model including versus one excluding a QTL at the test position. Such tests fit well into a multiple linear-regression framework, and the evolution of multiple-QTL modeling has progressed from regression on sets of markers (Cowen 1989), to "hybrid" models containing both QTL expectations at a test point and "background" cofactor markers at other places in the map (Jansen 1992; Jansen and Stam 1994; Zeng 1994), and finally to models in which all markers are replaced by QTL genotypes (Kao and Zeng 1999; Sen and Churchill 2001).

"Composite" methods are not interested in the cofactor markers per se, which are used only to absorb the approximate trait variation due to presumed QTLs outside the test interval. The QTL search is still a one-dimensional scan across the map. A variety of methods are used for selecting the cofactor markers, though the more added to increase QTL resolution, the lower the detection power and estimation precision. Error in the modeled QTL genotype can be minimized with weighted regression (Jansen and Stam 1994) or use of ML instead of LS (Zeng et al. 1999).

The "MQM" method of Jansen has been implemented in the commercial package MapQTL and the ML-CIM method of Zeng in QTL Cartographer. LS implementations are provided by MMQTX, PLABQTL (Utz and Melchinger 1996), and QTLMapper (Wang et al. 1999). MultiQTL, another commercial program, offers an elaborate suite of QTL analyses that includes CIM. A simpler version of CIM (sCIM) was provided by (Tinker and Mather 1995) in the program MQTL, which fits a multiple-regression model only once instead of at each QTL test position.

The "composite" feature of CIM can be incorporated into most variants of SIM, which may be thought of as CIM with no cofactors. Noteworthy examples are the logistic-regression method for mapping QTLs for binary diseases on the "threshold" model (Xu and Atchley 1996) and the "random model" algorithm of Xu and Atchley (1995). CIM has not yet been extended to autopolyploid models.

3.2.5
Significance Testing

A perennial problem in QTL mapping has been assessing the reliability of test statistics, e. g., LOD scores, from QTL scans. There simply is no single theoretical sampling distribution that applies to all variants (e. g., differing numbers of individuals, mating design, trait distributions, marker density and information content) of a QTL experiment. While theoretical significance thresholds were proposed by Lander and Botstein (1989), Lander and Kruglyak (1995), and Mangin et al. (1998), robust ones may be produced with shuffling (Churchill and Doerge

1994), simulation (van Ooijen 1999), or resampling (Visscher et al. 1996; Lebreton et al. 1998) methods. If the phenotype data are shuffled (randomized) with respect to the marker data numerous times and the QTL test is computed at each shuffle, the distribution of the resulting statistics represents that under the null hypothesis of no effect. A suitable (say, 5th) percentile of this distribution can then be used as a test threshold for the statistic computed on the unshuffled data. While these permutation tests are free of the distributional assumptions that cause parametric tests to fail, they can take a long time to run. Some as-yet unexploited ways of speeding them up are discussed below. A fast, general method relying only on the results of a QTL scan (Piepho 2001) appears to give the same results as permutation testing.

3.2.6 Interval Mapping: Multiple-QTL Model Building

3.2.6.1 Stepwise and Exhaustive-search Methods for Building Multiple-QTL Models

An early effort to construct multiple-QTL models employed multiple regression (Moreno-Gonzalez 1992a, 1992b) of the trait on dummy QTLs representing the midpoint of each of the marker intervals in the genetic map. It could then be refined by stepwise model-building methods to eliminate intervals of no influence and to refine the positions of the QTLs in the remaining interval. Elements of this method have reappeared in more recent efforts. Multiple interval mapping (MIM; Kao and Zeng 1999; Zeng et al. 1999), implemented in QTL Cartographer, uses a stepwise selection method to add and remove QTLs from a model first arrived at by CIM, then employs the EM algorithm to estimate simultaneously the QTL genotypes and their likelihoods, and finally searches for epistatic effects between modeled QTLs and each other or the unoccupied QTL positions on the map. A second method (Sen and Churchill 2001) replaces this "exact" computation method with Monte Carlo sampling. For a given dataset, many (10 to >200) independent sets of QTL genotypes called pseudomarkers are generated ("imputed") at densely spaced test positions in the map by sampling from their conditional distributions on the flanking markers in each individual, thus approximating the *prior* probability distributions of the QTL genotypes. The *posterior*, taking into account the phenotypes of the individuals, is produced by weighting each pseudomarker set by the fit statistic from regression of the phenotype on it. These operations are implemented in software (Broman et al. 2003). The method currently requires a q-dimensional search, where q is the number of QTLs modeled (so that $q > 3$ is impractical), but offers the advantage that only the weights, not the pseudomarkers, must be changed as the model is changed.

A method described by Wang et al. (1999) and implemented in QTLMapper 1.0 employs stepwise regression to select markers for inclusion in a mixed linear model of main plus epistatic and environmental interaction effects.

3.2.6.2 **Markov Chain Monte Carlo (MCMC) Methods**

MCMC (Hastings 1970) is a Bayesian method (one that estimates parameters by updating the analyst's prior suppositions about them with the information contained in the dataset). MCMC aims to generate, by probabilistic ("Monte Carlo") sampling, a *joint* (collective) posterior distribution of QTL parameters, given the observed QTL mapping data. The algorithm is related to EM in that it progressively refines each parameter estimate in turn based on the pretense that the current estimates of the other parameters are correct. Beginning with initial values for QTL genotypes, positions, and effects, MCMC (the analyst hopes) progresses towards a stable and explanatory model by updating each parameter in turn. Where its probability distribution (say, that of an additive QTL effect) with respect to the other parameters is known, a new value is created by random ("Gibbs") sampling from that distribution. Where the distribution is unknown (such as that of QTL position), a new "proposal" value is randomly (within certain constraints) generated and "accepted" with a probability corresponding to the change in likelihood that the new value would produce (a "Metropolis-Hastings" step). Thousands of iterations furnish *distributions* (in contrast to point estimates) of parameter values, providing built-in significance tests. Tuning MCMC is tricky, though diagnostics are available (http://www.statslab.cam.ac.uk/~mcmc/pages/list.html) to ensure that the chain of updating adequately samples the parameter distributions.

While a Monte Carlo EM method for evaluating QTL likelihoods during a QTL scan was described by Jansen (1996), the first application of MCMC – with QTL number and location themselves treated as random variables – to plant QTL mapping appears to have been that of Satagopan et al. (1996). A few of the many subsequent investigations of the method are due to Sillanpää and Arjas (1998, 1999), Bink et al. (2000), and Yi and Xu (2002). The only public software for MCMC QTL mapping appears to be Multimapper (Sillanpää 1998), which does not provide a full QTL model-building scheme of the type described in Yi and Xu (2002), involving so-called reversible-jump MCMC (Green 1995) for switching between models with different QTL numbers. MCMC analyses are computationally intensive and can consume many hours of computing time. A comparative study (Maliepaard et al. 2001) found general agreement between QTL findings of MQM and MCMC but also "some differences" – not very helpful for practical guidance!

3.2.6.3 **Genetic Algorithms**

GAs, described above in the context of map ordering, appears promising for QTL mapping. Carlborg (2000) addressed the problem of modeling epistasis, while Nakamichi (2001) and Nakamichi et al. (2003) published a Unix computer program. These studies reported speed gains of at least two orders of magnitude over exhaustive-search or MCMC methods for model building and testing. Like MCMC, GAs must be tuned for proper convergence.

3.2.7 Multiple-trait (MT) QTL Mapping

Often in QTL studies some of the measured traits are intercorrelated. The analyst would like to use this correlation structure to reduce the number of statistical tests and, if possible, to refine estimates of QTL position and to test for pleiotropy versus tight linkage of QTLs in the same map region affecting different traits. For the first goal, Weller et al. (1996) and Mangin et al. (1998) replaced correlated traits with their uncorrelated principal components, supposing that a QTL association with one of these "composite" traits would be evidence for pleiotropy. Jiang and Zeng (1995) and Korol et al. (1995) showed increased QTL detection power from MT interval mapping, and Knott and Haley (2000) and Hackett et al. (2001b) described a multivariate-regression method that is easy to program in statistical packages. QTL Cartographer provides an MT method, but the most comprehensive implementation is that of MultiQTL.

3.2.8 Multiple-cross (MC) QTL Mapping

Use of complex cross designs such as those developed for classical quantitative-genetics studies increases the power to detect QTLs that may not segregate in every cross. LS models described in Rebaï et al. (1997) and Rebaï and Goffinet (2000) have been implemented in QTL Express and MCQTL (Jourjon et al. 2004), and fixed- and random-model methods have been developed by Xu (1998b) and Xie et al. (1998). In species such as *Arabidopsis* (Syed and Chen 2004) and mouse (Threadgill et al. 2001; Hitzemann et al. 2001), appropriate crosses between well-genotyped inbred mapping lines lead to replicated heterozygous F_1 progenies with fully known genotypes, which are useful for testing interesting genetic hypotheses. Such designs can be analyzed with minor modification of standard QTL algorithms.

3.2.9 Computational Optimization Methods

Though the slowness of some QTL algorithms may be tolerated in comparison with the time and expense required for producing the data, it holds up both data exploration by the analyst and data processing that may be needed for production operations such as marker-assisted selection. Speed of execution is a concern of several recent papers, among them the GA articles cited above. Parallel computing (Carlborg et al. 2001), which involves distributing computing tasks to multiple processors, is applicable to several areas of QTL mapping, perhaps the most obvious being permutation testing.

Two further proposed improvements are in solving linear models and searching for QTL models. In one-dimensional QTL searches, even though much of the model matrix does not change from one test position to the next, naïve implementations expensively decompose the entire matrix anew into an intermediate format known as QR matrices. Yet, updating the QR form rather than regenerating can speed up CIM (by both LS and ML) by up to three orders of magnitude (Ljungberg et al. 2002). The same authors (Ljungberg et al. 2004) describe a method that, given a number of unlocalized QTLs and effects to be modeled, rapidly finds the global optima (as confirmed by comparison with exhaustive search) for their parameter values. The algorithm works by finding and progressively subdividing only those regions in which the objective function is mathematically expected to have an optimum. It is reported to be up to four orders of magnitude faster than exhaustive searching and one order faster than the (also fast) GA method.

For Bayesian algorithms, an acceleration method that also reduces the instability of MCMC was described by Berry (1998).

3.3 Future Directions in Mapping Methods and Tools

3.3.1 Future of Linkage and QTL Mapping

Genetic mapping will remain a vital research activity for years to come. Only a fraction of species are presently represented among mapped organisms. While it may be thought that the acquisition of full-genome sequence will remove the need for construction of these low-resolution maps, making crosses and acquiring marker data is still much cheaper than sequencing and gives, as physical mapping cannot, a view of the meiotic behavior of chromosomes.

Until recently, QTL analysis has been a search for correlations between genetic markers and phenotypic observations representing the summation, over time, of gene effects. These include interactions with other genes and with environment (Korol et al. 1998; Wang et al. 1999; Cao et al. 2001; Carlborg and Andersson 2002). An area of growing interest is QTL variation over plant developmental stages (Wu et al. 1999; Cao 2001; Wu et al. 2004); a WWW-interfaced computer program, FunMap (Ma et al. 2004), for identifying these dynamic QTLs has recently appeared. A second area is the use of RNA expression data as quantitative traits for QTL scans (Jansen and Nap 2001), giving us an unprecedented ability to exploit genome sequence and SNP data to identify *cis* and *trans* operating genetic regulators (Morley et al. 2004). It appears that the aims of QTL research are evolving from identification of individual genes at coarse resolution to elucidation of the full spectrum of gene action and interaction over time. A third trend is based on the recognition of linkage disequilibrium (LD) as the basis of causative inference in genetics, with naturally occurring LD carrying complementary information to LD created with a planned cross (Wu et al. 2001, 2002; Jannink et al. 2001; Bink et al. 2002).

The most active research in the coming years may be in *general* solutions, e. g., integration of sources of genetic information to predict phenotype, such that the experimental design is only one of the variables. Therefore, not only unified methods of handling different sources of LD are needed for this, but also meta-analysis to unite disparate experiments (Goffinet and Gerber 2000; Arcade et al. 2004) and methods for combining QTL models (Sillanpää and Corander 2002).

3.3.2 Adequacy of Software Tools for Plant Mapping

3.3.2.1 Software Merit Criteria

From software, as from any analytical tool, researchers want ease of data preparation, ease of learning the use of the tool, suitable statistical methods readily applied, and readily understanddable output in a form suitable for communication. Some may also wish for a richer set of features, e. g., ease of exploring and comparing data and the behaviors of algorithms, graphic visualization, or other sensory conversion; integration with other software and with databases; programmatic access to the software, or open availability of high-quality source code. Some may insist that software, at least for public research, be free of monetary cost.

3.3.2.2 Analytical Scope

The software packages described above vary widely with respect to features and ease of use. None covers more than a limited range of variations on mapping problems, though MMQTX and MultiQTL seem to be the most ambitious in offering a widening array of options. The disuse in practice of many interesting statistical advances described in the previous sections may stem less from lack of merit than from lack of software to apply them.

Systematic comparative studies of output from the various programs are lacking, though a common practice for publication of new theoretical methods or tools is to compare their results with those from MAPMAKER, JoinMap, QTL Cartographer, or other popular programs. While data preparation for one program may not be hard, reformatting it for multiple programs that offer needed complementary analyses can be burdensome. Some software adopts data formats close to those of popular predecessors. MMQTX offers conversion routines to different formats.

3.3.2.3 Ease of Learning and Use

The author, himself a software developer with strong opinions about what constitutes quality, finds fault with all packages he has tried. Indeed, the volume of publications still reporting the use of SIM and other one-dimensional approximations for QTL mapping, a dozen years after clear demonstration of the weakness in this approach, suggests that researchers (many, of course, students) choose familiarity and ease of use over statistical power.

A minimal list of software properties that would seem to remove most excuses for using outdated analytical methods would include an intuitive user interface (UI) that prevents or forgives user errors such as in data formatting; requirement of minimal need for typing; clear and well-written documentation, UI labels, and prompts; fast performance; and freedom from hanging or crashing. Higher-order features – which, rather than merely providing accepted analysis methods, actually help the analyst understand and communicate their behavior – include ease of changing parameters or algorithms, facility of comparison between different analyses (e. g., marker regression and CIM), means of assessing the significance of results, and effective and exportable graphic displays.

3.3.2.4 Accessibility and Extensibility

The only way for an analyst to extend the functionality of an executable software program is to write a script to direct its operation, where the capability is offered (e. g., MAPMAKER and CarthaGene). Where source code is open, an external developer can in principle modify and extend it to add new analyses and graphics; however, unless the code is of good quality and is written in the preferred language and style, it is easier to build an entirely new program. The result is the current soup of limited-purpose programs written in diverse languages and styles, with diverse UIs, and without interconnectivity or ready comparability. In an academic setting, developer turnover is apt to be high, so that a project may have uncertain continuity; errors are not fixed, new analyses are not added, and the software is frozen in time. High-quality, specialized analytical software is difficult and expensive for academics to write and maintain. Is there any way to do it while keeping the product freely available to the community?

3.3.3 A Development Model for Public Genetic Mapping Software

One way to reduce duplication of work, encourage the rapid and wide testing and adoption of new analytical methods, and accommodate the work of statisticians and developers without restrictions from closed code or licensing fees would be an open-architecture system accommodating plug-in features. Two models for such a system are ISYS (Siepel et al. 2001), a general platform for dynamically connecting bioinformatics data with software services, and Eclipse (http://www.eclipse.org/), a software-development environment, both written in the Java language. But a QTL platform need not be of the generality and complexity of these if all that is wanted is a way to incorporate new statistical analyses, graphical displays, and data import and export utilities. Programmers familiar with the language should not need to understand more than a few rules for "plugging in" a new feature without rebuilding the entire application, but they would still have a voice in larger design decisions because the code base is a community resource. If such a system comes into use – and there appear to be several institutions where similar ideas are being developed – implementing new QTL-analysis methods in software would be easier to fund from public sources than would developing more limited-scope, limited-UI, closed-source applications.

References

Arcade A, Labourdette A, Falque M, Mangin B, Chardon F, Charcosset A, Joets J. (2004) BioMercator: integrating genetic maps and QTL towards discovery of candidate genes. Bioinformatics, April 1, 2004, electronic preprint.

Berry CC. (1998) Computationally efficient Bayesian QTL mapping in experimental crosses. Proceedings Biometrics Section, American Statistical Association, pp. 164–169.

Bink MCAM, Janss LLG, Quaas RL. Markov chain Monte Carlo for mapping a quantitative trait locus in outbred populations. (2000) Genet. Res., 75:231–241.

Bink MCAM, Uimari P, Sillanpää MJ, Janss LLG, Jansen RC. (2002) Multiple QTL mapping in related plant populations via a pedigree-analysis approach. Theor. Appl. Genet., 104:751–762.

Broman K, Wu H, Sen S, Churchill GA. (2003) R/qtl: QTL mapping in experimental crosses. Bioinformatics, 19:889–890.

Cao G, Zhu J, He C, Gao Y, Yan J, Wu P. (2001) Impact of epistasis and QTL x environment interaction on the developmental behavior of plant height in rice (Oryza sativa L.). Theor. Appl. Genet., 103:153–160.

Carlborg Ö. (2000) The use of a genetic algorithm for simultaneous mapping of multiple interacting quantitative trait loci. Genetics, 155:2003–2010.

Carlborg Ö, Andersson L. (2002) Use of randomization testing to detect multiple epistatic QTLs. Genet. Res., 79:175–184.

Carlborg Ö, AnderssonEklund L, Andersson L. (2001) Parallel computing in interval mapping of quantitative trait loci. J. Hered., 92:449–451.

Charmet G, Cadalen T, Sourdille P, Bernard M. (1998) An extension of the 'marker regression' method to interactive QTL. Mol. Breed., 4:67–72.

Churchill GA, Doerge RW. (1994) Empirical threshold values for quantitative trait mapping. Genetics, 138:963–971.

Cowen NM. (1989) Multiple linear regression analysis of RFLP data sets used in mapping QTLs. In: Development and Application of Molecular Markers to Problems in Plant Genetics,; Helentjaris T, Burr B, eds. Cold Spring Harbor Laboratory, Cold Spring Harbor, NY. pp. 113–116.

Crane CF, Crane YM. (2004a) A nearest-neighboring-ends algorithm for genetic mapping. Poster 976, Plant Animal Genome XII, San Diego, CA.

Crane CF, Crane YM. (2004b) FLIPPER: A general, high-capacity genetic mapping program. Poster 993, Plant Animal Genome XII, San Diego, CA.

Darvasi A, Weinreb A, Minke V, Weller JI, Soller M. (1993) Detecting marker-QTL linkage and estimating QTL gene effect and map location using a saturated genetic map. Genetics, 134:943–951.

Doerge R. (1996) Constructing genetic maps by rapid chain delineation. J. Quant. Trait Loci, 2. http://www.cabi-publishing.org/gateways/jag/index.html

Doerge RW. (2002) Mapping and analysis of quantitative trait loci in experimental populations. Nature Rev. Genet., 3:43–52.

Ellis THN. (1997) Neighbour mapping as a method for ordering genetic markers. Genet. Res., 69:35–43.

Goffinet B, Gerber S. (2000) Quantitative trait loci: a meta-analysis. Genetics, 155:463–473.

Green PJ. (1995) Reversible jump Markov Chain Monte Carlo computation and Bayesian model determination. Biometrika, 82:711–732.

Hackett CA, Luo ZW. (2002) TetraploidMap, software suite for calculating linkage maps for autotetraploid populations. ftp://ftp.bioss.sari.ac.uk/pub/cah/

Hackett CA, Weller JI. (1995) Genetic mapping of quantitative trait loci for traits with ordinal distributions. Biometrics, 51:1252–1263.

Hackett CA, Bradshaw JE, Meyer RC, McNicol JW, Milbourne D, Waugh R. (1998) Linkage analysis in tetraploid species: a simulation study. Genet. Res., 71:143–154.

Hackett CA, Bradshaw JE, McNicol JW. (2001a) Interval mapping of quantitative trait loci in autotetraploid species. Genetics, 159:1819–1832.

Hackett CA, Meyer RC, Thomas WTB. (2001b) Multi-trait QTL mapping in barley using multivariate regression. Genet. Res., 77:95–106.

Hackett CA, Pande B, Bryan GJ. (2003) Constructing linkage maps in autotetraploid species using simulated annealing. Theor. Appl. Genet., 106:107–115.

Haley CS, Knott SA. (1992) A simple regression method for mapping quantitative trait loci in line crosses using flanking markers. Heredity, 69:315–324.

Haley CS, Knott SA, Elsen J-M. (1994) Mapping quantitative trait loci in crosses between outbred lines using least squares. Genetics, 136:1195–1207.

Hastings WK. (1970) Monte Carlo sampling methods using Markov chains and their applications. Biometrika, 57:97–109.

He Y, Xu X, Tobutt KR, Ridout MS. (2001) Polylink: to support two-point linkage analysis in autotetraploids. Bioinformatics, 17:740–741.

Hitzemann R, Malmanger B, Reed C, Lawler M, Hitzemann B, Coulombe S, Buck K, Rademacher B, Walter N, Polyakov Y, Sikela J, Gensler B, Burgers S, Williams RW, Manly K, Flint J, Talbot C. (2001) A strategy for the integration of QTL, gene expression, and sequence analyses. Mamm Genome, 14:733–747.

Jannink J-L, Bink MCAM, Jansen RC. (2001) Using complex plant pedigrees to map valuable genes. Trends Plant Sci., 6:337–342.

Jansen RC. (1992) A general mixture model for mapping quantitative trait loci by using molecular markers. Theor. Appl. Genet., 85:252–260.

Jansen RC. (1993) Interval mapping of multiple quantitative trait loci. Genetics, 135:205–211.

Jansen RC. (1996) A general Monte Carlo method for mapping multiple quantitative trait loci. Genetics, 142:305–311.

Jansen RC, Nap J-P. (2001) Genetical genomics: the added value from segregation. Trends Genet., 17:388–391.

Jansen RC, Stam P. (1994) High resolution of quantitative traits into multiple loci via interval mapping. Genetics, 136:1447–1455.

Jensen J, Jørgensen JH. (1975) The barley chromosome 5 linkage map. I. Literature survey and map estimation procedure. Hereditas, 80:5–16.

Jiang C, Zeng Z-B. (1995) Multiple trait analysis of genetic mapping for quantitative trait loci. Genetics, 140:1111–1127.

Jiang C, Zeng Z-B. (1997) Mapping quantitative trait loci with dominant and missing markers in various crosses from two inbred lines. Genetica, 101:47–58.

Jones N, Ougham H, Thomas H. (1997) Markers and mapping: we are all geneticists now. New Phytol., 137:165–177.

Jourjon M-F, Jasson S, Marcel J, Ngom B, Mangin B. (2004) MCQTL: Multi-allelic QTL mapping in multi-cross design. Bioinformatics, August 19 2004, electronic preprint.

Julier B, Flajoulot S, Barre P, Cardinet G, Santoni S, Huguet T, Huyghe C. (2003) Construction of two genetic linkage maps in cultivated tetraploid alfalfa (*Medicago sativa*) using microsatellite and AFLP markers. BMC Plant Biol. 3:9.

Kao C-H, Zeng Z-B. (1999) Multiple interval mapping for quantitative trait loci. Genetics, 152:1203–1216.

Kearsey MJ, Hyne V. (1994) QTL analysis: a simple 'marker regression' approach. Theor. Appl. Genet., 89:698–702.

Knapp SJ, Bridges WC, Liu BH. (1992) Mapping quantitative trait loci using nonsimultaneous and simultaneous estimators and hypothesis tests. Beckmann JS, Osborn TC. (eds.) Plant genomes: methods for genetic and physical mapping. Kluwer Academic Publishers, Dordrecht, Netherlands.

Knott SA, Haley CS. (2000) Multitrait least squares for quantitative trait loci detection. Genetics, 156:899–911.

Korol AB. (2004) MultiQTL: http://esti.haifa.ac.il/~poptheor/MultiQtl/MultiQtl.htm.

Korol AB, Ronin YI, Kirzhner VM. (1995) Interval mapping of quantitative trait loci employing correlated trait complexes. Genetics, 140:1137–1147.

Korol AB, Ronin YI, Nevo E. (1998) Approximate analysis of QTL-environment interaction with no limits on the number of environments. Genetics, 148:2015–2028.

Kruglyak L, Lander ES. (1995) A nonparametric approach for mapping quantitative trait loci. Genetics, 139:1421–1428.

Lander ES, Botstein B. (1989) Mapping Mendelian factors underlying quantitative traits using RFLP linkage maps. Genetics, 121:185–199.

Lander ES, Green P. (1987) Construction of multilocus genetic linkage maps in humans. Proc. Nat. Acad. Sci. USA, 84:2363–2367.

Lander ES, Kruglyak L. (1995) Genetic dissection of complex traits: guidelines for interpreting and reporting linkage results. Nature Genetics, 11:241–247.

Lander ES, Green P, Abrahamson J, Barlow A, Daly MJ, Lincoln SE, Newburg L. (1987) MAPMAKER: an interactive computer package for constructing primary genetic linkage maps of experimental and natural populations. Genomics, 1:174–181.

Lebreton CM, Visscher PM, Haley CS, Semikhodskii A, Quarrie SA. (1998) A nonparametric bootstrap method for testing close linkage vs. pleiotropy of coincident quantitative trait loci. Genetics, 150:931–943.

Lincoln S, Daly M, Lander E. (1992) Mapping genes controlling quantitative traits with MAPMAKER/QTL 1.1. 2nd ed. Whitehead Institute Technical Report, Cambridge, MA.

Liu BH. (1998) Statistical Genomics: Linkage Mapping and QTL Analysis. CRC Press, NY.

Liu BH, Knapp SJ. (1990) GMENDEL: A program for Mendelian segregation and linkage analysis of individual or multiple progeny populations using log-likelihood ratios. J. Hered., 81:407.

Ljungberg, Holmgren S, Carlborg Ö. (2002) Efficient algorithms for quantitative trait loci mapping problems. J. Comput. Biol., 9:793–804.

Ljungberg K, Holmgren S, Carlborg Ö. (2004) Simultaneous search for multiple QTL using the global optimization algorithm DIRECT. Bioinformatics, 20:1887–1895.

Luo ZW, Kearsey MJ. (1989) Maximum likelihood estimation of linkage between a marker gene and a quantitative trait locus. Heredity, 63:401–408.

Lynch M, Walsh B. (1998) Genetics, and Analysis of Quantitative Traits. Sinauer Associates, Inc.

Ma C-X, Wu R, Casella G. (2004) FunMap: functional mapping of complex traits. Bioinformatics, 20:1808–1811.

Maliepaard C, Jansen J, van Ooijen JW. (1997) Linkage analysis in a full-sib family of an outbreeding plant species: overview and consequences for applications. Genet. Res., 67:237–250.

Maliepaard C, Sillanpää MJ, van Ooijen JW, Jansen RC, Arjas E. (2001) Bayesian versus frequentist analysis of multiple quantitative trait loci with an application to an outbred apple cross. Theor. Appl. Genet., 103:1243–1253.

Mangin B, Thoquet P, Grimsley N. (1998) Pleiotropic QTL analysis. Biometrics, 54:88–99.

Manly KF, Olson JM. (1999) Overview of QTL mapping software and introduction to Map Manager QT. Mamm. Genome, 10.

Martínez O, Curnow RN. (1992) Estimating the locations and the sizes of the effects of quantitative trait loci using flanking markers. Theor. Appl. Genet., 85:480–488.

Martínez O, Curnow RN. (1994) Missing markers when estimating quantitative trait loci using regression mapping. Heredity, 73:198–206.

Mester D, Ronin Y, Minkov D, Nevo E, Korol A. (2003) Constructing large-scale genetic maps using an evolutionary strategy algorithm. Genetics, 165:2269–2282.

Moreno-Gonzalez J. (1992a) Estimates of marker-associated QTL effects in Monte Carlo backcross generations using multiple regression. Theor. Appl. Genet., 85:423–434.

Moreno-Gonzalez J. (1992b) Genetic models to estimate additive and non-additive effects of marker-associated QTL using multiple regression techniques. Theor. Appl. Genet., 85:435–444.

Morley M, Molony CM, Weber TM, Devlin JL, Ewens KG, Spielman RS, Cheung VG. (2004) Genetic analysis of genome-wide variation in human gene expression. Nature, 430:743–747.

Nakamichi R. (2003) Sample Software of GA for QTL Mapping. http://lbm.ab.a.u-tokyo.ac.jp/~naka/software.html

Nakamichi R, Ukai Y, Kishino H. (2001) Detection of closely linked multiple quantitative trait loci using a genetic algorithm. Genetics, 158:463–475.

Nelson JC. (1997) QGENE: software for marker-based genomic analysis and breeding. Mol. Breed., 3:239–245.

Newell WR, Mott R, Beck S, Lehrach H. (1995) Construction of genetic maps using distance geometry. Genomics, 30:59–70.

Paterson AH, ed. (1996) Genome Mapping in Plants. R. G. Landes, Austin, Texas.

Piepho H-P. (2001) A quick method for computing approximate thresholds for quantitative trait loci detection. Genetics, 157:425–432.

Rebaï A. (1997) Comparison of methods for regression interval mapping in QTL analysis with non-normal traits. Genet. Res., 69:69–74.

Rebaï A, Goffinet B. (2000) More about quantitative trait locus mapping with diallel designs. Genet. Res., 75:243–247.

Rebaï A, Blanchard P, Perret D, Vincourt P. (1997) Mapping quantitative trait loci controlling silking date in a diallel cross among four lines of maize. Theor. Appl. Genet., 95:451–459.

Ripol MI, Churchill GA, da Silva JAG, Sorrells M. (1999) Statistical aspects of genetic mapping in autopolyploids. Gene, 235:31–41.

Ritter E, Salamini F. (1996) The calculation of recombination frequencies in crosses of allogamous plant species with applications to linkage mapping. Genet. Res., 67:55–65.

Ritter E, Gebhardt C, Salamini F. (1990) Estimation of recombination frequencies and construction of RFLP linkage maps in plants from crosses between heterozygous parents. Genetics, 125:645–654.

Satagopan JM, Yandell BS, Newton MA, Osborn TC. (1996) A Bayesian approach to detect quantitative trait loci using Markov chain Monte Carlo. Genetics, 144:805–816.

Sax K. (1923) The association of size difference with seed-coat pattern and pigmentation in Phaseolus vulgaris. Genetics, 8:552–560.

Schiex T, Gaspin C.(1997) CARTHAGENE: constructing and joining maximum likelihood genetic maps. Proc. Int. Conf. ISMB, 5:258–267.

Seaton G, Haley CS, Knott SA, Kearsey M, Visscher PM. (2002) QTL Express: mapping quantitative trait loci in simple and complex pedigrees. Bioinformatics, 18:339–340.

Sen S, Churchill GA. (2001) A statistical framework for quantitative trait mapping. Genetics, 159:371–387.

Siepel A, Farmer A, Tolopko A, Zhuang M, Mendes P, Beavis W, Sobral B. (2001) ISYS: a decentralized, component-based approach to the integration of heterogeneous bioinformatics resources. Bioinformatics, 17:83–94.

Sillanpää MJ. (1998) Multimapper Reference Manual. http://www.rni.helsinki.fi/~mjs/

Sillanpää MJ, Arjas E. (1998) Bayesian mapping of multiple quantitative trait loci from incomplete data based on line crosses. Genetics, 148:1373–1388.

Sillanpää MJ, Arjas E. (1999) Bayesian mapping of multiple quantitative trait loci from incomplete outbred offspring data.Genetics, 151:1605–1619.

Sillanpää MJ, Corander J. (2002) Model choice in gene mapping: what and why. Trends Genet., 18:301–307.

Simpson SP. (1989) Detection of linkage between quantitative trait loci and restriction fragment length polymorphisms using inbred lines. Theor. Appl. Genet., 77:815–819.

Soller M, Brody T, Genizi A. (1979) The expected distribution of markerlinked quantitative effects in crosses between inbred lines. Heredity, 43:179–190.

Stam P. (1993) Construction of integrated genetic linkage maps by means of a new computer package: JoinMap. Plant J., 8:739–744.

Stam P, van Ooijen JW (1995) JoinMap™ version 2.0: Software for the calculation of genetic linkage maps. CPRO-DLO, Wageningen.

Syed NH, Chen ZJ. (2004) Molecular marker genotypes, heterozygosity and genetic interactions explain heterosis in *Arabidopsis thaliana*. Heredity, electronic preprint.

Threadgill D, Airey DC, Lu L, Manly KF, Williams RW. (2001) Recombinant inbred intercross (RIX) mapping: a new approach extending the power of existing mouse resources. The 15th International Mouse Genome Conference, Braunschweig, Germany. (http://imgs.org/abstracts/2001abstracts/threadgill.shtml)

Tinker NA, Mather DE. (1995) Methods for QTL analysis with progeny replicated in multiple environments. J. Quant. Trait Loci, 1, http://www.cabi-publishing.org/gateways/jag/index.html

Tinker NA, Mather DE. (1995) MQTL: software for simplified composite interval mapping of QTL in multiple environments J. Quant. Trait Loci, 1, http://www.cabi-publishing.org/gateways/jag/index.html

Ukai Y. (1997) http://peach.ab. a.u-tokyo.ac.jp/~ukai/

Ukai Y, Ohsawa R, Saito A. (1990) Automatic determination of the order of RFLPs in a linkage group by metric multi-dimensional scaling method. Jpn. J. Breed., 40 Suppl. 2:302–303.

Utz HF, Melchinger AE (1996) PLABQTL: a program for composite interval mapping of QTL. J. Quant. Trait Loci, 2, http://www.cabi-publishing.org/gateways/jag/index.html

van Berloo R, Kraakman A.. (2002) Quality analysis of linkage maps by means of a new software package: MapInspect. http://www.dpw.wau.nl/pv/pub/MapComp/Mapinspect_poster.pdf

van Ooijen JW. (1999) LOD significance thresholds for QTL analysis in experimental populations of diploid species. Heredity, 83:613–624.

Visscher PM, Thompson R, Haley CS. (1996) Confidence intervals in QTL mapping by bootstrapping. Genetics, 143:1013–1020.

Wang DL, Zhu J, Li ZK, Paterson AH. (1999) Mapping QTLs with epistatic effects and QTL x environment interactions by mixed linear model approaches. Theor. Appl. Genet., 99:1255–1264.

Wang S, Basten CJ, Zeng Z-B. (2004) Windows QTL Cartographer v2.0. http://statgen.ncsu.edu/qtlcart/WQTLCart.htm

Weller JI. (1986) Maximum likelihood techniques for the mapping and analysis of quantitative trait loci with the aid of genetic markers. Biometrics, 42:627–640.

Weller JI, Wiggans GR, VanRaden PM, Ron M. (1996) Application of a canonical transformation to detection of quantitative trait loci with the aid of genetic markers in a multi-trait experiment. Theor. Appl. Genet., 92:998–1002.

Whittaker JC, Thompson R, Visscher PM. (1996) On the mapping of QTL by regression of phenotype on marker-type. Heredity, 77:23–32.

Wu R, Zeng Z-B. (2001) Joint linkage and linkage disequilibrium mapping in natural populations. Genetics, 157:899–909.

Wu KK, Burnquist W, Sorrells ME, Tew TL, Moore PH, Tanksley SD. (1992) The detection and estimation of linkage in polyploids using single-dose restriction fragments. Theor. Appl. Genet., 83:294–300.

Wu W-R, Li W-M, Tang D-Z, Liu H-R, Worland AJ. (1999) Time-related mapping of quantitative trait loci underlying tiller number in rice. Genetics, 151:297–303.

Wu SS, Wu R, Ma C-X, Zeng Z-B, Yang MCK, Casella G. (2001) A multivalent pairing model of linkage analysis in autotetraploids. Genetics, 159:1339–1350.

Wu R, Ma C-X, Casella G. (2002) Joint linkage and linkage disequilibrium mapping of quantitative trait loci in natural populations. Genetics, 160:779–792.

Wu R, Ma C-X, Lin M, Casella G. (2004) A general framework for analyzing the genetic architecture of developmental characteristics. Genetics, 166:1541–1551.

Xie C, Xu S. (2000) Mapping quantitative trait loci in tetraploid populations. Genet. Res., 76:105–115.

Xie C, Gessler DDG, Xu S. (1998) Combining different line crosses for mapping quantitative trait loci using the identical by descent variance component method. Genetics, 49:1139–1146.

Xu S. (1998a) Further investigation on the regression method of mapping quantitative trait loci. Heredity, 80:364–373.

Xu S. (1998b) Mapping quantitative trait loci using multiple families of line crosses. Genetics, 148:517–524.

Xu S, Atchley WR. (1995) A random model approach to interval mapping of quantitative trait loci. Genetics, 141:1189–1197.

Xu S, Atchley WR. (1996) Mapping quantitative trait loci for complex binary diseases using line crosses. Genetics, 143:1417–1424.

Yi N, Xu S. (2002) Mapping quantitative trait loci with epistatic effects. Genet. Res., 79:185–198.

Zeng Z-B. (1994) Precision mapping of quantitative trait loci. Genetics, 136:1457–1468.

Zeng Z-B, Kao CH, Basten CJ. (1999) Estimating the genetic architecture of quantitative traits. Genet. Res., 74:279–289.

Zhuchenko AA, Korol AB, Andryushchenko VK. (1979) Linkage between loci of quantitative characters and marker loci. Genetica, 14:771–778.

4
Single nucleotide Polymorphisms: Detection Techniques and Their Potential for Genotyping and Genome Mapping

Günter Kahl, Andrea Mast, Nigel Tooke, Richard Shen, and Dirk van den Boom

4.1
Introduction

Among the many types of mutations that occur naturally in genomes, the single nucleotide exchanges ("point mutations," single nucleotide polymorphisms, SNPs, pronounced "snips") stand out by their sheer numbers per genome, their relatively low mutation rates (as opposed to, e. g., microsatellites of all types), their even distribution across the genomes, and their relative ease of detection. In essence, the term SNP describes single base pair positions in the genomes of two (or more) individuals, at which different sequence alternatives (alleles) exist in populations. As a threshold, the least frequent allele should have an abundance of 1% (or more).

SNPs are the most prevalent sequence variations between individuals of all species and are a consequence of either transition or transversion events:

Individual A	5′-ATGGATCGCCA-3′	Individual A	5′-CGTTACAGGTT-3′
Individual B	5′-ATGGACCGCCA-3′	Individual B	5′-CGTTCCAGGTT-3′
	⇑		⇑
	Transitional SNP		Transversional SNP

SNPs are di-, tri-, or tetra-allelic polymorphisms. However, in the human genome, di-allelic SNPs prevail massively. The higher level of, e. g., C↔T (G↔A) SNPs is at least partly a consequence of deamination of 5-methylcytosine, particularly at CpG dinucleotides.

Generally, SNPs are highly abundant, but their density differs substantially in different regions of a genome and from genome to genome in any species, and even more so from species to species. For example, the average number of SNPs in the human genome is two to three per kilobase. However, in non-coding HLA regions, SNP density (i. e., the frequency of SNPs per unit length of genomic DNA, usually expressed as SNPs per 100 kb) increases dramatically (Horton et al. 1998). SNP density is generally high in intergenic and intronic regions, as compared to the genic space. In the mouse genome, a series of SNP-rich segments

The Handbook of Plant Genome Mapping.
Genetic and Physical Mapping. Edited by Khalid Meksem, Günter Kahl

ISBN 3-527-31116-5

have been discovered in which 40 SNPs per 10 kb (as opposed to only 0.5 SNPs per 10 kb in the intermitting regions) can be located.

In plants, SNP research is still in its infancy. The total SNP content of plants is unknown, and in only a few cases have SNPs been rigorously searched for, although a high level of sequence polymorphism in many cultivated plants (e. g., maize) certainly facilitates SNP detection. In wheat, for example, SNP density is two SNPs per kilobase but could be much higher in genes encoding enzymes of starch biosynthesis (Bhattramakki and Rafalski 2001). In soybean, 1.64 SNPs per kilobase are found in coding regions and 4.9 SNPs per kilobase are found in non-coding regions. In wild soybean, repeated sequencing of distinct genomic regions detected 99 coding SNPs and 329 non-coding SNPs, whereas a certain SNP depression was found in cultivated soybean (from 33–59 coding SNPs and 127–214 non-coding SNPs; Rafalski 2002). A number of SNPs are available for *Arabidopsis* (http://www.arabidopsis.org/cereon/index.html), for which a first medium-density SNP map has been established (Cho et al. 1999). SNP genotyping with 12 nuclear and 13 chloroplast interspecific SNPs of different tree species discriminated black spruce (*Picea mariana*) from red (*P. rubens*) and white spruce (*P. glauca*; Germano and Klein 1999). Analysis of several hundred loci in eight maize inbred lines revealed an extremely high prevalence of SNPs (one SNP per 83 bp), probably as a consequence of open pollination in this species (Bhattramakki and Rafalski 2001). The flanking sequences of microsatellites in maize contain one SNP for every 40 bp, bringing the total number of SNPs per whole genome to 62 million (Edwards and Mogg 2001). Furthermore, rather sporadic reports on the use of SNPs in genotyping and genetic mapping in plants have appeared, but doubtless SNPs are not yet in common practice in the plant sciences. The tremendous cost of developing SNPs, especially the sequencing load, may be one reason to prefer more economic markers such as SSRs in plant genomics.

SNPs fall into several classes, depending on (1) their precise location in a genome and (2) the impact of their presence within coding or regulatory regions onto the encoded protein or phenotype. Since most of the genomic DNA of eukaryotic organisms is non-coding and mostly repetitive DNA, the majority of SNPs fall into this category. They are deceptively called non-coding SNPs (ncSNPs). Among these ncSNPs, many are located within introns (intronic SNPs). Fewer SNPs can be found in coding regions ("coding SNPs"), e. g., in exons (exonic SNPs; also gene-based SNPs or, misleadingly, copy SNPs or cSNPs). Exonic SNPs also appear in the corresponding cDNAs and are then called cDNA SNPs, or cSNPs. Any SNP in an exon of a gene that can be expected to have an impact on the function of the encoded protein is named a candidate SNP. Still others occur in promoters or other regulatory regions of the genome and are coined regulatory SNPs and promoter SNPs (pSNPs), respectively. It is obvious that any promoter SNP can dramatically influence the activity of the promoter-driven gene. If, for example, a pSNP prevents the binding of a transcription factor to its recognition sequence in the promoter, the promoter becomes partly dysfunctional. In contrast, intronic SNPs are regarded as more or less inert. However, many researchers value the extragenic SNPs for association studies and whole-genome linkage-disequilibrium mapping.

In this context, an anonymous SNP represents any one of the most frequently occurring SNPs that have no known effect on the function of a gene. Any SNP at a specific site of a genome (or part of a genome, e. g., a BAC clone) that serves as a reference point for the definition of other SNPs in its neighborhood is called a reference SNP (refSNP, rsSNP, rsID). An rsID number ("tag") is assigned to each refSNP at the time of its submission to the databanks (e. g., dbSNP, a public database maintained by the National Center for Biotechnology Information, NCBI; http://www.ncbi.nlm.nih.gov/SNP). As more and more SNPs are accumulating in the databases, they are labeled with the organism of origin (e. g., human SNP). Tapping the dramatically increasing source of genomic sequences, SNPs can be identified by mining overlapping sequences in expressed sequence tag (EST) or genomic databases and are coined *in silico* or electronic SNPs (*in silico* SNPs, isSNPs, eSNPs). Since isSNPs represent "virtual" polymorphisms, they have to be validated by resequencing the region in which they occur.

The appearance of a SNP in an exon may be totally neutral (i. e., it does not change the amino acid composition of the encoded domain or protein and is therefore without any effect on its function). In such a case, the SNP is called a synonymous SNP (synSNP). A non-synonymous SNP (nsSNP) will change the encoded amino acid. NsSNPs may therefore cause the synthesis of a nonfunctional protein and therefore be involved in diseases (diagnostic SNPs). Clearly, the detection of diagnostic SNPs is the aim of all SNP discovery projects. The underlying strategy is either a whole-genome linkage approach, which uses a common set of evenly spaced SNPs spanning the entire genome as a standard mapping panel to randomly search for susceptibility alleles through linkage disequilibrium (LD), or a more focused approach, in which only cSNPs of candidate genes for the targeted disease are tested in association studies. Both strategies have their advantages and disadvantages. In short, the former is extremely costly if one considers, for example, 1000 individuals (e. g., patients) and 500,000 high-frequency SNPs to be tested (adequate number of SNPs to find associations). This discovery would require some 200–400 million US$. The latter approach certainly is cheaper, but it excludes SNPs in regulatory domains outside the coding regions, which may be truly responsible for a disease and therefore be more diagnostic than cSNPs themselves.

The direct identification of diagnostic SNPs basically follows two approaches. First, genomic regions containing the identified candidate genes are screened for all SNPs by a series of techniques such as microchip hybridization (Wang et al. 1998; Cargill et al. 1999; Halushka et al. 1999), denaturing HPLC (Giordano et al. 1999), or enzyme cleavage (Barroso et al. 1999), to name only few. Second, SNPs are identified *in silico*. Presently, the resulting isSNPs are being discovered at dramatic speed (see the Incyte program at http://www.incyte.com/) and may soon represent the majority of newly discovered single-base exchanges.

Eventually, all common SNPs in a genome will be discovered and a comprehensive SNP map will be established, the first most likely being the complete human SNP map (Sachidanandam et al. 2001). For practical purposes, it is probably better to identify haplotypes, linear arrangements of relatively few specific SNPs on a chromosomal segment (in diploids, on the corresponding segments of homologous chromosomes). If, for example, the target sequence from different individuals is compared after repeated sequencing, the haplotypic organization becomes apparent (also called "block" or "haplotype block," "hapblock") (Daly et al. 2001; Gabriel et al. 2002). The coinheritance of SNP alleles in these haplotypes (so called haplotype-tagged SNPs) associates the alleles in a population (known as linkage disequilibrium, LD). Research into the levels of LD in human populations revealed that large regions of high LD extend over 100 kb and more (Goldstein 2001). Therefore, genotyping only a few carefully chosen SNPs in the target region provides enough information to predict the constitution of the remaining SNPs in the region. These selected SNPs are known as "tag" SNPs. Therefore, a rough calculation led to the assumption that most of the information about genetic variation represented by the 10 million common SNPs in the population is already provided by genotyping only 200,000–800,000 tag SNPs across the genome. Hence, determining haplotypes spanning a candidate gene region is a more directed approach to detecting a real association than is testing candidate SNPs individually. Haplotype analysis thus is a logic approach to establishing genetic risk profiles and the prediction of clinical reaction of an individual towards pharmaceutically active compounds. Much of the haplotyping effort is bundled in the so-called HapMap Project of the International HapMap Consortium (www.hapmap.org), which aims at establishing the haplotype map of the human genome and describing the common patterns of variation as well as the association between SNPs.

The SNP genotyping process requires high-throughput automated platforms, of which microarray- and bead-based techniques are most promising, though they suffer from being closed technologies (Green et al. 2001). Also, a variety of microtiter plate–based platforms are available. Eventually, the target SNPs have to be detected directly. Various assays have been developed for detection of known and unknown SNPs, some of which are relatively easy to perform and low in cost. Others manage high-volume screening and therefore are very costly. There is no one protocol that meets all needs; therefore, different protocols may have to be involved in a single core genotyping lab to provide flexibility and accurate validation. Basically, the major SNP genotyping techniques fall into five groups, though a series of novel technologies are in the pipeline: (1) sequencing, (2) primer extension, (3) allele-specific hybridization (allele-specific oligonucleotide hybridization), (4) allele-specific oligonucleotide ligation, and (5) *in silico* SNPs (SNP mining).

Sequencing is the method of choice for SNP discovery and is based on primer extension with either (1) fluorophore primers and unlabeled terminators or (2) unlabeled primers and fluorophore terminators. Reaction products are then separated by either slab gel or capillary electrophoresis. For single-base primer extension, the target region is first PCR-amplified. Then a single-base sequencing reaction is performed that uses a primer annealing exactly one base off the polymorphic site. A series of detection methods are available, one of which uses a labeled primer and separates the extension products by gel electrophoresis. Alternatively, the single-base extension product can be fragmented into smaller pieces, whose masses are then measured by mass spectrometry. Most current SNP projects still involve fluorophore-labeled dideoxynucleotide terminators that stop chain extension. Allele-specific oligonucleotide hybridization discriminates between two target DNAs differing by one base. Fluorophore-labeled PCR fragments are hybridized to immobilized oligonucleotides representing the SNP sequences. After stringent hybridization and washing, fluorescence intensity is measured for each SNP oligonucleotide. For an allele-specific oligonucleotide ligation, the genomic target sequence is first PCR-amplified. Then, allele-specific oligonucleotides complementary to the target sequence and with the allele-specific base at their 3′- or 5′ ends are ligated to the DNA adjacent to the polymorphic site. Only in the case of a complete match is ligation possible; mismatches are definitely excluded. SNP mining ("*in silico* SNPing") is becoming a very efficient SNP discovery strategy, especially since it exploits already-existing and ever-increasing sequence databases. It depends on effective bioinformatic tools for, e. g., base calling, assembly of called sequences, and editing of the assembled sequence contigs, some of which are already in use (e. g., PolyPhred; Nickerson et al. 1997). A major concern relates to the sequence information from uncharacterized regions of genomes, which may well contain sequencing errors that probably lead to false SNPs. Special software dealing with these problems is available (e. g., POLYBAYES; Marth et al. 1999). Presently, most of the SNPs are extracted from EST databases (Picoult-Newberg et al.1999; Beutow et al. 1999).

The choice of a particular detection technique depends on (1) the number of SNPs, (2) the number of samples to be screened, and (3) the number of simultaneous SNP profiling projects. Usually a decision has to be made as to whether outsourcing is more economic than SNPing in-house. In-house SNP discovery requires instrumentation, whose cost ranges from 25,000 to over 400,000 US$. Costs for reagents, plates, and other plastic-ware are calculated from $0.10 to $3.00 per single genotyping experiment. For example, if 100 SNPs have to be tested in 1000 individuals, then the 100,000 genotyping reactions would amount to at least 10,000 and a maximum of 300,000 US$.

Below we demonstrate the potential of four selected SNP discovery and genotyping technologies that provide the desirable criteria (high-throughput, automation, and miniaturization), are firmly established, and have already shown potential in SNPing. This selection is by no means complete but is restricted by space limitations.

4.2
Selected Techniques

4.2.1
SNP Analysis I: The Invader Technology

4.2.1.1 Introduction

The Invader platform is a genotyping method suitable for direct analysis of genomic DNA with or without target amplification. It is an accurate and specific detection method for single-base changes, insertions, deletions, gene copy number, infectious agents, and gene expression. Invader reactions can be performed directly on genomic DNA or on RNA-, PCR-, or RT-PCR-amplified products. In an Invader reaction, a target-specific signal is amplified, but not the target itself. This technology is founded on the Cleavase enzyme's ability to recognize and cleave a specific invasive nucleic acid structure. This structure is generated when two oligonucleotides (oligos) – the Invader oligo and primary probe – overlap on the nucleic acid target.

4.2.1.2 Cleavase Enzymes

Cleavase enzymes are a collection of both naturally occurring and engineered thermophilic structure-specific 5′ endonucleases. Invader DNA assays use Cleavase enzymes that are derived from members of the flap endonuclease (FEN-1) family, typically found in thermophilic archaebacteria (Kaiser et al. 1999). The Invader RNA format uses Cleavase enzymes that are engineered from the 5′exonuclease domain of DNA polymerase I of thermophilic eubacteria (Lyamichev et al. 1993). The optimal substrate for these nucleases is comprised of distinct upstream, downstream, and template strands, which mimic the replication fork formed during displacement synthesis. The enzymes cleave at the 5′ end of the downstream strand between the first two base pairs and remove the single-stranded 5′ arm or flap (Kaiser et al. 1999).

4.2.1.3 Oligonucleotides and Structure

The primary probe consists of two regions: a 3′ target-specific region (TSR) and a 5′ flap. The sequence of the 5′ flap is independent of the target and consists of a universal sequence used for detection. In the presence of the specific target, the 3′ base of the Invader oligo overlaps with the target-specific region of the primary probe at the base referred to as "position 1." If position 1 of the primary probe is complementary to the target, the primary probe and Invader oligo create an overlapping structure. The Cleavase enzyme then cleaves the primary probe on the 3′ end of position 1, releasing the 5′ flap plus position 1 and the 3′ TSR. If the base at position 1 is not complementary to the target, a nicked structure is formed instead of an invasive structure, and position 1 of the primary probe effectively becomes part of the non-complementary 5′ flap. Because the proper substrate has not been formed, the Cleavase enzyme does not recognize and cleave the nicked structure. The requirement for complementarity at position 1 provides the ability to detect single-base changes.

4.2.1.4 Probe Cycling and Signal Amplification

The TSR of the primary probe is designed to have a melting temperature that is close to the reaction temperature to allow for the exchange of cleaved and uncleaved primary probes on the target. Following cleavage, the 5′ flap and the 3′ TSR dissociate from the target and a new, uncleaved primary probe takes their place. Since the primary probe is present in excess, numerous primary probes can be cleaved for each target molecule present, resulting in a linear accumulation of cleaved 5′ flaps. Once released, the 5′ flap functions as an Invader oligo on a universal fluorescence resonance energy transfer (FRET) oligo, which contains a donor fluorophore and a quencher dye on each side of the cleavage site. The Cleavase enzyme separates the donor fluorophore from the quencher dye, generating a fluorescent signal. Two sequences can be detected in a single well by using two different 5′ flap sequences and corresponding FRET oligos with non-overlapping fluorophores. Cleavage of primary probes and FRET probes takes place simultaneously in the DNA format. In the RNA format, the two reactions are sequential. Endpoint fluorescence signal is detected using a standard fluorescence plate reader after the sample has been incubated for 4 h at 63 °C. A ratio is determined for each sample using the net signal from the wild-type and mutant probes. Ranges of wild type/mutant ratios specify the homozygous and heterozygous genotypes. Copy number is determined by comparing the specific signal generated from the gene of interest with that of a constitutively expressed gene such as β-tubulin.

4.2.1.5 DNA Format

In the DNA format, the 5′ flap forms a transient invasive duplex with a FRET oligo. The Cleavase enzyme recognizes the invasive structure and cleaves the 5′ fluorophore from the FRET oligo. After the fluorophore has been released, the 5′ flap dissociates and is able to form an invasive structure with a new, uncleaved FRET oligo. Signal amplification results from the combined effect of having multiple flaps cleaved for each target molecule and from each of those released flaps driving the cleavage of multiple FRET oligos. The cleavage of both the primary probes and the FRET probes occurs simultaneously at a single temperature near the melting temperature of the primary probe (Figure 4.1) (de Arruda et al. 2002; Kwiatkowski et al. 1999; Neville et al. 2002).

4.2.1.6 RNA Format

The FEN-type Cleavase enzymes used in the DNA format do not recognize RNA targets. Therefore, the RNA format uses Cleavase enzymes derived from DNA polymerases (pol-type), which recognize both DNA and RNA. The reaction used to

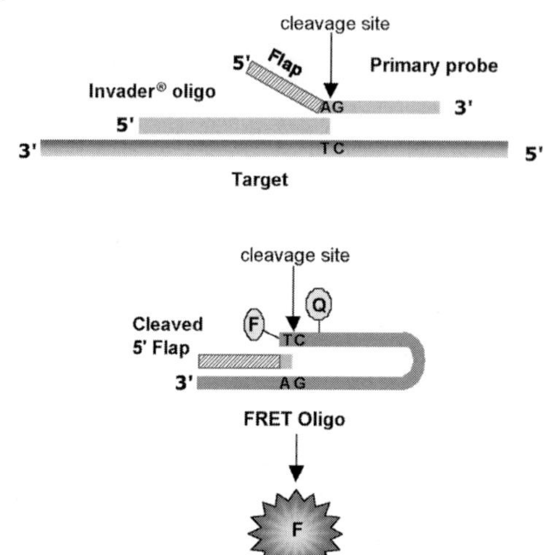

Figure 4.1. Schematic of the Invader assay.

cleave the FRET oligo is adapted to accommodate the slightly different substrate specificity of the pol-type Cleavase enzymes. The two sequential steps in the RNA format are the primary reaction and the secondary reaction. During the primary reaction, the Invader oligo and primary probe form an invasive structure on the RNA target. An additional oligo, the stacker oligo, coaxially stacks with the 3′ end of the primary probe and allows the primary probe to cycle on and off the target at a higher temperature. Stacker oligos increase the sensitivity of the assay.

In the separate secondary reaction, in contrast to the DNA format, the cleavage product of the primary reaction (the cleaved 5′ flap plus position 1 of the TSR) forms a one-base overlap structure with a secondary reaction template (SRT) and a FRET oligo. The enzymatic cleavage of the FRET oligo separates a fluorophore from a quencher molecule, as described for the DNA format, and fluorescent signal is generated. The released 5′ flap remains bound while the FRET oligos cycle on and off the SRT. Fluorescent signal accumulates linearly as a result of each 5′ flap causing the cleavage of multiple FRET oligos. An arrestor oligo, which hybridizes to the TSR and part of the 5′ flap of the uncleaved primary probes, is added to the secondary reaction. This allows for greater signal generation by preventing the uncleaved primary probes, but not the 5′ flaps, from hybridizing to the SRT during the secondary reaction (de Arruda et al. 2002; Eis et al. 2001; Kwiatkowski et al. 1999).

4.2.1.7 Alternate Detection Formats

The composition of the 5′ flap is independent of the target and can be detected in a variety of ways (e. g., by size, sequence, charge, or fluorescence). These properties enable the Invader technology to be used with numerous alternate detection formats, including mass spectrometry, capillary electrophoresis, microfluidics, universal array chips, capture, eSensor, fluorescence polarization, and time-release fluorescence.

ACLARA Biosciences (Mountain View, CA) produces small, electrophoretically distinct, fluorescent molecules called eTag reporter molecules that can be incorporated into primary probes as 5′ flaps. The released eTag flaps are resolved and quantitated using universal separation conditions on standard capillary DNA sequencing instruments. High-level multiplex capabilities are possible with the development of hundreds of eTaq reporters. ACLARA has demonstrated the ability to detect 26 Invader RNA assays from a single tube.

4.2.1.8 Specificity

Superior specificity has been achieved with the Invader technology by combining sequence identification with enzyme structure recognition. The key to the assay's specificity is the strict requirement of an invasive structure. If the sequence at the cleavage site is not complementary to the target, an invasive structure does not form, and the 5′ flap is not released for detection in the secondary reaction.

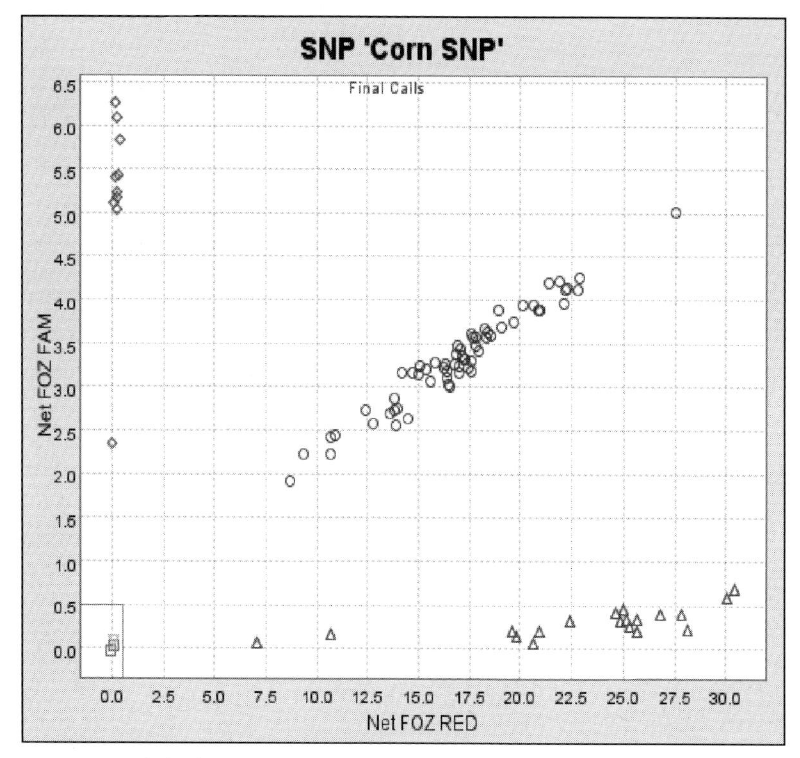

Figure 4.2. Results from a corn Invader SNP assay. Diamonds: homozygous wild-type samples; circles: heterozygote samples; triangles: homozygous mutant samples; squares: no target controls. The data were analyzed using Invader Analyzer software (Third Wave Technologies, Madison, WI).

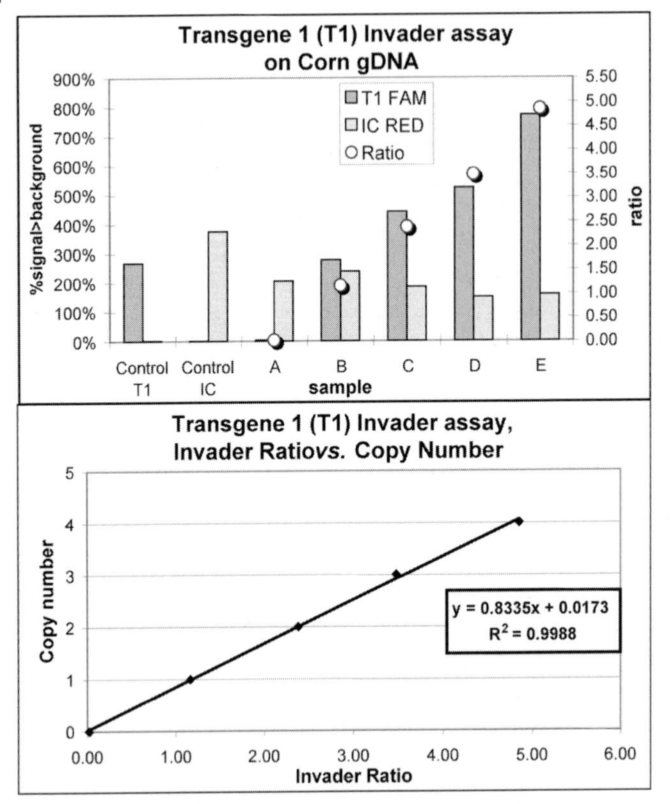

Figure 4.3. Results from a corn Invader copy number assay. Solid bars represent signal from a transgene inserted into corn DNA, and patterned bars represent signal from the internal β-tubulin control gene. The ratio is determined by the formula ((transgene signal/no target control signal) −1)/((internal control signal/no target control signal) −1). Plotting the sample ratios against a linear range of numbers identifies the relative copy number of the transgene.

For mutation detection applications, this discriminatory capability provides high accuracy, as demonstrated by its ability to classify the three possible genotypes (Figure 4.2).

The quantitative aspect of the Invader platform allows for the determination of chromosome and gene copy number directly from genomic DNA. This analysis may be accomplished by comparing the specific signal generated from the gene or chromosome of interest with that of a reference gene such as β-tubulin, which is not known to be polymorphic for either duplication or deletion. Careful design of the Invader oligos to regions with minimal homology to other regions within the genome enables accurate quantitation of the desired target sequence (Figure 4.3).

4.2.1.9 **Robustness**

Numerous sample preparation methods and sample sources are compatible with the Invader assay. The Invader assay provides an accurate genotype for both plant and animal samples using FTA-treated paper as a source for genomic DNA (Kozlowski et al. 2003). In addition, many homebrew and commercially available sample preparation methods can be used with the Invader assay.

Ease of use, flexibility, and scalability are some of the advantages that make the Invader platform ideally suited for automation. Assays using the Invader platform are completely homogeneous, require minimal hands-on time, and can be carried out in microtiter plates. Assays are readily adaptable from manual to semiautomated to fully automated for use in ultrahigh-throughput settings with a variety of standard plate formats. Most automated liquid handling systems can accomplish the assay setup, which requires only a few pipetting steps. Regardless of format (e.g., 96- or 384-well) or setup (e.g., manual or automated), the robustness of the platform remains constant (de Arruda et al. 2002).

4.2.1.10 **Invader Applications**

Applications of the Invader technology include the following: mutation detection, SNP scoring, high-throughput genotyping, marker-assisted selection, transgene detection and quantitation, copy number determination, genetic disease testing, and gene expression analysis. Custom assays have been developed for wheat, barley, oat, corn, soybean, cotton, rice, canola, tomato, pepper, lettuce, tobacco, cattle, swine, sheep, poultry, and mice.

4.2.1.11 **Conclusion**

The Invader platform is a homogeneous, isothermal DNA probe–based system for the quantitative detection of specific nucleic acid sequences. Invader assays can be performed directly on DNA-, RNA-, PCR-, or RT-PCR-amplified products. Based on the ability of the Cleavase enzymes to recognize and cleave specific nucleic acid structures, the Invader platform is a highly specific method for detecting SNPs, insertions, deletions, and gene copy numbers. The flexibility, ease of use, and robustness of the technology make it suitable for both small and high-throughput laboratories.

4.2.2 **SNP Analysis II: Pyrosequencing**

4.2.2.1 **Introduction**

Pyrosequencing technology is a method for sequencing by synthesis in real-time (Ronaghi et al. 1998). It is based on an indirect bioluminometric assay of the pyrophosphate (PPi) released from each deoxynucleotide (dNTP) upon DNA chain elongation (Ronaghi et al. 1998; Alderborn et al. 2000). A DNA template/primer complex is presented with a dNTP in the presence of exonuclease-deficient Klenow

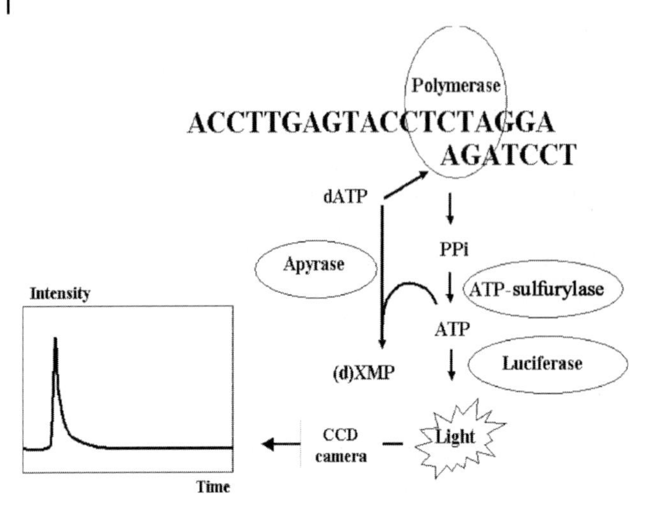

Figure 4.4. The principle of Pyrosequencing technology. Polymerase-mediated extension of a primer on a template releases pyrophosphate (PPi). The PPi is converted to ATP and then to a light signal via an enzyme cascade including ATP sulfurylase and luciferase. The light signal is detected by a CCD camera. Excess nucleotide and ATP are degraded by apyrase before addition of the next nucleotide.

DNA polymerase (see Figure 4.4). The four nucleotides are sequentially added to the reaction mixture in a predetermined order. If the nucleotide is complementary to the template base and thus incorporated, PPi is released and used as a substrate, together with adenosine 5′-phosphosulphate (APS) for ATP sulfurylase, which results in the formation of adenosine triphosphate (ATP). Luciferase then converts the ATP, together with luciferin, to oxyluciferin, AMP, Ppi, and visible light, which is detected by a luminometer or charge-coupled device. The light produced is proportional to the number of nucleotides added to the extended primer chain, thus giving Pyrosequencing technology a unique position as a truly quantitative sequencing method. Excess nucleotide is digested by apyrase present in the reaction mixture before the addition of the next nucleotide. Further improvements on the initial method have enabled extended and more robust read lengths, e. g., through the use of single-stranded DNA binding protein (SSB) to reduce secondary structure in DNA templates (Ronaghi 2000).

4.2.2.2 SNP Genotyping Using Pyrosequencing Technology

Pyrosequencing technology has been commercialized by Biotage AB (formerly Pyrosequencing AB) in Sweden, which provides a range of systems, software, and reagent kits for processing up to 96 post-PCR samples in parallel on solid phase (see http://www.biotage.com). Methods are included for rapid sample preparation (Dunker et al. 2003). The method also lends itself to automation, since it involves a homogeneous assay. The software supports multiplexing of SNP or mutation detection in different templates or positions, detection of multiple

SNPs in one template, analysis of insertions and deletions, quantification of allele frequency, and sequencing of short stretches of typically 20–40 bases. Pyrosequencing technology compares favorably to other techniques for genotyping of SNPs (see, e. g., Nordfors et al. 2002).

This technology provides the option of analyzing neighboring, or even contiguous, SNPs with the same primer, as well as SNPs on different templates, in the same assay (multiplexing). In addition, primer placement is somewhat flexible, since the 3′ end does not need to be juxtaposed to the SNP, which can be an advantage in highly variable genomes such as those of certain plant species where variation might compromise the placement of sequencing primers. An additional aspect that is especially relevant in plant genomes is the ability to provide single assays to deliver accurate, quantitative SNP genotyping data from genomes with a high level of ploidy, in which allele ratios vary. Pyrosequencing technology has been applied to several areas of research, such as human genetics, bacterial typing, plant genetics, and animal breeding (see reviews by Ronaghi 2001; Berg et al. 2002).

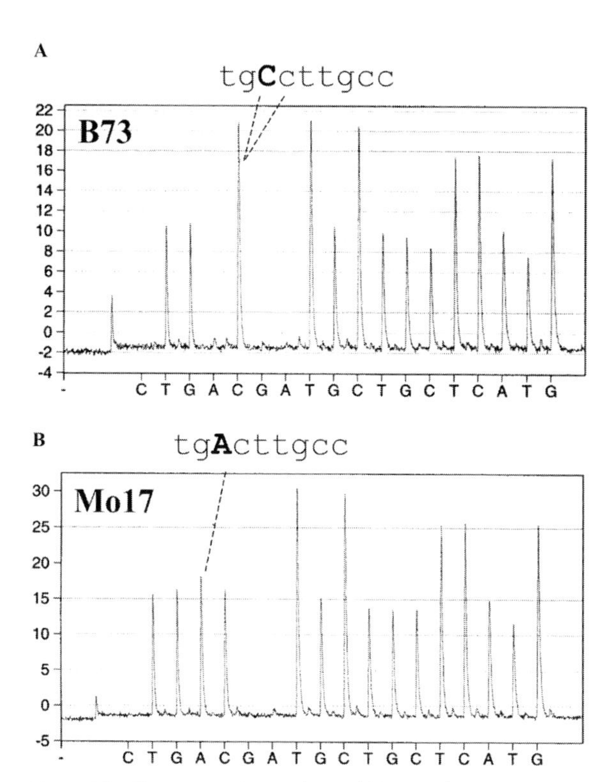

Figure 4.5. Pyrosequencing analysis of SNPs in three genes in maize. The polymorphic positions are marked in bold type. (A) The B73 allele of nucleoside diphosphate kinase. (B) The Mo17 allele of nucleoside diphosphate kinase (Ching and Rafalski 2002; reproduced with kind permission of the authors).

An illustrative example of the use of Pyrosequencing technology is the genetic mapping of ESTs in maize using SNPs (Ching and Rafalski 2002). Sequences obtained from mapping parents identified SNPs in three genes, and Pyrosequencing systems were used to analyze these loci in 94 individuals of a maize recombinant-inbred population. The results of SNP genotyping of two alleles of nucleoside diphosphate kinase are shown in Figure 4.5. The sequencing primer was located two nucleotides from the SNP. The Pyrogram results show the appearance of peaks, where the correct nucleotide was presented and incorporated and peak height is related to the number of nucleotides. The method also provides information on the sequence around the analyzed SNP, which provides a quality assessment of the data. In this example, the three loci were successfully mapped with a success rate for individual genotyping of greater than 95 % per locus. Failures were attributed to poor PCR yield.

4.2.2.3 **Allele Frequency Quantification**

Three significant issues in SNP-based mapping are the cost of large-scale genotyping itself, the accuracy of quantification, and the possibility of employing direct (molecular) haplotyping to increase predictive power. Linkage disequilibrium studies on individual samples require high-throughput genotyping, which can be cost-prohibitive. An alternative approach is to measure SNP allele frequencies in DNA pools, which in turn places high demands on the accuracy of quantitative genotyping. Pyrosequencing analysis is quantitative in that relative peak sizes indicate the number of incorporated bases. This permits accurate detection of heterozygotes and even quantification of allele frequencies in pools and sample mixtures (Neve et al. 2002; Wasson et al. 2002; Gruber et al. 2002). This is illustrated in Figure 4.6, which shows the relationship between allele frequencies and Pyrogram peak heights in a mixing experiment between two individuals homozygous for a

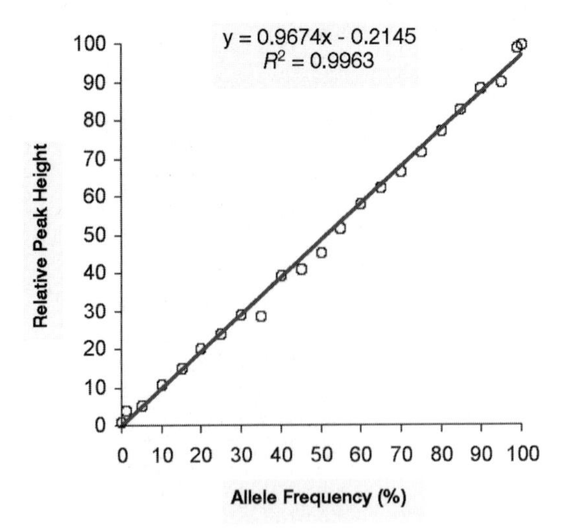

Figure 4.6. Regression line between allele frequencies and Pyrogram peak heights in a mixing experiment between two individuals homozygous for a particular SNP. Each point is the mean of duplicate determinations (Wasson et al. 2002; reproduced with kind permission of the authors and Eaton Publishing Co, USA).

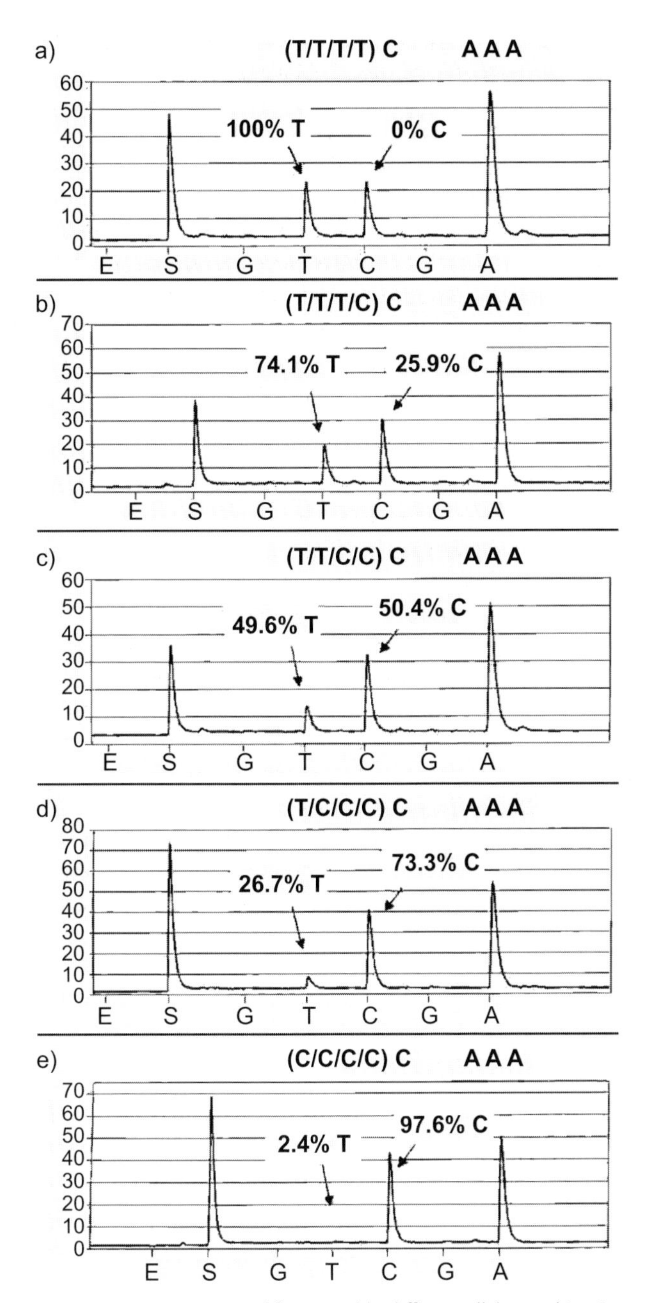

Figure 4.7. Genotyping of five possible different allele combinations of SNP-44/114 in the tetraploid genome of potato. The sequence to analyze was T/CCAAA. Panels (a) and (e) show the two homozygotes, whereas panels (b–d) show the three heterozygous allelic states, together with the percentage distribution of the Tand C allelic bases as calculated by the software (Rickert et al. 2002; reproduced with kind permission of the authors and Eaton Publishing Co, USA).

SNP (Wasson et al. 2002). The R^2 for the regression line was 0.996, with a significance of $P<0.0001$, which means that the authors could determine allele frequencies from DNA pools with confidence.

Another example of the application of quantitative Pyrosequencing analysis is shown in Figure 4.7 (from Rickert et al. 2002). In this example, five possible different allele combinations of a SNP (SNP-44/114) in the tetraploid potato genome were determined. The accuracy enabled all five possible allelic states to be correctly identified. Over 80 % of polymorphic sites could be genotyped successfully, which favorably compared with other studies. It is likely that recent developments in Pyrosequencing technology – including improved use of SSB, software-assisted primer design, and downscaling of reaction volumes – will result in an even higher success rate at lower cost.

4.2.2.4 **Haplotyping**

The power of SNP genotyping in genome mapping can be extended to the construction of haplotypes, which increase the information content. Sequencing methods, including Pyrosequencing technology, are particularly suitable for the construction of molecular haplotypes, since they facilitate identification of several adjacent SNP alleles (Bhattramakki and Rafalski 2001). One method of haplotype determination exploits allele-specific polymerase chain reaction (AS-PCR) in combination with Pyrosequencing technology. For example, DNA pools and allele-specific, long-range haplotype amplification have been used in combination with genotyping using Pyrosequencing technology to directly measure haplotype frequencies (Inbar et al. 2002). Similarly, Pyrosequencing has been used to analyze multiple SNPs on amplified fragments, including introduction of a mismatch at the second base from the 3′ end of the PCR primer to dramatically improve allele specificity (Pettersson et al. 2003). Genotyping of heterozygote samples after AS-PCR gave a typical monoallelic pattern at each SNP, in which the identity of the present allele depended on the allele-specific initial amplification. The results obtained by Pyrosequencing technology are truly quantitative, enabling detection of any nonspecific allele amplification (Pettersson et al. 2003).

Haplotyping in multi-genomes, such as the allopolyploid genome of wheat (AABBDD) is particularly challenging. Mochida and coworkers have successfully tested the application of Pyrosequencing for generating haplotypes for polymorphic regions of purothionin genes in the wheat multi-genome, using normal wheat and a nullisomic-tetrasomic series. Their method involved the use of a single sequencing primer to simultaneously determine sequences, and thus allelic frequencies, for the three genomes A, B, and D. The heights of the peaks were used to calculate the ratios of the polymorphic nucleotides in the nullisomic-tetrasomic series and enabled assignment of contigs (haplotypes) to their homoeologous chromosomes. The results from Pyrosequencing analysis fully agreed with the chromosomal assignment of the purothionin gene previously reported. The authors found Pyrosequencing technology to be the most suitable method for SNP analysis among cultivars or strains of polyploid plants due to the linear response that provided quantitative SNP data in the hexaploid background, and also because of the ability to produce sequence data that facilitated haplotype and linkage analysis.

A natural extension of Pyrosequencing technology, which provides means for genotyping at different positions along a template, would be to develop methods for direct molecular haplotyping without AS-PCR. The possibility of designing assay-specific orders for nucleotide addition means that Pyrosequencing technology provides the possibility of primer extension over a polymorphic site such that the two alleles become out of phase, thus providing patterns that are specific for each haplotype (Odeberg et al. 2002). Future improvements in read length are expected to further extend the limits of direct haplotyping based on Pyrosequencing technology.

In conclusion, Pyrosequencing technology has proved itself in a large number of applications and offers unique features that make it suitable for gene mapping based on SNP genotyping, particularly in the complex genomes of plants.

4.2.3 SNP Analysis III: A Scalable High-multiplex SNP Genotyping Platform

4.2.3.1 Introduction

It is estimated that there are over 10 million SNPs present at a population frequency greater than 1% in the human genome (Kruglyak and Nickerson 2001). Some of these SNPs, along with environmental factors, may be responsible for individual predisposition to common diseases, such as cardiovascular heart disease, diabetes, and cancer (Goldstein and Weale 2001). In order to build and collect the resources for large-scale genetic disease association studies, the International HapMap project is determining a minimal set of approximately 500,000 tag SNPs across a diverse population of individuals that will enable whole-genome studies of complex genetic diseases. A need clearly exists for a high-throughput method to determine the genotypes for a large set of SNPs across many individuals in an efficient and cost-effective way. The Illumina genotyping platform enables these large-scale genomic studies by combining several technologies. These include a miniaturized array of individual arrays (Sentrix Array Matrix), a high-resolution confocal scanner (BeadArray Reader) with which the arrays are read, and a highly multiplexed genotyping assay (GoldenGate assay). These combined technologies maximize ease of use, improve data quality, and reduce the cost per data point. This section will provide an overview of the Illumina platform, which has been used for high-throughput SNP genotyping studies.

4.2.3.2 The Illumina Genotyping Platform

The Sentrix Array Matrix consists of 96 miniature arrays, each 1.2 mm in diameter, formatted in an 8×12 matrix suitable for mating to microtiter plates. Each of these miniature arrays contains approximately 50,000 fiber optic strands, fused together in a bundle to form the array (Michael et al. 1998). This array is chemically etched

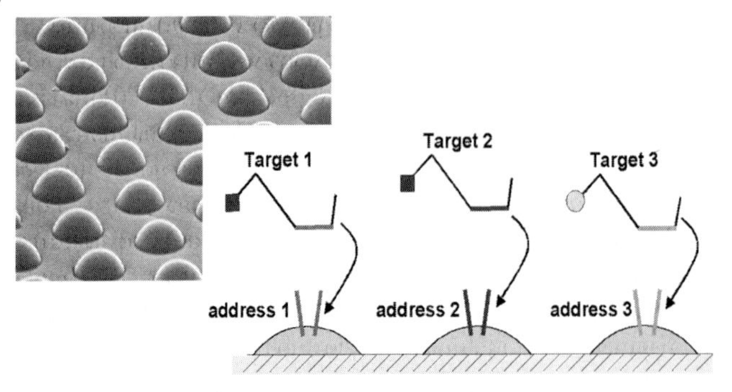

Figure 4.8. Universal arrays.

to form wells at the end of each fiber, each measuring 3 μm in diameter, with center-to-center spacing of approximately 5 μm. Into these ~50,000 wells are randomly self-assembled 3-μm beads that have covalently attached oligonucleotides (Figure 4.8). In our universal array product, we use 1624 unique oligonucleotide sequences. A different oligo sequence is associated with each bead type. Each bead type forms an address sequence, which is designed to hybridize to an assay product (Figure 4.8). The number of 3-μm wells in the miniature array outnumbers the types of beads, so there are on average 30 beads for each bead type. This redundancy of measurement improves genotyping assay results. Since these beads are randomly assembled into the miniature array, a hybridization-based decoding process is performed on the array to determine the type of bead that is in every position on the array. This decoding process also enables a quality-control check of every bead in every array.

The small feature size of the Sentrix Array Matrix necessitated the design and manufacture of a laser confocal scanner that is uniquely tailored to read these high-density arrays. Illumina's BeadArray Reader has a resolution of approximately 0.8 μm (Barker et al. 2003). It has two excitation lasers (532 nm and 635 nm) and is able to simultaneously capture images in two color channels. The BeadArray Reader software automatically images all 96 miniature arrays in a Sentrix Array Matrix, registers the images, and extracts and saves the intensity information, which is used by the genotyping software to determine genotype calls.

The GoldenGate assay is designed to easily multiplex many loci in a single reaction while maintaining the ability to target any particular SNP of interest. Previous efforts on multiplexing assay development attempted to amplify the targets of interest prior to allelic discrimination or attempted to perform allelic discrimination during amplification. Those prior efforts had limited success because of site-specificity and/or allelic discrimination problems in the amplification reaction. We have greatly improved the methods to successfully multiplex to high levels by separating the site-specificity and the allelic discrimination process from the signal amplification step. Separation of these steps further allows us to optimize each step independently of the other. We optimized the assay oligo hybridization process under conditions that are independent of the allelic discrimination step. Likewise, we were able to optimize the allelic discrimination condition independently of the assay oligo hybridization condition or the signal amplification condition.

The DNA in the GoldenGate assay is first attached to paramagnetic particles and acts as a purification scaffold on which properly hybridized assay oligos may be retained (Figure 4.9). Three assay oligos are synthesized for each SNP locus: two allele-specific oligos (ASOs) and one locus-specific oligo (LSO). The two ASOs have a 3′ region complementary to the SNP genomic region, and each ASO has a 3′ base complementary to one of the two SNP alleles and a 5′ region complementary to one of two universal PCR primers (P1 and P2, Figure 4.8). The LSO has a 5′ region complementary to the SNP genomic region, a middle portion later used to target one of 1624 capture sequences, and a 3′ portion complementary to universal PCR primer P3. The LSO is phosphorylated on the 5′ end. The GoldenGate assay multiplexes at a level greater than 1536 SNP loci. All three assay oligos for all 1536 SNP loci are pooled and hybridized to a sample of genomic DNA. The hybridization process is performed, and properly hybridized assay oligos are preferentially retained, while improperly hybridized or non-hybridized oligos are washed away.

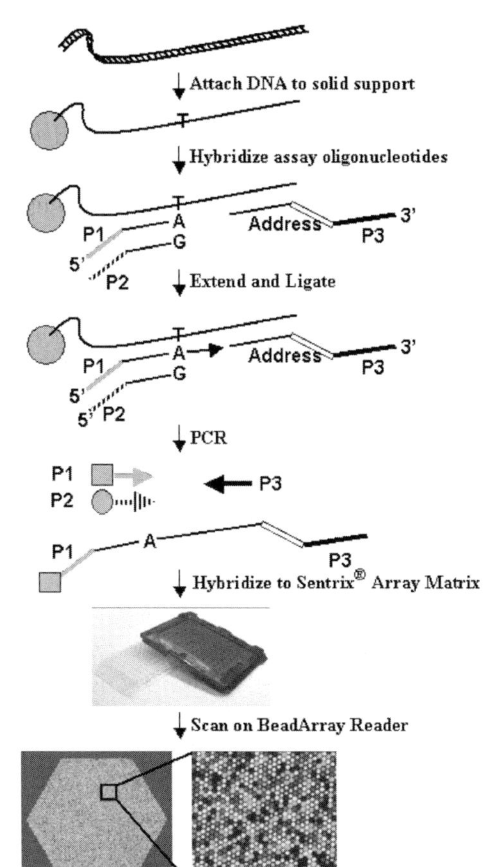

Figure 4.9. The Illumina genotyping platform.

DNA polymerase is added to the hybridized assay oligos, and allelic discrimination is carried out directly on the solid-phase DNA. The 3′ ends of ASOs that properly base-pair with the SNP site on the genomic DNA are extended to the 5′ ends of the LSO. A ligase is then added and the extended ASO is joined with the nearby LSO to form a PCR template. The requirements that the ASO hybridize independently of the LSO and that an extension reaction be performed prior to a ligation reaction serve two purposes. An extra degree of freedom is added to the positioning of the LSO, thus providing better hybridization targets, and mis-hybridized oligos are unlikely to be adjacent to each other and thus are unlikely to be extended and joined.

PCR primers P1, P2, and P3 are added to the extended and ligated products. P1 and P2 are fluorescently labeled with Cy3 and Cy5, respectively. In the example diagrammed in Figure 4.9, P1-A-P3 product would be produced for the SNP loci gen-

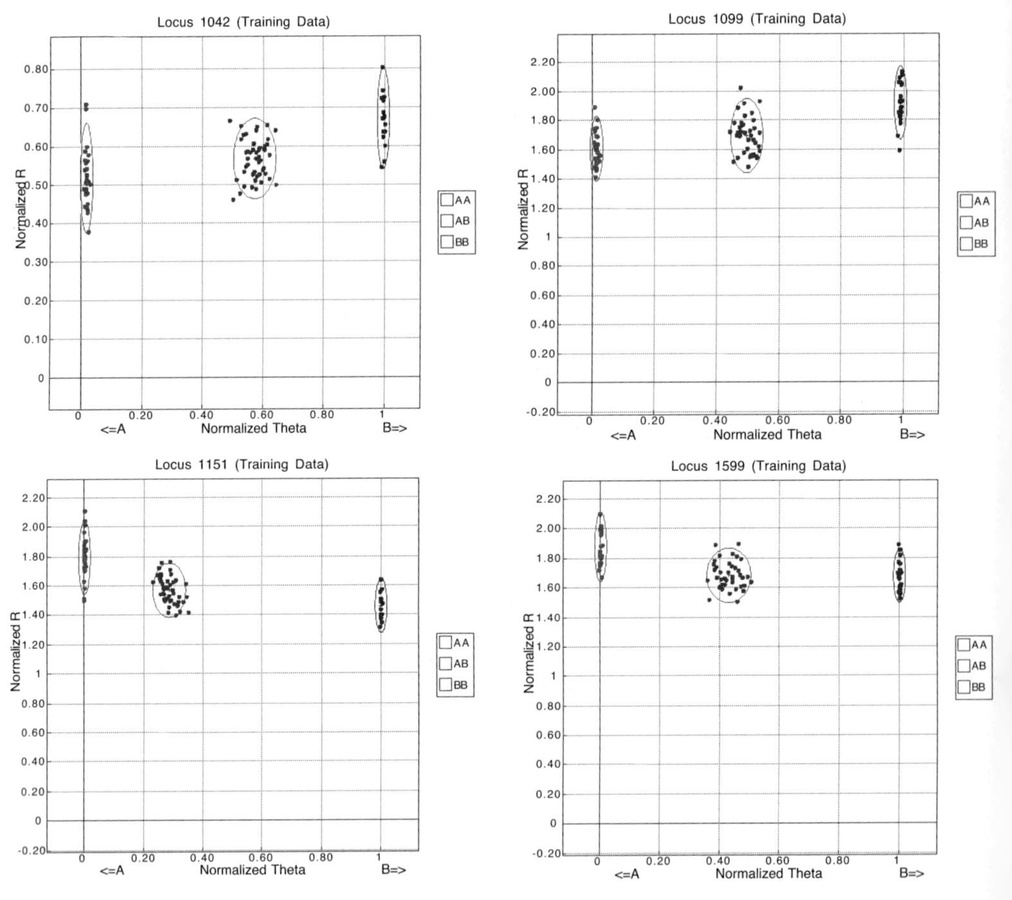

Figure 4.10. Results from a 1152-loci multiplex reaction on 96 DNA samples.

otyped. Likewise, if the SNP in the figure were a heterozygote, both P1-A-P3 and P2-G-P3 products would be produced. If it were a homozygote G, then only P2-G-P3 product would be produced. The PCR primers P1, P2, and P3 amplify the signal from all 1536 loci simultaneously. The fluorescently labeled PCR products are made single-stranded and hybridized to the Sentrix Array Matrix. The address sequence on the LSO is unique for each interrogated SNP locus and will hybridize to the complementary oligo sequences on the beads in the array. A reaction that is highly multiplexed in solution is de-multiplexed onto an array for readout. The picture at the bottom of Figure 4.9 is a composite image of a single array, scanned in two colors. The white beads have Cy3 signal, the dark gray beads have Cy5 signal, and light gray beads have both Cy3 and Cy5 signal. Intensity information in both color channels for each bead is determined. Since there are on average 30 beads per bead type, the trimmed mean of the intensity in each color channel is calculated for each bead type.

Four representative plots of genotyping data for a 1152-loci multiplex reaction on 96 DNAs are shown in Figure 4.10. Each dot represents the signal intensity for a single DNA sample. The y-axis of these graphs represent the normalized intensities, and the x-axis represents the normalized theta angles (theta=$[2/\pi]\mathrm{Tan}^{-1}$[Cy5 intensity/Cy3 intensity]). The ovals enclosing the samples are the two standard deviation boundaries for each genotype cluster. The genotype clusters are well separated from each other and the sample-to-sample variation within each cluster is low.

4.2.3.3 **Genotyping Data**

In a recent genotyping study, 17,280 human SNPs were run through the GoldenGate assay (Fan et al. 2003). These SNPs were selected from a collection of approximately 124,000 high-quality SNPs that have relatively high minor allele frequency, and each allele was independently observed by at least two sequence reads. These 17,280 SNPs were chosen because they were predicted by the assay design software to have a high likelihood of success. The SNPs were grouped into 15 sets of 1152 loci multiplex pools. Assays for 10×1152 SNPs were developed on both strands and 5×1152 SNPs were developed on only one strand. The results of this study are represented in Table 4.1. Assay conversion rates ranged from 96 % to 99 % when assays were designed to both strands, and data are from the better-performing strand. The assay conversion rate drops only slightly to a range between 94 % and 96 % when assays are designed for only one strand. The very high assay conversion rates observed in this study are due to the selection for the highest-quality SNPs. In another study, in which SNPs were chosen randomly from dbSNP and assays were designed to both strands, an assay conversion rate of 81 % was obtained (Fan et al. 2003).

Table 4.1. Assay development conversion rate and genotyping call rate for 17,280 loci (modified from Fan et al. 2003).

Bundle	*Multiplex*	*Successful assays*	*Conversion rate (%)*	*Called genotypes*	*Possible genotypes*	*Call rate (%)*
DS_1	1152	1136	99	107,882	107,920	99.96
DS_2	1152	1129	98	107,232	107,255	99.98
DS_3	1152	1132	98	107,515	107,540	99.98
DS_4	1152	1134	98	107,704	107,730	99.98
DS_5	1152	1121	97	106,464	106,495	99.97
DS_6	1152	1112	97	105,619	105,640	99.98
DS_7	1152	1112	97	105,575	105,640	99.94
DS_8	1152	1120	97	106,376	106,400	99.98
DS_9	1152	1114	97	105,794	105,830	99.97
DS_10	1152	1107	96	105,139	105,165	99.98
SS_1	1152	1101	96	104,540	104,595	99.95
SS_2	1152	1107	96	105,111	105,165	99.95
SS_3	1152	1097	95	104,189	104,215	99.96
SS_4	1152	1086	94	103,132	103,170	99.96
SS_5	1152	1084	94	102,947	102,980	99.97
Total	17,280	16,692	97	1,585,219	1,585,740	99.97

4.2.3.4 Conclusion

The Illumina genotyping platform is an efficient, accurate, and cost-effective system for high-throughput SNP genotyping. The GoldenGate assay coupled with the Sentrix Array Matrix is a flexible platform that allows the genotyping of a large number of user-definable SNPs across a large number of samples. The throughput and scalability of the assay are well suited for large-scale association studies to discover the genetic basis of complex diseases.

4.2.4 SNP Analysis IV: High-throughput SNP Analysis by MALDI-TOF MS

4.2.4.1 Introduction

The rapidly expanding field of biotechnology has fostered the development of several high-throughput technologies. Although widely applied in analytical settings for several decades, mass spectrometry found more widespread use in biotechnology only after the "soft" ionization and desorption technologies matrix-assisted laser desorption (MALDI) and electrospray ionization (ESI) were developed. These new techniques of mass spectrometry enabled analysis of large biomolecules such as proteins and nucleic acids.

The core process of MALDI-TOF (time of flight) mass spectrometry is the laser-induced desorption of biomolecules mediated by an excess of co-crystallized matrix substance. Matrix substances are organic molecules selected for their energy absorption spectrum (matching the selected laser wavelength). They are also chemically inert towards the molecule of interest. The matrix absorbs the laser energy and evaporates into the vacuum of the mass spectrometer. Analytes are co-desorbed with the matrix and are ionized by proton transfer. The ions are then accelerated in an electric field, and their flight time (TOF) through a field-free drift region is determined. The flight time is proportional to the molecular mass of the analytes: molecules of small masses travel through the drift region faster than molecules of large masses.

The matrix-mediated desorption process yields more intact analyte molecules than do other types of mass spectrometry (with the exception of ESI-TOF) and is the preferred method when nonvolatile substances such as nucleic acids and proteins need to be detected with appropriate sensitivity.

Analysis of DNA by MALDI-TOF MS offers some significant advantages. For example, the analyte is analyzed directly rather than through surrogate markers such as fluorescent labels. Determination of an intrinsic molecular property of the analyte, the molecular mass, is more accurate and allows discrimination between artifacts and desired reaction products. Determination of the molecular mass is also independent of structural features. Hence, misinterpretations due to secondary structures of the analyte – as observed, for example, in gel electrophoresis – are avoided. Additionally, the MALDI process is extremely fast. Today, data acquisition can be achieved on the order of a few hundred milliseconds and allows real-time analysis of samples.

Within recent years, MALDI-TOF MS has been developed into a viable platform for high-throughput genotyping. Several further qualitative and quantitative applications have now been added to the nucleic acid analysis portfolio. These include the relative quantitation of genetic information in DNA pools and sample mixtures (Bansal et al. 2002; Buetow et al. 2001; Mohlke et al. 2002; Ross et al. 2000; Werner et al. 2002) and resequencing methods that allow the rapid discovery of SNPs, the screening for mutations or signature sequence-based identification of organisms such as pathogens (Hartmer et al. 2003; Lefmann et al. 2004; Stanssens et al. 2004), as well as relative and absolute quantitation of gene expression (Ding and Cantor 2003). With these, MALDI-TOF MS has matured into a high-performance nucleic acid analysis method. A recent review summarizes these developments (Jurinke et al. 2004). The focus of the present contribution will be the analysis of SNPs by MALDI-TOF MS.

4.2.4.2 **SNP Analysis Using the Massarray Platform**

The Homogeneous MassEXTEND (hME) Assay

The most common mass spectrometry–based methods for genotyping of SNPs or known mutations employ primer extension (Braun et al. 1997). The homogeneous MassEXTEND assay (originally published as PRimerOligoBaseExtension) is a primer extension assay specifically developed for high-throughput processing (Rodi et al. 2002). It is usually performed in 384-well microtiter plates.

Figure 4.11 depicts the basic steps of MassEXTEND. The region containing the SNP is first amplified in a PCR reaction. Unreacted dNTPs are then dephosphory-

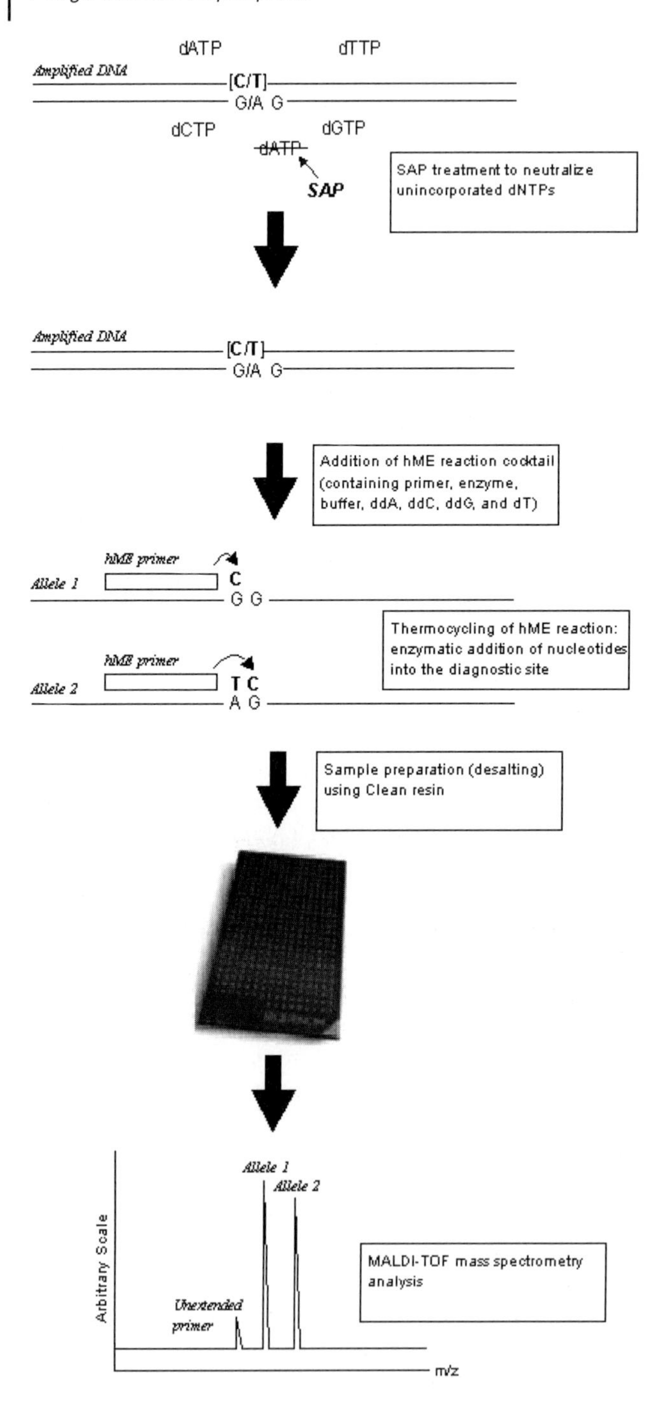

Figure 4.11. Schematic overview of the MassEXTEND process. The SNP-specific target region is amplified by PCR. Unincorporated dNTPs are dephosphorylated. In a post-PCR primer extension reaction, allele-specific termination products are generated. Samples are transferred to a chip array and automatically analyzed by MALDI-TOF MS.

lated by shrimp alkaline phosphatase (SAP) treatment. A sequence-specific hME primer is annealed immediately adjacent to the polymorphic site. The hME primer is extended in the presence of three dideoxyribonucleotide triphosphates (e.g., ddA, ddC, and ddG) and one deoxyribonucleotide. The selection of the deoxyribonucleotide is driven by the nature of the polymorphism. The dNTP should correspond to one of the alleles (in the depicted case the dTTP corresponds to the A allele). Analysis of a SNP leads to three possible masses: a mass signal for unextended primer, a mass signal corresponding to a single nucleotide extension specific for one allele (allele 1), and a mass signal corresponding to a primer extended by two or more nucleotides specific for the other allele (allele 2). The masses generated in this reaction differ by at least the mass of a single nucleotide (around 300 Da). This mass difference is about two orders of magnitude more than the actual resolving power of instruments currently employed.

Upon completion of the cycled primer extension reaction, ion-exchange resin is added to the reaction vessels. This step removes sodium and potassium from the phosphate backbone of the nucleic acid products and conditions the sample for MALDI-TOF MS.

Following conditioning, samples are dispensed onto chip arrays (SpectroCHIP). The chip arrays are highly precise and uniform silicon chips pre-applied with matrix. Only a minute volume of about 10 nL analyte is required for analysis. The dispensing step is carried out by a specialized Pintool device. Sample measurement from miniaturized chip arrays allows fully automated assignment of genotypes. During the measurement, algorithms can evaluate the data quality and decide within microseconds whether a genotype call can be made accurately or whether more data should be acquired from a sample spot. The user is presented with the genotyping results in real time. Databasing allows visualization of mass spectra and post-acquisition data handling such as application of statistical tools.

Rapid SNP Validation and Genome-wide Association Studies in DNA Pools

As described above, MALDI-TOF MS efficiently genotypes sequence variations using primer extension. The same platform also allows relative quantification of primer extension products (Buetow et al. 2001; Ross et al. 2000). For each sample, the ratio of analytes that are co-crystallized with matrix and subsequently ionized after laser desorption is proportional to the peak area of the corresponding mass signals. This allows for relative quantitation of DNA analytes and thus facilitates estimation of the relative abundance of alleles in sample mixtures or DNA pools. Technically, the relative quantitation is achieved by normalizing the baseline of each mass spectrum. A Gaussian fit is then applied to specified mass signals (corresponding to targeted alleles of the hME reaction) and the area under the curve is integrated. Allele frequencies of the primer extension products are estimated as the ratio of the area of one allelic mass signal to the total area of all expected allelic mass signals.

The process of allele frequency estimation (allelotyping) by MALDI-TOF MS was implemented originally as a rapid means to validate SNPs. The analysis of each SNP marker in a sample pool provides a fast way to separate true SNPs from sequencing errors or false positives of the *in silico* selection process and allows the characterization of the allele frequency in selected populations without the experimental burden and cost of individual genotyping (Buetow et al. 2001). The technical characteristics of this process have been evaluated in several studies (e. g., Bansal et al. 2002; Mohlke et al. 2002; Werner et al. 2002). Using MALDI-TOF MS from miniaturized chip arrays, the limit of detection (LOD) was determined at 2% minor allele frequency, whereby the limit of quantitation (LOQ) was determined at 5–10% minor allele frequency. The precision of the process (including sample pooling) allows the estimation of allele frequencies in DNA pools with a standard deviation of ±3.0%, whereby the majority of deviation is introduced in the PCR amplification. MALDI-TOF MS contributes only around 1% standard deviation (SD) to the overall process.

SEQUENOM has used allelotyping to validate over 400,000 SNPs and to generate a human genome gene-based SNP map consisting of over 100,000 SNP markers with experimentally validated allele frequencies. The assay database www.realsnp.com contains 400,000 working assays with associated frequency information for polymorphic SNPs (assays are specific to the human genome).

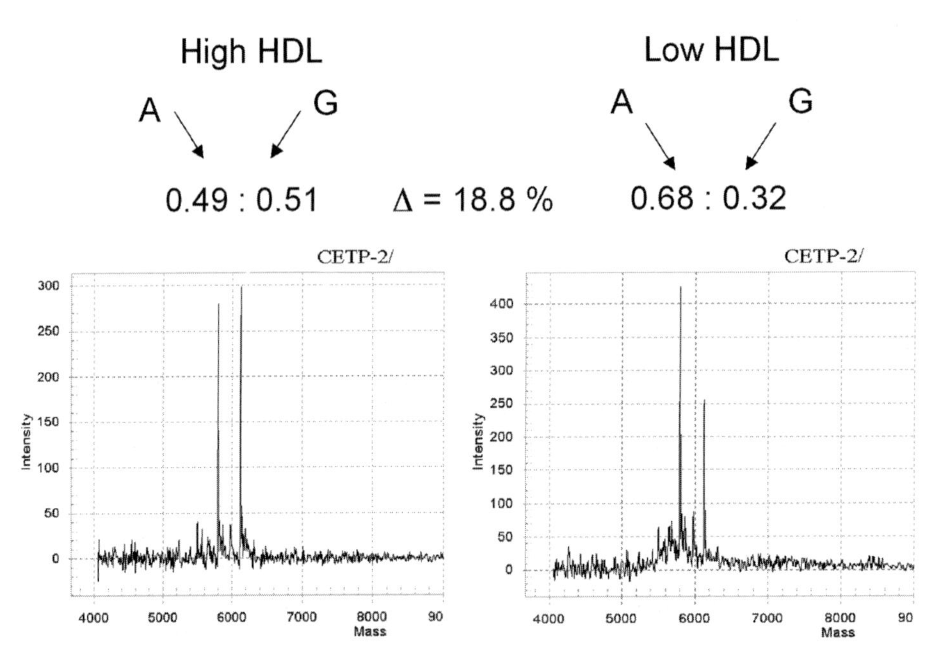

Figure 4.12. Allelotyping results of a SNP in the CETP gene obtained from DNA pools of individuals with high HDL levels and low HDL levels, respectively. The two predominant mass signals represent the primer extension products for the A and G alleles of the CETP SNP. As can be seen, the relative peak intensity (peak area) of the two alleles changes drastically between the two pools. This indicates that the G allele is less prevalent in individuals with low HDL. The example demonstrates how the combination of DNA pooling and a technology providing precise assessment of the relative abundance of alleles can be used to minimize processing efforts and cost in association studies.

Although an established SNP assay portfolio exists, analysis of large numbers of SNPs by individual genotyping of sample numbers large enough to provide sufficient statistical power in linkage studies or association studies is still costly and time-consuming. Several groups have therefore proposed the use of sample pools rather than individual genotyping. The estimation of allele frequentcies from case and control DNA sample pools allows for a significant reduction of experimental workload without major compromises in statistical power. For comparison, the analysis of 100,000 SNPs over two pools of 300 samples each requires only 200,000 reactions, whereas individual genotyping requires 6×10^7 reactions. Several groups have used allelotyping by MassARRAY successfully for candidate gene approaches or genome-wide association studies (Bansal et al. 2002; Mohlke et al. 2002). Figure 4.12 exemplifies the principle of association testing by DNA pooling using the MassARRAY system. The allele frequency of a SNP in the CETP gene changes significantly between a sample pool of individuals with high HDL values and a sample pool of individuals with low HDL values. This allows the prioritization of particular genomic regions for more in-depth analysis.

Multiplexing SNP Genotyping

A simple way to increase throughput and reduce cost per genotype is to perform multiple PCR/hME reactions in a single reaction vessel (multiplexing). The available mass window for analysis of primer extension products ranges from 5000 Da to 9000 Da (equivalent to products of 17–30 nucleotides). If primers targeting different SNPs or mutations are designed such that the masses of unextended primers and their corresponding extension products do not overlap, a single mass spectrum allows for unambiguous genotyping of multiple SNPs.

Although not all aspects of multiplexed SNP genotyping can be discussed here, a few points need to be highlighted. The most important aspect of multiplexing is the efficient but balanced PCR amplification of multiple SNP-specific target regions. As a first consideration, PCR primers should be checked for cross-loci amplification and, if sufficient genome sequence information is available, should also be blasted against the chosen genome to assure that each PCR primer pair amplifies only the desired region. We have found that a balanced primer T_m (around 60 °C), a balanced GC content, and a balanced amplicon length of roughly 100 bp usually provide good performance. Despite necessary adjustments in dNTP, enzyme amount, and primer concentrations, the two most critical factors for amplification are the ion strength of the PCR buffer and the final concentration of "free" Mg^{2+} ions. We found a final PCR buffer concentration of 1.25× (compared to the usual 1×) and a free magnesium concentration between 1.3 mM and 1.7 mM to be essential for efficient multiplexed amplification.

Multiplexed hME reactions follow considerations similar to those of the PCR step. hME primers should exhibit low hairpin formation and should not form primer-dimers. They should anneal specifically to only one amplicon in the multiplexed reaction. Primers and their extension products should be intercalated in such a way that each mass signal allows unambiguous assignment of alleles. Potential mass signal conflicts with byproducts, such as those generated by polymerase pausing or salt adducts, need to be avoided.

On a larger scale, the design of multiplexed PCR, the selection of appropriate termination mixes, and the efficient intercalation of assays are best performed with a computer-aided assay-design program (such as MassARRAY Assay Design available from SEQUENOM). Using these programs, up to 15 plexed MassEXTEND reactions can be designed, implemented, and analyzed in a fully automated manner. Using standardized reaction conditions, current MALDI-MS multiplexing (12-plex level) provides a first-pass call rate of 90 % (first-pass is defined as assays run straight from *in silico* design without prior knowledge of assay performance). The accuracy of the genotype data (automated calls, first-pass) averages to 99.7 %. Removal of those assays, which repeatedly did not PCR-amplify, increases the call rate to over 97 % and the averaged accuracy to 99.8 %. These are remarkable numbers considering that no prior assay optimization is necessary.

Figure 4.13 depicts a mass spectrum of a successfully called 12-plex reaction. All expected signals are marked with a dotted line. Signals with the descriptor “P” are unextended primers. Signals with the descriptor “Pausing Peak” mark polymerase artifacts (unspecific termination of the primer extension reaction). All termination mixtures are selected such that mass signals of pausing products do not interfere with allele calling.

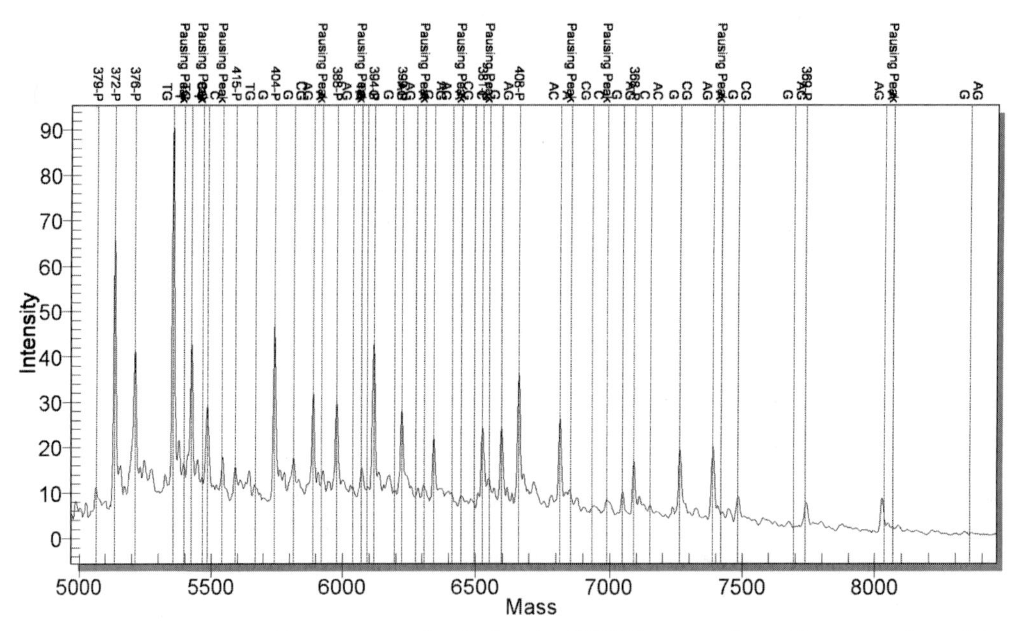

Figure 4.13. Mass spectrum of a 12-plex MassEXTEND reaction. Dotted lines mark all possible extension products of the multiplexed reaction, including unextended primer, alleles, and potential pausing signals. All mass signals (indicated by the dotted lines) are sufficiently separated to ensure that each genotype can be assigned unambiguously.

Performance Summary

The following features characterize the performance of MALDI-TOF MS-based SNP genotyping using the MassARRAY system.

- Efficient assay design: The simplicity of the MassEXTEND assay design and MALDI-MS readout combined with in-house experience comprising hundreds of thousands of hME reactions has generated a specific software tool (MassARRAY Assay Design) that matches the speed of the MALDI process. The designs include both PCR and hME primers; determination of optimal terminator mixes; reaction plate confirmation; primer purchase requisitions; and allow multiplexing of reactions. The use of ordinary oligodeoxyribonucleotide primers without any modification keeps the assay design fast, simple, and inexpensive. Implementation of *in silico*-designed multiplexed assays provides first-pass call rates of 90% without assay optimization. This allows rapid generation of SNP assay panels for genetic studies with minimal experimental effort prior to the initiation of a study.
- Simple assay processing: Multiplexed MassEXTEND assays are designed as addition-only assays. Both PCR and hME steps are designed as single-tube reactions, which enables automated assay setup on standard liquid handling equipment and simplifies sample tracking.
- Automated data collection and genotype reporting: Current MALDI-TOF MS instruments generate mass spectra in less than 1.5 s per chip array element, including transit time to the next element. Using recently released software (MassARRAY TYPER), genotypes can be analyzed in real-time during these same 1.5 s.
- MassARRAY throughput: The described combination of factors enables the processing and determination of thousands of genotypes in a single day using automated assay setup and a MassARRAY system. Acquisition/ real-time analysis of a 384-element chip array proceeds in $\approx$30 min. This translates to an analytical speed of about 150 genotypes per minute (9000 genotypes per hour). This throughput coupled with a cost per genotype of less than $0.10 renders whole-genome scans feasible and affordable (see www.sequenom.com).

4.3 Conclusions and Perspectives

The portrayed technical platforms in this chapter, though being only four among several others, are witness to the advance of technologies since the advent of the SNP era some 10 years ago. A continuing effort to improve the available techniques and the design of novel methods for SNP discovery and SNP genotyping will expand the repertoire of technologies markedly. Up to now the SNP technology has almost exclusively been used in animal, particularly human, genomics. And it is obvious that the major breakthroughs will occur in mammalian systems, be it forensic analyses, comparative and evolutionary genetics, or the use of SNPs in large-scale association studies to identify disease-susceptibility genes for common human disorders such as type II diabetes, hypertension, and cancer. Once the culprit genes are identified, the encoded proteins can be targeted by novel diagnostic tests or therapeutic drugs. Comprehensive SNP maps will be available for the human genome, and haplotype blocks for the detection of real associations with candidate disease genes will be identified. All these techniques are already miniaturized and designed for high-throughput (in some cases ultrahigh-throughput) settings. All these exciting developments somehow bypassed plant genomics. Though the value of SNPs is highly praised and appreciated, factual research on SNPs in plant genomes is rare and is still restricted to powerful breeding companies. We hypothesize that the costs for SNPing are presently still prohibitive for the average plant genomics laboratory. However, with the availability of SNP chips and affordable technologies Lemieux 2001, SNPs will arrive on the plant (especially crop) market as tools for DNA fingerprinting, genomic mapping, and linkage analyses, at least for the more economically important crop plants.

Acknowledgments

The authors highly appreciate the significant contributions of many individuals to the development of the portrayed genotyping systems and apologize that space restrictions do not allow us to list the individual names. Günter Kahl's research is supported by BMZ and GTZ (Germany) grant no. 2001.7860.8-001.00.

References

Alderborn, A., Kristofferson, A., Hammerling, U. Determination of single nucleotide polymorphisms by real-time pyrophosphate DNA sequencing Genome Res. 10:1249–1258 (2000).

Bansal A., van den Boom D., Kemmerer S., Honisch C., Adam G., Cantor C. R., Kleyn P., Braun A.: Association testing by DNA pooling: an effective initial screen. Proc. Natl. Acad. Sci. USA 99:16871–16874 (2002).

Barker D. L., Therault G., Che D., Dickinson T., Shen R. and Kain R. Self-assembled random arrays: high-performance imaging and genomics applications on a high-density microarray platform. Proc. SPIE 4966:1 (2003).

Barroso et al., Dominant negative mutations in human PPARg associated with severe insulin resistance, diabetes mellitus and hypertension. Nature 402:880–883 (1999)

Berg, L. M., Sanders, R., Alderborn, A. Pyrosequencing technology and the need for versatile solutions in molecular clinical research. Expert Rev. Mol. Diagn. 2:361–369 (2002).

Beutow K. H., Edmonson M. N., Cassidy A. B. Reliable identification of large numbers of candidate SNPs from public EST data. Nature Genet. 21:323–325 (1999).

Bhattramakki, D. & Rafalski, A. Discovery and application of single nucleotide polymorphism markers in plants. In 'Plant Genotyping: the DNA Fingerprinting of Plants'(ed. R. J. Henry), pp. 179–192. CAB International (2001).

Braun A., Little D. P., Koster H.: Detecting CFTR gene mutations by using primer oligo base extension and mass spectrometry. Clin. Chem. 43:1151–1158 (1997).

Buetow K. H., Edmonson M., MacDonald R., Clifford R., Yip, P., Kelley J., Little D. P., Strausberg R., Koester H., Cantor C. R., et al: High-throughput development and characterization of a genome-wide collection of gene-based single nucleotide polymorphism markers by chip-based matrix-assisted laser desorption/ionization time-of-flight mass spectrometry. Proc. Natl. Acad. Sci. USA 98:581–584 (2001).

Cargill, M., Daley, G. Mining for SNPs: putting the common variants-common disease hypothesis to the test. Pharmacogenomics 1: 27–37 (1999).

Ching, A, Rafalski, A. Rapid genetic mapping of ESTs using SNP Pyrosequencing and indel analysis Cell. Mol. Biol. Lett. 7:803–810 (2002).

Cho R. J., et al. Genome-wide mapping with bialleleic markers in Arabidopsis thaliana. Nature Genet. 23:203–207 (1999).

Daly M. J., Rioux J. D., Schaffner S. F., Hudson T. J., Lander E. S. High-resolution haplotype structure in the human genome. Nature Genet. 29:229–232 (2001).

de Arruda, M., Lyamichev, V. I., Eis, P. S., Iszczyszyn, W., Kwiatkowski, R. W., Law, S. M., Olson, M. C., Rasmussen, E. B. Invader Technology for DNA and RNA analysis: principles and applications. Expert Rev. Mol. Diagn. 2:487–496 (2002).

Ding C., Cantor C. R.: A high-throughput gene expression analysis technique using competitive PCR and matrix-assisted laser desorption ionization time-of-flight MS. Proc. Natl. Acad. Sci. USA, 100:3059–3064 (2003).

Dunker J, Larsson U., Petersson D, Forsell J., Schiller A.-L,. Alderborn A,. Berg, L. M. Parallel DNA template preparation using a vacuum filtration sample transfer device. BioTechniques 34:862–868 (2003).

Edwards K. J., Mogg R. Plant genotyping by analysis of single nucleotide polymorphisms. In: Plant Genotyping: The DNA Fingerprinting of Plants. Ed. H. R. J. Wallingford. CABI Publishing (2001).

Eis, P. S., Olson, M. C., Takova, T., Curtis, M. L., Olson, S. M., Vener, T. I., Ip, H. S., Vedvik, K. L., Bartholomay, C. T., Allawi, H. T., Ma, W.-P., Hall, J. G., Morin, M. D., Rushmore, T. H., Lyamichev, V. I., Kwiatkowski, R. W. An invasive cleavage assay for direct quantitation of specific RNAs. Nat. Biotechnol. 19:673–676 (2001).

Fan, J. B., Oliphant, A., Shen, R., Kermani, B., Garcia, F., Gunderson, K. L., Hansen, M., Steemers, F., Butler, B. L., Deloukas, P., Chee, M. S. Highly parallel SNP genotyping. Cold Spring Harbor Symposium Biology 68: (2003).

Gabriel S. B., Schaffner, S. F., Nguyen H., Moore J. M., Roy J., Blumenstiel B., Higgins J., DeFelice M., Lochner A., Faggart M. The structure of haplotype blocks in the human genome. Science 296:2225–2229 (2002).

Germano J., Klein A. S. Species-specific nuclear and chloroplast single nucleotide polymorphisms to distinguish *Picea glauca, P. mariana* and *P. rubens*. Theor. Appl. Genet. 99:37–49 (1999).

Giordano M. et al., Identification by denaturing high-performance liquid chromatography of numerous polymorphisms in a candidate region for multiple sclerosis susceptibility. Genomics 56:247–253 (1999).

Goldstein D. B. Islands of linkage disequilibrium. Nature Genet. 29:109–111 (2001).

Goldstein D. B. and Weale M. E. Population genomics: Linkage disequilibrium holds the key. Current Biology 11:R576 (2001).

Griffin, T. J., Smith L. M. Single nucleotide polymorphism analysis by MALDI-TOF mass spectrometry. Trends in Biotech. 18:77–84 (2000).

Green C. D., Simons J. F., Taillon B. E., Lewin D. A. Open systems: panoramic views of gene expression. J. Immunol. Methods 250:67–79 (2001).

Gruber, J. D., Colligan, P. B., Wolford, J. K. Estimation of single nucleotide polymorphism allele frequency in DNA pools using Pyrosequencing. Hum Genet. 110:395–401 (2002).

Halushka M. K. et al., Patterns of single nucleotide polymorphisms in candidate genes for blood-pressure homeostasis. Nat. Genet. 22:239–247 (1999).

Hartmer R., Storm N., Boecker S., Rodi C. P., Hillenkamp F., Jurinke C., van den Boom D.: RNase T1 mediated base-specific cleavage and MALDI-TOF MS for high-throughput comparative sequence analysis. Nucleic Acids Res. 31:e47 (2003).

Horton, R., Niblett D., Milne S., Palmer S., Tubby B., Trowsdale J., Beck S. Large-scale sequence comparisons reveal unusually high levels of variation in the HLA-DQB1 locus in the class II region of the human MHC. J. Mol. Biol. 282:71–97 (1998).

Inbar E, Yakir B, Darvasi A. An efficient haplotyping method with DNA pools. Nucleic Acids Res. 1;30(15):e76 (2002).

The International HapMap Consortium. The International HapMap Project. Nature 426: 789–796 (2003).

Jurinke C., Oeth P., Van Den Boom D.: MALDI-TOF mass spectrometry: a versatile tool for high-performance DNA analysis. Mol. Biotechnol. 26:147–164 (2004).

Kaiser, M. W., Lyamicheva, N., Ma, W.-P., Miller, C., Neri, B., Fors, L., Lyamichev, V. I. A comparison of eubacterial and archaeal structure-specific 5′-exonucleases. J. Biol.-Chem. 274:21387–21394 (1999).

Kozlowski, R. G., Donner, A. L., Roeven, R. SNP and transgene analysis directly from sample collection paper using the Invader DNA assay. In: Plant and Animal Genome Conference, San Diego, CA ,USA (2003).

Kruglyak L. and Nickerson D. A. Variation is the spice of life. Nature Genetics 27:234 (2001).

Kwiatkowski, R. W., Lyamichev, V., de Arruda, M., Neri, B. Clinical, genetic, and pharmacogenetic applications of the Invader assay. Mol. Diagn. 4:353–364 (1999).

Lefmann M., Honisch C., Bocker S., Storm N., von Wintzingerode F., Schlotelburg C., Moter A., van den Boom D., Gobel U. B.: A novel mass spectrometry based tool for genotypic dentification of mycobacteria. Journal of Clin. Microbiol. 42:339–346 (2004).

Lemieux B. Plant genotyping based on analysis of single nucleotide polymorphisms using microarrays. In: Plant Genotyping: The DNA Fingerprinting of Plants. Ed. H. R. J. Wallingford. CABI Publishing (2001).

Lyamichev, V., Brow, M. A.D., Dahlberg, J. E. Structure-specific endonucleolytic cleavage of nucleic acids by eubacterial DNA polymerases. Science 260:778–783 (1993).

Marth G. T., Korf I., Yandell M. D., Yeh R. T., Gu Z., Zakeri H., Stitziel N. O., Hillier L., Kwok P. Y., Gish W. R. A general approach to single nucleotide polymorphism discovery. Nature Genet. 23:452–456 (1999).

Michael K. L., Taylor L. C., Schultz S. L., Walt D. R. Randomly ordered addressable high-density optical sensor arrays. Anal. Chem. 70: 1242 (1998).

Mohlke K. L., Erdos M. R., Scott L. J., Fingerlin T. E., Jackson A. U., Silander K., Hollstein, P., Boehnke M., Collins F. S.: High-throughput screening for evidence of association by using mass spectrometry genotyping on DNA pools. Proc. Natl. Acad. Sci. USA 99:16928–16933 (2002).

Neve, B., Froguel, P., Corset, L., Vaillant, E., Vatin, V., Boutin, P., Rapid SNP allele frequency determination in genomic DNA pools by Pyrosequencing. BioTechniques 32:1138–1142 (2002).

Neville, M., Selzer, R., Aizenstein, B., Maguire, M., Hogan, K., Walton, R., Welsh, K., Neri, B., de Arruda, M. Characterization of cytochrome P450 2D6 alleles using the Invader system. BioTechniques 32:S34–S43 (2002).

Nickerson D. A., Tobe V. O., Taylor S. L. PolyPhred: automating the detection and genotyping of single nucleotide substitutions using fluorescence-based resequencing. Nucl. Acids Res. 25:2745–2751 (1997).

Nordfors, L., Jansson, M., Sandberg, G., Lavebratt, C., Sengul, S., Schalling, M., Arner, P., Large-scale genotyping of single nucleotide polymorphisms by Pyrosequencing and validation against the 5′nuclease (TaqMan®) assay. Human Mutation 19:395–401 (2002).

Odeberg J, Holmberg K, Eriksson P, Uhlén M. Molecular haplotyping by Pyrosequencing. BioTechniques 33:1104–1108 (2002).

Pettersson M, Bylund M, Alderborn A. Molecular haplotype determination using allele-specific PCR and Pyrosequencing technology. Genomics 82:390–396 (2003).

Picoult-Newberg L., Ideker R., Pohl M., Taylor S., Addelston M. B., Nickerson D. A., Boyce-Jacino M. Mining single nucleotide polymorphisms (SNPs) from expressed sequence tag (EST) databases. Genome Res. 9:167–174 (1999).

Rafalski A. Applications of single nucleotide polymorphisms in crop genetics. Curr. Opinion Plant Biol. 5:94–100 (2002).

Rickert, A., Premstaller, A., Gemhardt, C., Oefner, P. J. Genotyping of SNPs in a polyploid genome by Pyrosequencing. BioTechniques 32:592–603 (2002).

Rodi C. P., Darnhofer-Patel B., Stanssens P., Zabeau M., van den Boom D.: A strategy for the rapid discovery of disease markers using the MassARRAY system. BioTechniques Suppl: 62–66, 68–69 (2002).

Ronaghi M. Improved performance of Pyrosequencing using single-stranded DNA-binding protein. Anal Biochem. 286:282–288 (2000).

Ronaghi, M. Pyrosequencing sheds light on DNA sequencing. Genome Res. 11: 3-1-1 (2001).

Ronaghi, M., Uhlen, M., Nyrén, P. A sequencing method based on real-time pyrophosphate. Science 281:363–365 (1998).

Ross P., Hall L., Haff L. A. Quantitative approach to single-nucleotide polymorphism analysis using MALDI-TOF mass spectrometry. BioTechniques 29:620–6, 628–629 (2000).

Sachidanandam R., Weissmann D., Schmidt S. C., Kakol J. M., Stein L. D., Marth G., Sherry S., Mullikin J. C., Mortimore B. J., Wiiley D. L. A map of human genome sequence variation containing 1.42 million single nucleotide polymorphisms. Nature 409:928–933 (2001).

Stanssens M., Zabeau M., Meersseman G., Remes G., Gansemans Y., Storm N., Hartmer R., Honisch C., Rodi C. P., Bocker S., et al: High-throughput discovery of genomic sequence polymorphisms. Genome Res 14:126–133 (2004).

Wang D. G., et al., Large-scale identification, mapping and genotyping of single nucleotide polymorphisms in the human genome. Science 280:1077–1082 (1998).

Wasson, J., Skolnick, G., Love-Gregory, L., Permutt, M. A. Assessing allele frequencies of single nucleotide polymorphisms in DNA pools by Pyrosequencing Technology. BioTechniques 32:1144–1152 (2002).

Werner M., Sych M., Herbon N., Illig T., Konig I. R., Wjst M.: Large-scale determination of SNP allele frequencies in DNA pools using MALDI-TOF mass spectrometry. Hum. Mutat. 20:57–64 (2002).

5
Breeding By Design: Exploiting Genetic Maps and Molecular Markers Through Marker-assisted Selection

Johan D. Peleman, Anker P. Sørensen, and Jeroen Rouppe van der Voort

Abstract

The potential value of genetic markers, linkage maps, and indirect selection in plant breeding has been known for over 80 years. However, it was not until the advent of DNA marker technology in the 1980s that a large enough number of environmentally insensitive genetic markers could be generated to adequately follow the inheritance of important agronomic traits. Since then, DNA marker technology has dramatically enhanced the efficiency of plant breeding. In the past decade, a number of breeding companies have, to varying degrees, started using markers to increase the effectiveness of selection in breeding and to significantly shorten the development time of novel varieties. Now, advances in automated technology enable a new approach in marker-assisted breeding: "Breeding by Design." The advances in applied genomics and the possibility of generating large-scale marker datasets provide us with the tools to determine the genetic basis for all traits of agronomical importance. Methods for assessing the allelic variation at these agronomically important loci are also now available. This combined knowledge will eventually allow the breeder to combine favorable alleles at all these loci in a controlled manner, leading to superior varieties. Changing concepts and molecular approaches provide opportunities to develop rational and refined breeding strategies. Here it is argued that knowledge about map position and allelic variation at agronomically important loci in concert with available, easy-to-assay molecular markers allows the design of superior varieties. Depending on the crop-specific generation time, controlled marker-assisted selection strategies can lead to the development of superior varieties within 5 to 10 years.

The Handbook of Plant Genome Mapping.
Genetic and Physical Mapping. Edited by Khalid Meksem, Günter Kahl

ISBN 3-527-31116-5

5.1 Introduction

The purpose of plant breeding has always been to adapt the growth and production of the plant to the needs of man. The earliest advance in meeting this purpose is now called crop domestication and, for cereals and pulses, took place ca. 10,000 years ago. [1]. The first selections in natural populations were aimed at preventing seed dispersal at maturity, leading to the selection of the non-brittle spikes in cereals. As a result of this domestication, today's crops must rely on human intervention for their reproduction.

After the rediscovery of the Mendelian principles of heredity, the methodologies of modern plant breeding through directed crosses and selection for the desired recombinants were developed, leading to an enormous and steady increase in crop productivity and quality. The progress in plant breeding is based entirely on the availability of genetic variation. In principle, exploitation of genetic variability is the task of the modern plant breeder. With conventional breeding schemes, the genetic variation of breeding populations is estimated (and selected) by means of the phenotypic performance only. Even though this process has proven to be very effective, a selection procedure directly at the genotype level would greatly increase the efficiency of breeding efforts [2]. This is due to environmental influence on the phenotypic measurements, resulting in a biased measure of the true genetic potential of an individual. A prerequisite for the use of selection based on genotype is that the relative value of the different genotypes be well known and predictable.

After the discovery of the DNA molecule as the carrier of genetic and heritable information [3], followed 27 years later by the first description of the use of molecular (RFLP) markers for the construction of genetic maps [4], the possibility of factually describing the genotype of individuals, and thus using this information through selection, became theoretically feasible. The first molecular technique to address this challenge in plants, the RFLP technique, was reviewed by Tanksley et al. in 1989 [5]. However, it was only after the development of PCR-based molecular marker technologies such as RAPD [6] and AFLP [7] that the technologies could be applied at an acceptable cost to marker-assisted selection in plant breeding. The challenge since then has been to develop the methodologies needed for molecular marker assays and screens, which can link the genotypic scores to phenotypic performance in a repeatable, robust, and affordable manner [8]. This challenge has been a major focus of Keygene over the last 14 years.

The following two sections will summarize the knowledge and experience that has been generated through many different marker-assisted selection projects. The section on Breeding by Design (Section 5.4) presents our current vision on the optimal utilization of technologies and methodologies, with the ultimate aim of maximizing the exploitation of genetic variation in plants.

In our view, different stages of impact can be distinguished in the use of molecular markers in plant breeding. These stages are described below in more detail.

5.2 Marker-assisted Selection

DNA markers are reliable selection tools because they are stable, are not influenced by environmental conditions, and are relatively easy to score in an experienced laboratory. Compared to phenotypic assays, DNA markers offer great advantages in accelerating variety development time as a result of:

1. Increased reliability: the outcome of phenotypic assays is affected by factors such as environmental factors, the heritability of the trait, the number of genes involved, the magnitude of their effects, and the way these loci interact. Hence, error margins on the measurement of phenotypes tend to be significantly larger than those of genotyping scores based on DNA markers.
2. Increased efficiency: DNA markers can be scored at the seedling stage. This is especially advantageous when selecting for traits that are expressed only at later stages of development, such as flower, fruit, and seed characteristics. By selecting at the seedling stage, considerable amounts of time and space can be saved.
3. Reduced cost: there are ample traits where the determination of the phenotype costs more than the performance of a PCR assay. In a high-throughput setting, the material and consumable cost for a PCR assay will typically not exceed 2 Euro. In comparison, the growth of a tomato or pepper plant to full maturity in a heated greenhouse will cost approximately 20 Euro. Every plant that can be rejected before planting will in such settings save a considerable amount of money.

Before deciding to follow DNA marker-assisted approaches, practical concerns and cost-benefit analyses need to be addressed. Leaders of breeding programs must address a multifaceted evaluation of DNA marker-assisted approaches before committing to such new endeavors. Where none of the above arguments apply, the utilization of markers does not make any sense, at least for the above-described purposes.

The use of DNA markers for indirect selection offers the greatest benefits for quantitative traits with low heritability, as these are the most difficult characters to assess in field experiments. Obviously, the development of marker-assisted assays for such traits is difficult and costly due to the extensive phenotypic assays. However, once the knowledge exists to estimate the parameters that determine the trait of interest, a well-designed experimental setup will result in the availability of marker-assisted selection tools that can reduce the future application of phenotypical assays to a major extent.

The most straightforward applications of DNA markers in marker-assisted selection include genetic distance analysis, variety identification, identification of markers tightly linked to specific genes, and marker-assisted backcrossing. These applications are discussed in more detail in the following sections.

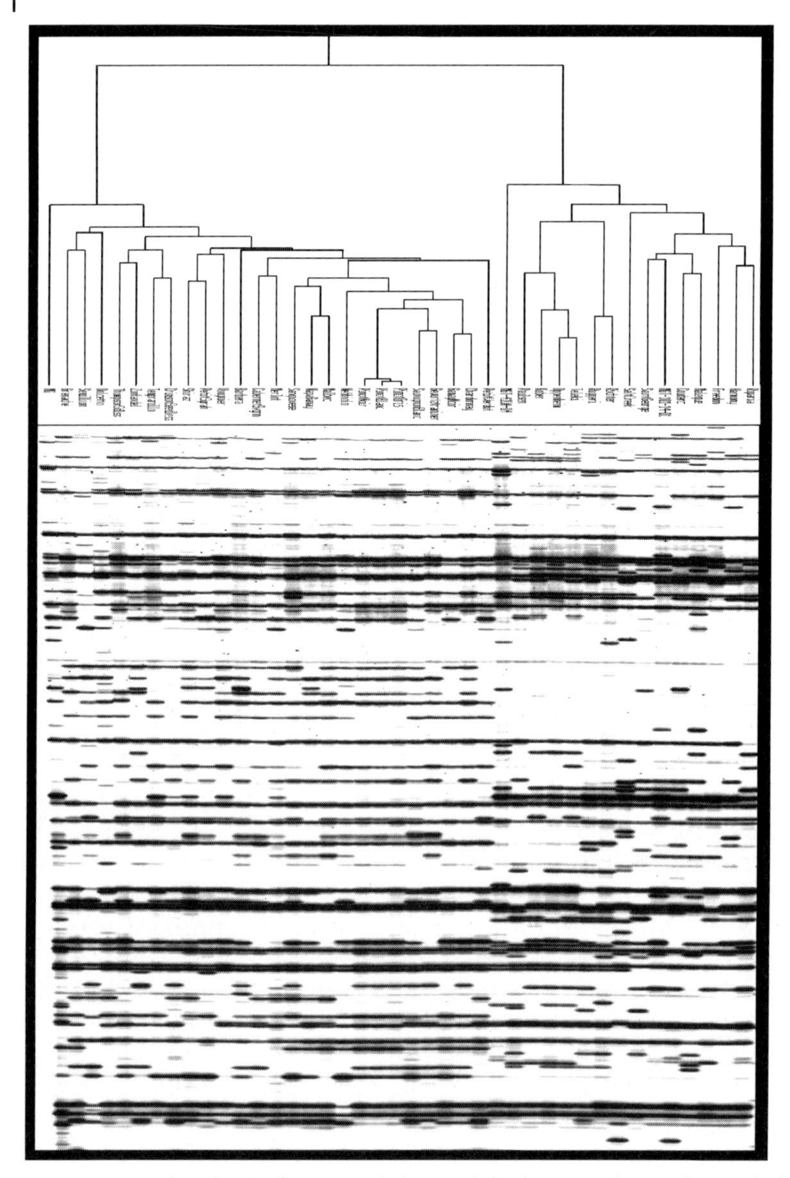

Figure 5.1. Combined AFLP fingerprint (below) and dendrogram (above) of a panel of grape varieties and rootstocks. Based on the AFLP fingerprint data, the genetic relatedness of all the varieties was determined and displayed by using a dendrogram. Subsequently, the fingerprints of the varieties were reordered according to the order of the dendrogram.

5.2.1 Genetic Distance Analysis, Variety Identification, and Seed Purity Analysis

Genotyping using DNA markers can be considered the most reliable method for the identification of lines and varieties. Therefore, the DNA fingerprinting methods can be used to analyze the purity of seed lots. Genetic distance analysis also provides a powerful tool for breeders to identify different heterotic groups and to increase the efficiency of finding crosses with good combinability. To determine the genetic distance between lines and groups of lines, the lines are fingerprinted and the markers are scored for each line. Based on the obtained score table, similarity indices can be calculated for all combinations of lines. Subsequently, the relatedness amongst the lines can be visualized using a dendrogram display or PCA plots (Figure 5.1).

5.2.2 Indirect Selection

Indirect selection can be a powerful method of selection in plant breeding. Especially for traits for which the phenotypic tests are unreliable or expensive, markers can offer a solution. Before indirect selection can be applied, the genetic basis of the trait of interest needs to be elucidated and markers linked to the gene(s) of interest have to be identified. The methods for identifying linked markers are described below. The AFLP marker technology has been used at Keygene for this purpose in a great variety of species for numerous traits. Once linked markers have been identified, the AFLP markers can be converted into simple PCR assays, which allow screening of large numbers of plants for the trait of interest in a cost-effective manner. A suitable linked DNA marker should allow prediction of the phenotype in a broad range of the germ plasm. To ensure reliable implementation of the marker in the practical breeding schemes, the identified marker must typically be located within 1–2 cM from the trait of interest. The occurrence of multiple alleles in the germ plasm for a desired locus may often complicate the identification of one single marker that will predict the phenotype in the entire breeding germ plasm. This hurdle can be overcome by the method of "locus haplotyping" (this method is explained further in Section 5.3.5).

5.2.2.1 Monogenic Traits

For the identification of markers linked with monogenic traits, different approaches can be followed. The preferred approaches are all based on the screening of a limited number of samples with a relatively large number of markers. In this way, many loci can be screened with limited effort. The number of samples that need to be fingerprinted can be limited by screening on set(s) of near-isogenic lines (NILs), if these are available. Candidate markers that are identified in this way are then screened on a panel of phenotypically well-characterized germ plasm lines to confirm their linkage and to determine the predictive value of the markers on the germ plasm.

An extremely powerful approach to identifying linked markers consists of the bulked segregant analysis (BSA) method [9]. For this type of screening, individuals from a segregating population are pooled on the basis of their phenotype, and the pools are then fingerprinted until a sufficient number of markers emerge. This method can be used for both dominant and recessive monogenic traits. For dominant genes, *cis* markers (linked with the trait of interest) will emerge from the screening, whereas *trans* markers (linked with the opposite allele) will be identified for recessive traits.

The BSA approach has proved to be useful for the identification of linked markers for oligogenic traits as well. An example where this is demonstrated is in an AFLP marker screening for disease resistance in tomato. The BSA screening yielded *trans* markers for a recessive gene involved in the resistance. When screening the *trans* markers on individuals of the segregating population, the resistance could be predicted in only ≈75 % of individuals, and the remaining 25 % of the individuals could not be explained as a result of recombination events. The results suggested that an additional (dominant) gene is needed for full resistance. By reconstructing the plant pools among the remaining 25 % of the individuals, an additional locus involved in the resistance could be identified.

5.2.2.2 **Polygenic (Quantitative) Traits**

The classical approach to the identification of loci involved in complex polygenic traits consists of screening a large number of individuals from a segregating population with a set of markers that are evenly distributed throughout the genome. Subsequently, statistical analysis is performed to identify regions in the genome that are involved in the trait. The laborious nature of this approach makes it unrealistic to screen sufficiently large populations to precisely locate the quantitative trait loci (QTLs). As a consequence, the QTLs cannot be localized precisely on the map, and sufficiently closely linked markers cannot be obtained, thereby preventing the broad-scale application of indirect selection for quantitative traits.

To increase the number of markers linked with the QTLs, a new approach based on the BSA principle can be applied. As an example, an oilseed rape F_2 population of ≈2500 individuals segregating for two quantitative traits, glucosinolate and erucic acid content, has been used. Based on the phenotypic scores, bulks were composed and approximately 2000 loci were screened on those bulks using the AFLP fingerprinting technique. Candidate markers that were identified on the bulk screening were subsequently analyzed on ≈200 randomly chosen individuals of the F_2 population. This screening demonstrated that the candidate markers identified using the BSA screening approach were derived from three different loci involved in glucosinolate content and two QTLs involved in erucic acid content. The results have been confirmed by independent studies in which the same map positions have been identified to be involved in the respective traits [10, 11]. The results demonstrate that a BSA strategy may be useful even for the identification of markers for quantitative traits.

To efficiently improve the precision of the interval in which oligogenic traits can be mapped, we have also devised a new method called reverse QTL mapping (RQM) [12, 13]. This method enables precise mapping of QTLs within very narrow genetic intervals. The principle of the approach relies on the selective phenotyping and genotyping of only those plants yielding information on the map position of the QTL. In the first step, a classical QTL analysis is performed on a fraction of a segregating population (e. g., 200 F_2 individuals) to identify the major QTL for the trait of interest. Subsequently, markers flanking the major QTL are used to screen the entire population (e. g., 2000 F_2 plants) to identify QTL isogenic recombinants (QNIRs): plants that carry a recombination at one QTL region and bear identical homozygous genotypes at the other QTL. These QNIRs are then geno-

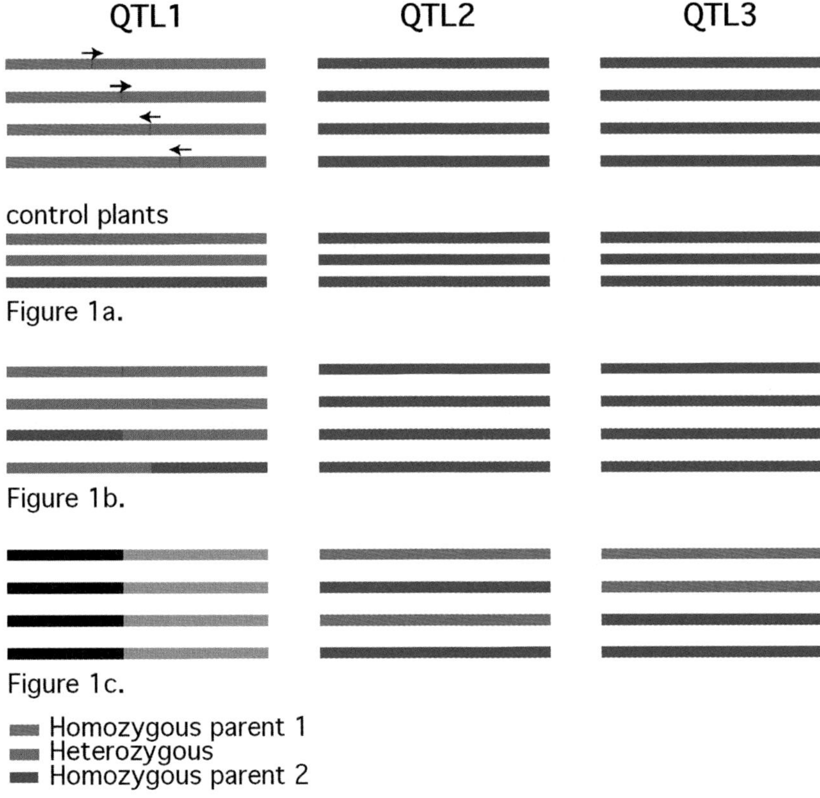

Figure 5.2. Principle of reverse QTL mapping demonstrated for three QTLs segregating in an F_2 population. (a) A stepwise reduction of the interval for QTL 1 for a single QTL near-isogenic recombinant (QNIR) set of plants. The position of the gene is determined based on the phenotypic values of the recombinants relative to the values of the control plants. (b) The different types of recombinants at QTL 1 that can be found for a single QNIR set. Such recombinants for QTL 1 also exist for other types of homozygotes at QTL 2 and QTL 3 (c). In a similar way, other QNIR sets exist for QTL2 and QTL3.

typed with sufficient markers at the recombinant QTL region to precisely map the recombination events. By growing progeny or clones of the QNIRs, highly accurate phenotyping data can be obtained on those plants. Comparison of the (averaged) phenotypic values of the QNIRs will lead to precise localization of the QTL gene within a sub-centimorgan interval (Figure 5.2).

The principle of RQM was demonstrated by using a subset of 1000 individuals derived from the same F_2 population as described above, which enabled the mapping of a QTL involved in erucic acid content within an interval smaller than 1 cM.

The RQM method has now been used successfully in our lab for several oligogenic traits. RQM is considerably more cost-effective than other approaches for fine mapping of oligogenic QTL traits. Because of the high resolution that can be obtained, the approach should also greatly facilitate map-based cloning of QTLs.

5.2.2.3 Marker-assisted Backcrossing

With the cost reductions that can be achieved using DNA markers, marker-assisted backcrossing (MABC) has become an important application in modern plant breeding. Two different aspects can be distinguished in MABC:

1. Selection for high recurrent parent genomic content: in this application, the DNA fingerprints are used to calculate the percentage of recurrent parent genome in each backcross individual, thereby taking the genome representation of the markers into account. In the case that genetic maps are constructed of the BC progeny, the analysis allows for a selection based on the number and position of remaining donor segments.
2. Selection against linkage drag: when negative characteristics are linked with the trait that needs to be introgressed, molecular markers can be used to select for recombinants in the genomic region surrounding the trait of interest. After phenotypic testing of these recombinants, individuals may be selected in which the region responsible for the linkage drag has been removed from the locus of interest.

MABC has been successfully implemented in a number of crops. For example, in maize the number of generations needed for recovery of the recurrent parent genome is reduced by MABC from eight to three [14].

5.3
The Creation of Novel Varieties (Marker-assisted Breeding)

The application of DNA markers in the breeding process can create substantially more added value than just improving the quality or cost of existing selection programs. By applying markers in a creative manner, new traits can be introduced that either could not be obtained or could be obtained only with great difficulty by classical breeding. Therefore, the application of markers in breeding can create a major competitive advantage to those breeders/companies that have integrated markers as part of their working tools. Where breeding goals cannot be achieved through traditional approaches, there is considerable scope for the use of molecular markers at almost any stage of development of new varieties. Here the limitation is not the facilitating technology but the imagination and motivation of the (marker-assisted) breeder. Below, a number of examples are provided where creative applications of markers clearly provide a major benefit over classical breeding.

5.3.1 Removal of Linkage Drag

In the mid-1990s, Keygene was involved in a marker-assisted breeding approach that led to the development of a novel lettuce variety resistant to the aphid *Nasonovia ribisnigri* [15]. This aphid is a major problem in areas in Europe and California, causing reduced and abnormal growth in addition to the spread of viral diseases. Resistance to this aphid could be introgressed from a wild relative of lettuce, *Lactuca virosa*, by repeated backcrossing. However, despite many rounds of backcrossing, the new product was of extremely poor quality, bearing yellow leaves and a greatly reduced head. This could have been caused either by a pleiotropic effect of the resistance gene or by so-called linkage drag, a negative trait closely linked to the positive trait of interest. Marker analysis eventually demonstrated that the reduced quality was indeed caused by linkage drag. In this case, the linkage drag was recessive, only visible in the homozygous state, thereby seriously increasing the difficulty in selecting for recombinations based on the phenotype. It was decided to use DNA markers flanking the introgression to pre-select for individuals that were recombinant in the vicinity of the gene. More than 1000 F_2 plants were screened this way, leading to the selection of some 100 individuals bearing a recombination or even double recombinations in the vicinity of the gene. Only those individuals needed to be phenotyped for both the resistance and, at the F_3 level, the absence of the negative characteristics. This approach eventually led to the selection of an individual bearing recombination events very close to each side of the gene, thereby removing the linkage drag. The results demonstrated that the (recessive) linkage drag was located on both sides of the resistance gene, in addition to being tightly linked. This result would have been very hard to obtain by classical selection methods (Figure 5.3).

5.3.2 Pyramiding Resistance Genes

Another example using markers for the creation of novel varieties is the pyramiding of disease-resistance genes. This approach can offer great financial rewards through extending the lifespan of new varieties.

Such an approach has been used for the backcross transfer of QTL for downy mildew resistance in pearl millet [16]. Here, a limited number of RFLP markers have been used for marker-assisted selection to improve disease resistance in both parent lines of a popular hybrid variety. Despite the labor-intensive nature

Figure 5.3. Example of a novel variety created by marker-assisted breeding.

of this approach and the resulting limitation in population size in a given generation, good progress has been made and field evaluation of the finished projects is underway just four years after initiation of the project.

As resistance genes tend to reside in resistance gene clusters, interesting alleles of different resistance genes may be located in tandem but be present in different accessions. In such a case, it is of paramount importance to precisely fine-map the alleles of the different genes with respect to one another. This goal can be most easily achieved using DNA markers. Subsequently, the linked markers can be utilized to select for the rare recombinants that combine the favorable alleles in tandem.

5.3.3
Marker-assisted Breeding of Polygenic Traits

Keeping track of all the genes involved in complex traits during a breeding program is an enormous challenge, if not an impossible task. In our experience, we have observed the loss of minor QTL when a round of marker-assisted selection was replaced by a round of phenotypic selection. The utilization of markers can obviously prevent such loss of QTL.

Simulation studies, although based on some oversimplified scenarios, provide some interesting insight into the optimum number of locations, replications, and population sizes in molecular breeding programs [17]. In these simulation studies, marker-assisted approaches remained efficient for QTLs with even very low heritability (0.15).

DNA markers allow us to unravel the genetic basis of traits expressing continuous phenotypic variations, as they are abundant and scattered throughout the genome. By using dense genetic marker maps, the contributions of separate regions of the genome on the trait values can be estimated once the mapping population is sufficiently large. Agronomically important traits such as nutritional quality, yield, flowering time, and "durable" resistance are all traits that appear to follow complex, polygenic inheritance patterns, with multiple genes having small effects on the trait value. Nevertheless, various lines of evidence obtained from various crops indicate that even such complex traits appear to be determined by only a few major factors [18–21].

We therefore argue that simplification of these complex analyses can offer an important key to success in mapping the loci involved in these traits. Such simplifications can be obtained at several levels.

1. Simplification of the phenotype: division of a complex phenotype into its separate genetic components. For example, an extremely important phenotype such as yield is determined by a vast array of component characters, e. g., root size, plant size, number of fruit, size of fruit, fruit contents, etc. Mapping the genes involved in these separate components provides a better understanding of the complex trait and a higher chance of success.

2. Simplification of the mapping: separating the effect of each QTL by generating NILs or by using the technique of introgression line libraries [22, 23] or RQM [12, 13] enables the more precise measurement of the effect of the QTL and thereby the precise/fine mapping of the QTL. Fine mapping of a QTL is an essential step in exploiting the QTL by marker-assisted selection.

These approaches significantly aid in unraveling the complexity of agronomically important traits. It is with such traits that, in the long term, the biggest benefits of marker-assisted selection can be obtained.

5.3.4 Introduction of Novel Characteristics

An innovative approach for introducing novel polygenic characteristics from wild germ plasm in a controlled manner is the application of advanced backcross breeding. This approach involves the simultaneous discovery and transfer of important QTL from unadapted germ plasm into elite breeding lines [24]. The advanced backcross breeding strategy postpones QTL mapping until the BC2 or BC3 generation. During the generation of BC2 or BC3 populations, negative selection is exercised to minimize the occurrence of unfavorable donor alleles. The advantage of focusing on BC2 or BC3 populations is that they offer sufficient statistical power for QTL identification on the one hand and provide sufficient similarity to the recurrent parent to select for QTL-NILs in a short time span (within 1–2 years) on the other. By use of QTL-NILs, the QTLs discovered can be verified and the NILs may serve directly either as improved varieties or as a parent variety in the case of hybrid crops.

The advanced backcross approach has been successfully used to identify markers for QTLs contributing to fruit size, shape, color, and firmness together with the content of soluble solids and total yield in tomato. On this basis, QTL marker associations were identified in one backcross generation and immediately applied in the subsequent backcross generation some six months later [25].

5.3.5 Effective Exploitation of (Exotic) Germ Plasm

Breeders have traditionally been reluctant to use wild germ plasm in their breeding programs due to complex, long-term, and unpredictable outcomes, particularly in crops where quality traits are important market criteria. This is a pity, because in most crops the cultured germ plasm represents only a small section of the vast diversity available in the species. Tanksley and coworkers have clearly demonstrated that wild relatives of tomato contain genes that contribute to interesting culture characteristics generally not expected to reside in those species [19, 26]. Marker-assisted backcrossing now enables the breeders to precisely introgress small sectors of wild/exotic accessions, thereby providing breeders with the tools to effectively unleash the vast resources held in germ plasm collections.

DNA marker-based diversity analysis enables gene banks to define core collections, which will provide a user-friendly entry point for breeders to access large and varied germ plasm collections. A large-scale genetic distance analysis of the complete CGN gene bank of lettuce in the Netherlands has been performed using AFLP markers. The analysis involved more than 6800 samples, and an enormous dataset of more than 1.35 million data points was produced in this study [27].

This type of analysis also greatly aids in selection of genotypes for broadening the genetic base of breeding populations and for the development of heterotic populations for breeding F_1 hybrid varieties.

By using markers tightly linked to a gene of interest, so-called locus haplotyping can be performed on accessions of germ plasm to identify those samples that bear different alleles at the locus of interest. It enables identification of accessions/lines bearing different alleles at a single locus, which can then be evaluated in further detail with respect to performance. This enables the breeders to efficiently identify new traits or better versions of existing traits that can then be quickly introgressed into their breeding lines. This approach allows the effective exploitation of germ plasm without the impossible task of having to phenotype all accessions.

5.4 Breeding by Design

The examples above show that the application of markers in breeding not only improves existing selection processes but also can aid in creating novel varieties bearing new characteristics of agronomical importance. Extending on these capabilities, an understanding of the genetic basis of *all* agronomically important characters *and* the allelic variation at those loci would enable the breeder to design superior genotypes *in silico*. We call this concept Breeding by Design. This goal can be reached by following a three-step approach.

5.4.1 Mapping Loci Involved in All Agronomically Relevant Traits

To elucidate the genetic basis of agronomically important traits, mapping populations are needed in which those traits are segregating. To map all traits that are relevant in breeding a crop, we strongly prefer the use of introgression line (IL) libraries. An IL library consists of a series of lines harboring a single homozygous donor segment introgressed into a uniform, cultivated background (reviewed in [23]) (Figure 5.4).

The following are great advantages of IL libraries in comparison with other mapping approaches.

- IL libraries consist of homozygous "immortal" lines and therefore can be phenotyped repeatedly and used for the simultaneous mapping of many traits.

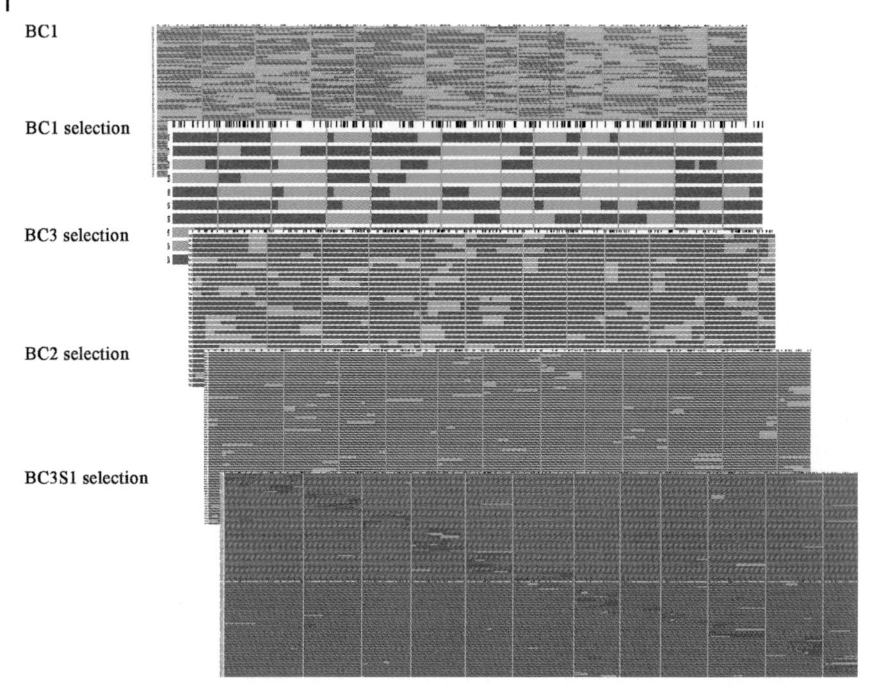

Figure 5.4. Example of an IL library construction process in tomato (12 chromosomes each separated by a gray bar) by marker-assisted backcrossing. Each horizontal bar represents an individual (best visible in the BC1 selection). Homozygous recurrent parent segments are shown in red. The homozygous donor segments are indicated in blue. Green bars represent heterozygous segments. The black bars on top show the relative positions of the markers used to assist the selection of the appropriate introgression lines. The construction and use of IL libraries are extensively reviewed by Zamir [23].

- IL libraries contain homogenous genetic backgrounds, only differing from one another by the introgressed donor segment. Thus, epistatic effects from the donor parent are eliminated.
- QTLs are dissected into separate monogenic components, which increases the reliability of measuring phenotypic traits.
- ILs containing interesting QTLs can be backcrossed to various lines to investigate interactive effects.
- Although dependent on the resolution of the IL library (i. e., average introgression segment size), QTLs are typically mapped into smaller intervals than by classical QTL mapping.
- IL libraries provide optimal starting material for the fine mapping of the mapped loci and for generating improved breeding lines.
- A secondary bonus from using IL libraries is that often new "exotic" alleles can be found that have a positive effect in the culture crop germ plasm.

To study interaction (epistasis) between genes, reciprocal IL libraries can be constructed. In such cases, IL libraries from line A into B and vice versa are constructed. By doing so, phenotypes that cannot be detected because they are mediated by interacting loci in the A×B library will be measured as a knocked-out phenotype in the B×A library. Subsequently, crosses between individual introgression lines, each bearing one of the interacting alleles, can be made to investigate the extent of the interaction [28].

To map loci contributing to heterosis, the IL library can be crossed to a tester parent. This will create an F_1 IL library in which each introgression segment is present in the heterozygous state. This F_1 IL library is then phenotyped to detect heterotic effects caused by specific introgression segments (Figure 5.3). In the past five years, considerable progress has been made for a number of different crops in the construction of IL libraries [23]. In the quest for mapping the loci underlying all agronomically important traits, IL libraries provide an extremely powerful tool.

In order to identify the allelic variation at each locus of interest by using chromosome haplotyping (see below), it is essential to determine a precise position for the loci of interest. IL libraries also provide perfect starting material for this purpose: each line containing a locus of interest can be backcrossed to the recurrent parent (and, if necessary, selfed) to create a large segregating population. This population can be used to identify recombinants within the introgression segment using flanking markers. Phenotyping these recombinants will enable the mapping of the locus at high resolution.

Another powerful fine-mapping approach is provided by the wealth of sequence information available from model plant species in combination with the rapidly expanding understanding of the function of genes. This knowledge can be linked to commercial species by exploiting synteny between both species, thereby opening the possibility for developing candidate-gene mapping approaches and markers [29, 30]. In our view, the candidate gene approach can provide a shortcut in the fine mapping or cloning of loci that are roughly mapped using an IL library.

An alternative but not yet sufficiently validated approach to fine-map loci of interest is provided by linkage disequilibrium mapping (LD mapping). LD mapping relies on associations between the phenotype and neutral polymorphisms located near the trait locus [31]. The predictive value of a marker for a trait of interest in a broad germ plasm is basically determined by the mode of inheritance of the trait, the recombination distance between the marker and the trait locus, and the mutation rate of the marker. Instead of single-marker-based tests, combinations of linked marker alleles at the same chromosome (haplotypes) can be used to predict phenotypic trait values in a panel of lines. Haplotypes of multiple markers have higher information content compared to single markers. Therefore, an increased predictive power is achieved by using haplotypes. We have successfully applied this approach in different crop species by using AFLP markers to identify and predict different alleles of agronomically important loci.

Figure 5.5 shows an example of three allelic variants of a hypothetical gene. Three tightly linked SNP markers were identified in a segregating population derived from parental lines G1 and G2. However, once the linked markers are used in a germ plasm panel (G1–G6), typically the association between (most of) the markers is lost, which is due to the independent origin of the SNPs. In this example,

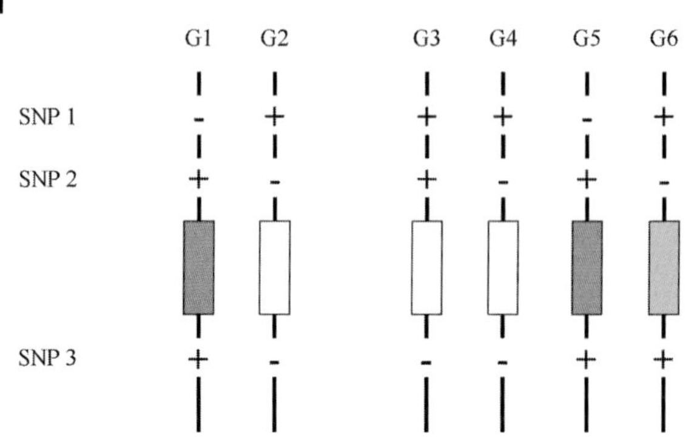

Figure 5.5. Predictive value of haplotypes. G1 and G2 represent parents of a segregating population, while G3–G6 represent germ plasm lines. Three gene alleles are distinguished: orange, white, and green. For clarity reasons, the genotypes of three SNP markers are indicated as "+/–" scores. Two different haplotypes are indicative of the white allele, whereas the orange and green alleles correspond to unique haplotypes.

the haplotypes composed of the three SNPs are required to discriminate the three different alleles.

Haplotype-based analysis of SNPs is currently the focus of intensive research as result of the effectiveness of haplotypes in (fine) mapping of complex traits and scanning for allelic variants (e. g., [32, 33], see chapter 4). Because of the complexity of this approach and the associated uncertainties, our preference is to use "targeted LD mapping." Once a rough position of a locus is known, LD mapping can be applied using markers or haplotypes (Figure 5.5) in that area of that locus, leading to the identification of markers or haplotypes that show a strong association with the phenotype under investigation, thereby fine-mapping the locus of interest [20].

5.4.2
Assessment of the Allelic Variation at the Loci Associated With Agronomically Relevant Traits

Although several strategies based on different population structures exist to unravel the genetic basis of complex traits, it will still be impossible to predict the phenotypes generated by those genes in the germ plasm. This is caused by the allelic variation that exists in the germ plasm at those loci. Typically, in a segregating population (F_2, BC, RIL, DH, IL library) only two alleles are segregating per locus. Mapping of the genes involved in a desired phenotype will enable prediction of only the phenotype within the same population or within populations in which the same alleles are segregating at that locus. To obtain broad predictive power, it is of paramount importance to be able to distinguish all alleles at the locus of interest and to assign phenotypic values to the different alleles.

As mentioned above, a potential shortcut to mapping genes and allelic variation simultaneously could be by using LD mapping with single markers [31]. However, a better method for assessing allelic variation at loci of interest is by using marker

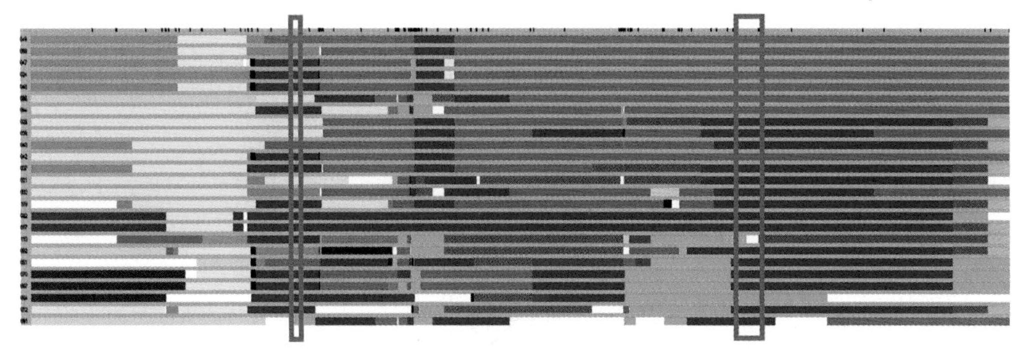

Figure 5.6. Chromosome haplotypes of a set of maize lines for one chromosome. Horizontal lines represent the same chromosome for a panel of lines. Identical stretches of marker scores (haplotypes) are indicated by the same color. The boxes indicate genome positions of interest.

haplotypes. By using a set of tightly linked markers at a candidate locus, the combined scores of all markers in principle can distinguish all the different alleles that occur at that locus in a set of accessions (Figure 5.5). For example, a high correlation was found between resistance phenotypes and marker haplotypes derived for the largest resistance gene cluster in lettuce [34].

By extending this technique to the whole genome, complete chromosome haplotypes can be generated (Figure 5.6). Assuming a high level of saturation with markers, these chromosome haplotypes enable the determination of the allelic variation at any position in the genome. After having fine-mapped the genes of agronomic importance, e.g., by using IL libraries combined with targeted LD mapping, these chromosome haplotypes can be used to assess the allelic variation at those loci.

Once the allelic variation at the loci of interest is identified, it is essential to attribute phenotypic values to the different alleles. For this purpose, the inbred lines bearing different alleles at the loci of interest need to be thoroughly phenotyped. In the case of polygenic traits, it is possible to pre-select those lines that bear different combinations of alleles at the different loci for phenotyping. Currently, we are developing algorithms that can combine those data with the phenotypic data on the lines to determine the contributing value of each allele to the phenotype.

In our view, the mapping of loci using simple mapping populations together with allele assessment using chromosome haplotyping of lines in combination with phenotyping provides the most powerful route to optimally exploiting the germ plasm of a species.

5.4.3
Breeding by Design

Eventually, the knowledge of the map positions of all loci of agronomic interest, the allelic variation at those loci, and their contribution to the phenotype will enable the breeder to design superior genotypes consisting of a combination of favorable alleles at all loci. Since the positions of all loci of importance are mapped precisely, recombination events can be accurately selected using flanking markers to collate the different favorable alleles next to each other (Figure 5.7). Software tools will enable determination of the optimal route to generate those mosaic genotypes by crossing lines and using markers to select for the specific recombinants that will eventually combine all those alleles. Since this is a precisely defined process, selection by phenotyping can be omitted. Only the eventually obtained superior varieties will have to be evaluated for field performance.

The above-described approach contains a number of prerequisites that require more discussion. Firstly, extremely saturated marker maps must be available to enable the generation of high-resolution chromosome haplotypes. Preferentially, a few markers are needed per window of LD to ensure reliable high-resolution chromosome haplotypes. Obviously, the extent of LD is strongly dependent on a number of factors, including, the crop species, the germ plasm selection, the genome region of interest, etc. [31]. Taking the rice genome as an example, let us assume that the average size of an LD window in rice is 100 kb and that the rice genome comprises 450 Mb, this means that 10,000–20,000 mapped markers need to be scored on a panel of inbred lines. With the currently available automation tools, this is a feasible task. Especially when using a multiplex PCR marker system such as AFLP [7, 35], which generates 20–100 markers per PCR, such a job can be performed reasonably within one year with a few well-trained people and the

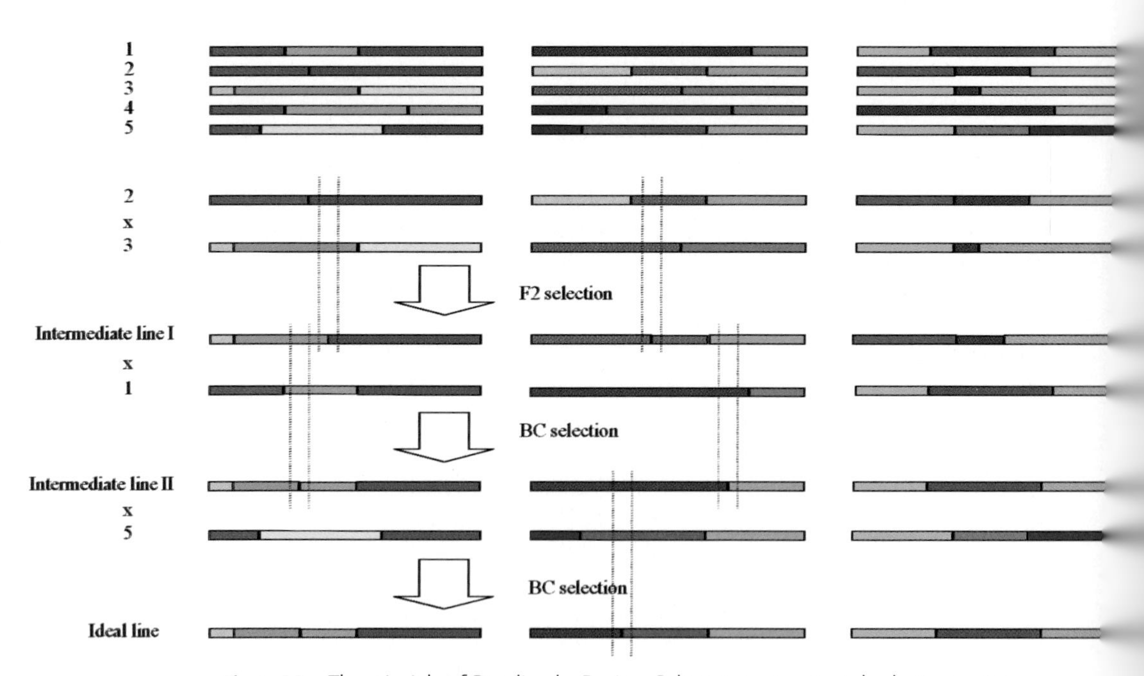

Figure 5.7. The principle of Breeding by Design. Subsequent crosses and selections using markers lead to the desired superior elite line genotype. Dotted lines indicate marker positions used to select for the desired recombinants.

appropriate equipment. High-resolution mapping of AFLP markers can be performed by using physical mapping strategies such as HAPPY mapping and radiation hybrid mapping [36] or the application of BAC pooling strategies [37].

Secondly, the described approach requires extensive phenotyping of all agronomical traits on both the mapping populations and the inbred lines that are used for chromosome haplotyping and allele assessment. Moreover, to map loci involved in heterosis, it is also necessary to phenotype hybrids derived from selected combinations of haplotyped lines. The most important factor determining success will be the precision of the phenotyping. Dissection of the phenotypes into components aids in objectively measuring phenotypes: for example, a complex trait such as taste can be dissected into sugar, acid, and volatile flavor molecules, which can be measured separately by using biochemical analysis equipment (e. g., gas chromatographers, etc.). Extending on this approach for all traits of agronomical importance will require great organization skills and will represent a major effort. In our view, phenotyping is the most critical and challenging factor in achieving the goals set by Breeding by Design.

5.4.4 Future of Breeding by Design

Breeding by Design contains an integrative, complementary application of the technological tools and materials currently available to develop superior varieties. During this process, enormous resources of knowledge are generated that will enable breeders to deploy more rational and refined breeding strategies in the near future. The recent developments in technology and statistical methodology have now brought this strategy within reach. The optimal exploitation of the naturally available genetic resources can create unsurpassed possibilities to generate new traits and crop performance. In our view, Breeding by Design will reach as much crop improvement potential as GMO strategies while requiring fewer investments and without challenging public acceptance.

Acknowledgments

The authors gratefully acknowledge Dr. Maarten Koornneef (Wageningen University and Research Center), Dr. Jonathan Crouch (ICRISAT), and Dr. Tom Gerats (University of Nijmegen) for their contributions and discussions during the development of this paper. The AFLP technology is covered by patents and/or patent applications of Keygene N. V.; AFLP is a registered trademark of Keygene N. V. A trademark registration for Breeding by Design has been filed by Keygene N. V. Figures 5.4–5.7, as well as parts of the section on Breeding by Design by Johan Peleman and Jeroen Rouppe van der Voort, published in TRENDS in Plant Science, Vol. 8, No. 7, pp. 301–352, have been reprinted with permission from Elsevier.

References

[1] Lev-Yadun, S., A. Gopher, and S. Abbo (2000) "The cradle of agriculture." Science 288:1602–1603.

[2] Dekkers, J. C.M. and F. Hospital (2002) "The use of molecular genetics in the improvement of agricultural populations." Nat. Rev. Genetics 3:22–32.

[3] Watson J. and F. Crick (1953) "A structure for Deoxyribose Nucleic Acid." Nature 4356:737–738.

[4] Botstein, D., R. L. White, M. Skonick, and R. W. Davis (1980) "Construction of a genetic linkage map in man using restriction length polymorphisms". Am J. Hum. Genet. 32:314–331.

[5] Tanksley, S.D, N. D. Young, A. H. Paterson, and M. W. Bonierblade (1989) "RFLP mapping in plant breeding: New tools for an old science." BIO/Technology 7:257–264.

[6] Williams, J., A. Kubelik, K. Livak, J. Rafalski, and S. Tingey (1990) "DNA polymorphisms amplified by arbitrary primers are useful as genetic markers." Nucleic Acids Research 18:6531–6535.

[7] Vos, P., R. Hogers, M. Bleeker, M. Reijans, T. Van der Lee, M. Hornes, A. Frijters, J. Pot, J. Peleman, M. Kuiper, and M. Zabeau (1995) "AFLP, a new technique for DNA fingerprinting." Nucleic Acids Research 23:4407–4414.

[8] Young, N. D. (1995) "A cautiously optimistic vision for marker assisted breeding." Molecular breeding 5:505–510.

[9] Michelmore, R. W., I. Paran, and R. V. Kesseli (1991) "Identification of markers linked to disease-resistance genes by bulked segregant analysis: A rapid method to detect markers in specific genomic regions by using segregating populations." Proc. Natl. Acad. Sci. 88:9828–9832.

[10] Toroser, D., C. E. Thormann, T. C. Osborn, and R. Mithen (1995) "RFLP mapping of quantitative trait loci controlling seed aliphatic-glucosinolate content in oilseed rape (*Brassica napus L.*)." Theor. Appl. Genet. 91:802–808.

[11] Jourdren, C., P. Barret, R. Horvais, N. Foisset, R. Delourme, and M. Renard (1996) "Identification of RAPD markers linked to the loci controlling erucic acid level in rapeseed." Molecular Breeding 2: 61–71.

[12] Wye, C., J., Rouppe van der Voort, and J. Peleman (2000) Poster # 538, Plant & Animal Genome Conference VIII, San Diego.

[13] Peleman, J.; C. Wye, J. Zethof, and J. Rouppe van der Voort: in preparation.

[14] Frisch, M., M. Bohn, and A. E. Melchinger (1999) "Comparison of selection strategies for marker-assisted backcrossing of a gene." Crop Sci. 39:1295–301.

[15] Jansen, J. P. (1996) Aphid resistance in composites. International application published under the patent cooperation treaty (PCT) No. WO 97/46080.

[16] Witcombe, J. R., and C. T. Hash (2000) "Resistance gene deployment strategies in cereal hybrids using maker-assisted selection: gene pyramiding, three-way hybrids and synthetic parent populations." Euphytica 112:175–186.

[17] Moreau, L., S. Lemarie, A. Charcosset, and A. Gallais (2000) "Economic efficiency on one cycle of marker-assisted selection." Crop Science 40:329–337.

[18] Young, N. (1996) "QTL mapping and quantitative disease resistance in plants." Rev. Ann. Phytopathol 34:479–501.

[19] Frary, A., T. Clint Nesbitt, S. Grandillo, E. Van der Knaap, B. Cong, J. Liu, J. Meller, R. Elber, K. B. Alpert, and S. D. Tanksley (2000) "A quanititative trait locus key to the evolution of tomato fruit size." Science 289:85–88.

[20] Thornsberry, J. M., M. M. Goodman, J. Doebley, S. Kresovich, D. Nielsen, and E. S. Buckler (2001) "Dwarf8 polymorphisms associate with variation in flowering." Nat. Genet. 28:286–289.

[21] Rouppe van der Voort, J., E. Van der Vossen, E. Bakker, H. Overmars, P. Van Zandvoort, R. Hutten, R. Klein Lankhorst, and J. Bakker (2000) "Two additive QTLs conferring broad-spectrum resistance in potato to *Globodera pallida* are localized on resistance gene clusters." Theor. Appl. Genet. 101:1122–1130.

[22] Eshed, Y., and D. Zamir (1995) "An introgression line population of *Lycopersicon*

pennellii in the cultivated tomato enables the identification and fine mapping of yield-associated QTL." Genetics 141:1147–1162.

[23] Zamir, D. (2001) "Improving plant breeding with exotic genetic libraries." Nature Reviews Genetics 2:983–989.

[24] Tanksley, S. D., and J. C. Nelson (1996) "Advanced backcross QTL analysis: a method for the simultaneous discovery and transfer of valuable QTLs from unadapted germplasm into elite breeding lines." Theor. Appl. Genet. 92:191–203.

[25] Tanksley, S. D., S. Grandillo, T. M. Fulton, D. Zamir, Y. Eshed, V. Petiard, J. Lopez, and T. Beck-Bunn (1996) "Advanced backcross QTL analysis in a cross between an elite processing line of tomato and its wild relative *L. pimpinellifolium*." Theor. Appl. Genet. 92:213–224.

[26] Tanksley, S. D. and S. R. McCouch (1997) "Seed banks and molecular maps: unlocking genetic potential from the wild." Science 277:1063–1066.

[27] Van Hintum, Th. J.L., H. Verbakel, I. W. Boukema, E. C. De Groot, and J. Peleman, submitted.

[28] Eshed, Y. and D. Zamir (1996) "Less than additive epistatic interactions of QTL in tomato." Genetics 143:1807.

[29] Fulton, T. M., R. Van der Hoeven, N. T. Eannetta, and S. D. Tanksley (2002) "Identification, analysis, and utilization of conserved ortholog set markers for comparative genomics in higher plants." The Plant Cell 14:1457–1467.

[30] Oh, K., K. Hardeman, M. G. Ivanchenko, M. Ellard-Ivey, A. Nebenfuhr, T. J. White, and T. L. Lomax (2002) "Fine mapping in tomato using microsynteny with Arabidopsis genome: the Diageotropica (Dgt) locus." Genome Biology 3 (9):0049.I–0049.II.

[31] Weir, S. (1996) "Genetic data analysis II." Genetic Data Analysis Sinauer.

[32] Johnson, G. C.L., L. Esposito, B. J. Barratt, and A. N. Smith, (2001) "Haplotype tagging for the identification of common disease genes." Nat. Genet. 29:233–237.

[33] Cardon, L. R. and G. R. Abecasis (2003) "Using haplotype blocks to map human complex trait loci." Trends Genet. 19:135–140.

[34] Sicard, D., S. S. Woo, R. Arroyo-Garcia, O. Ochoa, D. Nguyen, A. Korol, E. Nevo, and R. Michelmore (1999) "Molecular diversity at the major cluster of disease resistance genes in cultivated and wild *Lactuca* spp." Theor. Appl. Genet. 99:405–418.

[35] Zabeau, M., and P. Vos (1993) Selective restriction fragment amplification: a general method for DNA fingerprinting, European Patent 0 534 858.

[36] Waugh, R., P. H. Dear, W. Powell, and G. C. Machray (2002) "Physical education - new technologies for mapping plant genomes." Trends Plant Sci. 7:521–3.

37 Klein, P. E., R. R. Klein, S. W. Cartinhour, P. E. Ulanch, J. Dong, J. A. Obert, D. T. Morishige, S. D. Schlueter, K. L. Childs, M. Ale, and J. E. Mullet (2000) "A high-throughput AFLP-based method for constructing integrated genetic and physical maps: progress toward a sorghum genome map." Genome Res. 10 (6):789–807.

Part II
Physical Mapping

The Handbook of Plant Genome Mapping.
Genetic and Physical Mapping. Edited by Khalid Meksem, Günter Kahl

ISBN 3-527-31116-5

6
Physical Mapping of Plant Chromosomes

Barbara Hass-Jacobus and Scott A. Jackson

6.1
Introduction

All organisms contain DNA, which in higher organisms such as plants is organized into chromosomes. As the field of genetics moves forward into areas such as genomics and proteomics, it is increasingly important to investigate how the linear organization of genes and other sequences is involved in the structural organization of chromosomes and how genomes of different organisms are similar to and different from one another. Recombination has been used to order genes and molecular markers into linkage groups on genetic maps, but this has been done with the understanding that genetic distance (usually expressed in terms of centimorgans and based on the frequency of recombination between genetic markers) on a genetic map does not necessarily reflect the physical distance in base pairs between markers. Physical mapping – the ordering of genetic components on chromosome maps in terms of physical distance – is becoming increasingly important for our understanding of how genomes are regulated and their evolution. Physical mapping allows one not only to determine where markers are on a chromosome in relation to the centromere and telomeres but also to detect phenomena such as insertions, deletions, and translocations. Using physical mapping techniques, it is possible to differentiate between chromosomal segments from two different species in a hybrid organism. As the sequencing of genomes from all kingdoms of life continues to accelerate, physical mapping is being used to verify sequence data and to allow the comparison of sequenced genomes to genomes that have not yet been sequenced, a field of study called comparative genomics.

This chapter will examine physical mapping, with an emphasis on the work being done in plant systems. We will begin with a discussion of classical physical mapping methods, followed by a discussion of more-modern molecular physical mapping techniques. We will examine the reasons for physical mapping and look at which areas of the genome are investigated in physical mapping studies. Finally, we will briefly discuss the future direction of physical mapping as more genomes are fully sequenced.

The Handbook of Plant Genome Mapping.
Genetic and Physical Mapping. Edited by Khalid Meksem, Günter Kahl

ISBN 3-527-31116-5

6.2 Classical Physical Mapping Techniques

6.2.1 Knob Mapping in Maize

Knob heterochromatin in maize represents a unique chromatin feature that can be used for physical mapping. Knob heterochromatin was first observed by Barbara McClintock in 1929. Her observations state, "In the fourth from the smallest chromosome there is a deeply staining body which becomes very conspicuous during late prophase. Other chromosomes have less conspicuous bodies of this kind, but their exact position requires further study." By 1930, McClintock was using a terminal knob to map translocations in maize. Knobs are used to distinguish between each of the 10 maize chromosomes and can be used as markers to determine genetic distances (Mohammed and Shoeib 1965). While the number, size, shape, and distribution of knobs vary among lines of maize, these traits are genetically inheritable and can be used to differentiate between maize strains (Chughtai and Steffensen 1987). Also of cytological importance, it has been observed that knobbed chromosomes, under certain conditions, undergo chromatin loss; the larger the knob, the more frequently the arm of the knobbed chromosome is lost, whereas knobless arms are stable (Rhoades and Dempsey 1973).

Knobs have been used to study maize evolution and are thought to have played a significant role in the adaptation and evolution of maize and its relatives. Maize lines isolated from the regions where cultivated maize was thought to have originated, namely, Guatemala and Mexico, have the greatest number of knobs, while maize lines with the lowest number of knobs are found in outlying areas (Longley 1952; Moll et al. 1972). A large portion of knob heterochromatin contains a 180-bp repeat, a 27-bp region of which is conserved among maize and the relatives of maize teosinte and *Tripsacum* (Chughtai and Steffensen 1987). Another 350-bp repeat, TR-1, has also been found in knob heterochromatin. These arrays of tandem repeats have been found to be interrupted by retrotransposon insertions, and it has been hypothesized that knobs may be complex megatransposons (Ananiev et al. 1998). Although knob mapping in maize could be considered outdated due to the availability of molecular mapping and other physical mapping techniques that have been developed, the implications of knob heterochromatin in cell division and in maize genetics will continue to be studied and will likely play a role in the comparative genetics and evolutionary studies involving maize and its relatives in the future.

6.2.2 Deletion/Aneuploid/Substitution Mapping

Deletion mapping has been used in bacteria to analyze gene mutations and to determine gene structure. Deletions may be detected and mapped in a number of ways, including by an observed lack of reverse mutations, pseudodominance, and recessive lethality and by the appearance of deletion loops in chromosome preparations. In humans, deletion lines have been used to find recessive oncogenes that have been inactivated by a homozygous deletion. In plants, deletion lines have been used in wheat to place sequences on the chromosome map (Qi and Gill 2001).

Cytogenetic and DNA-based maps can be related to one another through the use of aneuploid and substitution lines (Young 2000). Aneuploids are organisms that have an increase or decrease in the normal chromosome number. The change in chromosome number can involve as little as a portion of a chromosome (i. e., telosomics) or may involve more than one chromosome (i. e., nullisomic-tetrasomics). Aneuploids of tetraploid and hexaploid wheat have been used to locate genes on chromosomes, to map gene-to-centromere distances, and to identify chromosome homoeologies (Joppa 1987). The hybridization of a genetically mapped clone to digested DNA from aneuploid lines that results in the loss of a band (nullisomics) or a change in the signal on an autoradiogram indicates the chromosomal location of the clone (Young 2000). Monosomic lines of maize and oat and trisomic lines of tomato are just three examples of species in which aneuploids have been used to physically map genetic markers to specific chromosomes (Young et al. 1987; Rooney et al. 1994). A major disadvantage of this technique is the availability of viable aneuploids, as few species are able to tolerate aneuploidy very well, making maintenance of the genetic lines difficult. Aneuploid mapping, nevertheless, has been widely used in plant physical mapping. In general, plants are much more tolerant of aneuploidy than are animal systems.

Substitution mapping is similar to deletion mapping, except that substitution mapping uses meiotic recombinants instead of deletion events (Paterson et al. 1990). In substitution mapping, lines of a species are developed where chromosomes or chromosome arms have been substituted with homoeologous chromosomal segments from another species (Young 2000). Traditionally, before genetic engineering was so widely used to introduce genes into lines, substitution lines were used in plants to develop varieties with traits of interest, such as disease resistance. More recently, these lines have been used to physically map genetic markers. Restriction fragment length polymorphisms (RFLP) are identified that lie near the gene or QTL of interest. Different chromosomal segments carrying the RFLP are identified, and their overlap is determined as precisely as possible using mapped genetic markers (Paterson 1990). The DNA from the substitution lines is restriction-digested and blotted onto a membrane, which is then probed with a DNA clone. The substitution line corresponding to the clone's location shows a different restriction-digest fingerprint (RFLP) compared to the other substitution lines, identifying the location of that clone (Young 2000; Sharp et al. 1989). The resolution of substitution mapping is limited to the resolution of existing genetic maps, because the number and spacing of genetic markers is used to determine the overlap of chromosomal segments (Paterson et al. 1990) One example of substitution mapping is that of mapping QTLs in tomato. Paterson et al. (1990) used substitution mapping to increase the resolution of the tomato linkage map from ~20-cM intervals to intervals of 3 cM, allowing finer mapping of their QTLs of interest.

Aneuploid, substitution, and deletion mapping represent three classical physical mapping techniques that use cytogenetic stocks to confirm results obtained with molecular mapping projects in plant systems. The ability to correlate the genetic and physical maps is becoming crucial to plant mapping, especially as more and more laboratories become interested in comparative genomics. The major disadvantages of these techniques are the difficulty in isolating a large number of lines and the fact that the cytogenetic lines cannot be developed in all plant species due to issues such as viability and chromosome identification (Cheng et al. 2001).

6.2.3 Polytene Chromosomes in *Drosophila*

Polytene chromosomes represent one of the earliest tools used by geneticists to physically map chromosomes. First discovered in the salivary glands of the fly *Bibio hortulanus*, polytene chromosomes consist of a chromosome that has been replicated many times without the replicas separating into chromatids. Instead, the replicas remain aligned and bundled together and have characteristic banding patterns as well as regions that appear swollen. These chromosomal landmarks are excellent for mapping regions of the chromosome. Maps of polytene chromosomes of *Drosophila melanogaster* have been around since the 1930s (reviewed in Rubin and Lewis 2000). Polytene chromosomes are also useful for visualizing mutation events such as deletions. A bulge, or looped configuration, can be seen in the polytene chromosomes of deletion heterozygotes, allowing the mapping of genes and markers in the deletion zone. Polytene chromosomes are especially good for physical mapping techniques such as FISH, where the hybridization signal is very clear compared to background hybridization, as it appears as a nice, distinct line across the width of the chromosome as compared to a small spot normally observed in FISH in plant mitotic spreads.

In plants, polytene chromosomes are somewhat different from those found in insects. The banding patterns are not as well developed and the degree of condensation of the chromosomes varies. While they have been found in some plants, such as *Phaseolus* (http://www.ba.cnr.it/Beanref/polytene.htm), polytene chromosomes have not been used extensively for physical mapping in plants, primarily due to the difficulty in obtaining them and the low resolution compared to other available methods.

6.2.4 Chromosome Banding

The analysis of longitudinal banding patterns that are produced when chromosomes are stained with various dyes is a classical tool for the identification of individual chromosomes and chromosome segments, as banding patterns provide unique fingerprints for individual chromosomes. There is a wide variety of chromosome banding techniques, many of which are described below, that have been used for physical mapping, mostly in animal systems. Some are widely used in plant systems while others are not, and many of the techniques have required modification for use with plants and seldom give the resolution obtained with animal chromosomes.

C-banding is a technique used widely in physical mapping in plants. It was first reported for use with animal chromosomes (Hsu 1973). C-banding stains the centromeric regions and other constitutive (C) heterochromatic regions (Knight 2003). The DNA is denatured under alkaline conditions, leading to depurination of the DNA. The DNA is then renatured (Sengbusch 2002). The degree to which the denatured DNA reanneals and the time it takes to do so (renaturation kinetics) is an index of the degree of repeats present on the chromosome (Sharma and Sharma 1999). After a wash step, the chromosomes are stained with Giemsa solution, which binds primarily to the repetitive DNA, or heterochromatin (Sengbusch 2002). These stained repetitive DNA bands are called C bands. DAPI/distamycin A can be used to label a subset of C bands to identify pericentromeric breakpoints and to identify chromosomes that are too small for standard banding techniques (Knight 2003). C-banding has been used extensively in the physical mapping of crop plants such as wheat. C bands are polymorphic among wheat cultivars and therefore provide cytological markers along the entire lengths of wheat chromosome arms. The bands can be mapped in relation to other genetic markers and are used to build cytogenetic maps (Jiang and Gill 1993). Studies of repetitive DNA sequences in rye (*Secale cereale*) showed that sequences of the 480-bp family are located mainly at the telomeres on all seven pairs of rye chromosomes. Later studies found that the locations of these repetitive sequences correspond to most of the major C-banded regions in rye, revealing a relationship between heterochromatic bands and repetitive sequences in plants (reviewed in Jiang and Gill 1994c).

N-banding is another Giemsa-staining method. Whereas C-banding stains all heterochromatin, N-banding stains only chromatin containing polypyrimidine sequences (Gill 1987). Staining of chromosomal regions containing (GAA)m(GAG)n satellite DNA results in a banding pattern that is conserved between the genomes of different accessions of the same species (Zoller et al. 2001). N-banding is used in nucleolar regions to differentiate heterochromatin from C-banded centromeric sites on the chromosome (Sharma and Sharma 1999). While N-banding is less universal than C-banding for chromosome identification because it stains only a select portion of the heterochromatin, it is rapid and reproducible and provides excellent band resolution. Modifications of C-banding methods can produce N bands, and modification of N-banding methods can be used to produce C bands. In fact, N- and C-banding can be used on the same chromosome preparations to show heterochromatin heterogeneity (reviewed in Gill 1987).

Q-banding was first described by Caspersson et al. (1971). In this technique, chromosomes are stained with the fluorescent quinacrine mustard solution to identify specific chromosomes and structural rearrangements. Most of the stained DNA is heterochromatin (Knight 2003). The stain binds AT- and GC-rich regions of the DNA, although only the AT-rich regions, which are more common in heterochromatin than in euchromatin, fluoresce (Sengbusch 2002).

The R (reverse)-banding technique is used to look for deletions or translocations at the telomeres of chromosomes. Chromosomes are incubated in hot phosphate buffer followed by Giemsa staining, resulting in a pattern of dark R bands and pale G bands. When the chromosomes are denatured in the hot buffer, the AT-rich regions of the chromosome are readily denatured, while the GC-rich regions remain annealed, resulting in the staining of the R bands (Knight 2003). R-banding gives bands opposite to Q- and G-banding (Sharma and Sharma 1999).

Hy-banding is a method reported in plant cells that is not used as commonly as the techniques described thus far. In this method, the cells are treated with hot hydrochloric acid and then stained with acetocarmine. The binding of proteins to the DNA affects the binding ability of the stain (Sengbusch 2002).

In plants, banding analysis of paired chromosomes in hybrids during meiosis revealed aberrancies in pairing that were not detected using conventional staining techniques. Chromosome banding has also been used in plant breeding to assay the quality of cytogenetic stocks and to identify interspecific hybrids, addition lines, substitution lines, and translocation lines (reviewed in Gill 1987). In terms of physical mapping, banding is used to map genes in relation to cytological markers on the chromosome (reviewed in Gill 1987).

Chromosome banding can be used in conjunction with *in situ* hybridization (ISH) techniques to map specific DNA sequences to chromosomal regions. In plants, this was first tried by Hutchinson and Seal (1983), who used an isotopic ISH technique followed by a C-banding procedure to locate a repetitive DNA sequence on rye chromosomes (reviewed in Jiang and Gill 1994c). The C-banding resolution of this technique was poor, and use of this technique in plants was not reported again for 10 years, at which time Jiang and Gill (1993, 1994a, 1994b) used a modified N-banding/ISH procedure to localize repetitive DNA sequences and the breakpoints of intergenomic translocation chromosomes in wheat (reviewed in Jiang and Gill 1994c).

Recently, a fluorescent banding technique termed comparative genomic *in situ* hybridization (cGISH) was developed by Zoller et al. (2001). In this protocol, total genomic *Arabidopsis* DNA (labeled with biotin or digoxigenin) was hybridized to chromosomes of various plant species and was found to produce species-specific banding patterns, depending on the hybridization conditions and on the conservation and organization of repetitive DNA relative to *Arabidopsis*. The genomic probe detected signals in regions containing rDNA, telomere, and satellite sequences. The detection of these conserved sequences generated banding patterns that were specific enough to allow the identification of each chromosome tested in the study. The cGISH bands corresponded to some of the C bands and most of the N bands of rye. The technique also gave bands at previously unmarked positions on the chromosome, resulting in new markers for mapping. Zoller et al. (2001) recommend the technique for use especially in plant systems where C-banding has not been successful.

6.3
Molecular Physical Mapping Techniques

6.3.1
CHEF Gel Mapping

When physically mapping a genome, it is often desirable to separate DNA fragments of 100 or more kilobases in size on an agarose gel. However, standard gel electrophoresis can resolve fragments up to only ~40 kb in length (Sambrook and Russell 2001). Fortunately, this limitation was overcome by the use of pulsed-field gel electrophoresis (PFGE) (Schwartz et al. 1982), which allows the separation of fragments up to 5 Mb in length. In PFGE, the electrical field is switched between two different directions every 0.1 to 1000 s or more, causing the DNA molecules to reorient. Smaller molecules are able to reorient themselves more rapidly than larger molecules and thus wiggle their way through the gel more quickly as the electrical field switches, allowing the DNA fragments to be fractionated according to size (Sambrook and Russell 2001; Carle and Olson 1984; Chu et al. 1986; Smith et al. 1987).

There are many applications for CHEF gels in the physical mapping process. When large segments of DNA, such as BAC clones, are digested with rare-cutting restriction enzymes, CHEF gel electrophoresis allows the determination of insert sizes in high-capacity vectors used in physical mapping projects. Size fractionation of clones on CHEF gels can also be used to determine the sizes of deletions and insertions in chromosomal regions of interest. The DNA from a CHEF gel can also be blotted onto a membrane and used in Southern hybridization experiments to give physical estimates of distances between markers for comparison with genetic maps, such as RFLP maps. Sadowski and Quiros (1998) used PFGE to analyze and determine the chromosomal location of a gene complex of *Arabidopsis thaliana* in the *Brassica nigra* genome, powerfully demonstrating the importance of PFGE as a physical mapping tool.

6.3.2
Radiation Hybrid Mapping

Radiation hybrid mapping is a tool for mapping specific chromosomal regions. This technique is currently being used to map chromosome segments in a number of mammalian systems, including horse (Chowdhary et al. 2003), mouse (Rowe et al. 2003), dog (Guyon et al. 2003), and human (Deloukas et al. 1998; http://www.ncbi.nih.gov/mapview/static/humansearch.html#Radiation HybridMaps), and is just beginning to be used as a mapping tool in plants (Ananiev et al. 1997; Kynast et al. 2002).

The use of radiation hybrid somatic cell lines to map chromosomes was first developed by Cox et al. (1990) for the construction of high-resolution maps in mammalian systems. In this technique, X-rays are used to break the chromosome of interest into fragments, which are recovered in cells of another species. The fragments are then analyzed for the presence or absence of specific DNA markers, allowing an estimate of the frequency of breakage, or genetic distance, between two markers. Once the distance between the markers has been determined, the markers can be placed in order on a chromosome map. This method is similar to traditional meiotic mapping, which uses polymorphisms between two copies of the same chromosome to calculate the recombination frequency (genetic distance) between markers. A major advantage of recombinant hybrid mapping is that only one copy of the chromosome of interest is being analyzed, so DNA markers need not be polymorphic.

A variation of the radiation hybrid technique was first proposed as a physical mapping tool in plants by Ananiev et al. (1997). Disomic addition lines for each maize chromosome except for chromosome 10 have been established in an oat background; chromosome 10 is available in a haploid oat background. Radiation is then used to induce translocations of maize chromatin into oat, creating radiation hybrid lines containing fragments of maize chromosomes in an oat background. The maize genome is highly rearranged and duplicated. The removal of DNA segments from the native background into an alien background simplifies the physical mapping of duplicated sequences and gene families to their chromosomal location, enabling the structural and functional analysis of duplicated genomes (Kynast et al. 2002).

It is important to remember that radiation hybrid mapping is a statistical method; thus, the best map obtained using the technique does not necessarily reflect the actual order of markers on the chromosome. The method is most useful when combined with other physical mapping methods, such as CHEF gel analysis, to confirm the results obtained from the radiation hybrid analysis (Cox et al. 1990). The advantage of this system over conventional bacterial, yeast, or phage cloning systems is that all of the alien chromosome fragments (maize) would be derived from a known chromosome.

6.3.4
Large-insert Clone Libraries (YACs, BACs, Cosmids)

(This topic is discussed in more detail in Chapter 8.) Genomic libraries, prepared by digesting genomic DNA with restriction enzymes and then ligating the fragments into high-capacity vectors, have become staple ingredients for physical mapping projects. Digesting large-insert clones with restriction enzymes and hybridizing the fragments to the clone libraries, then end-sequencing the clones and using computer software to align the sequences, allow the construction of overlapping arrays of clones (contigs). These contigs are widely used in sequencing and in physical mapping projects in both plants and animals. Each type of large-insert vector has advantages and disadvantages; therefore, physical maps are often constructed from a combination of clone types (Sambrook and Russell 2001), the three most common of which are discussed here. The choice of vector depends on a number of factors, including the size of the insert, the ease of screening and maintaining the libraries, the ability to recover cloned sequences, and the stability of the cloned sequences in the chosen vector (Sambrook and Russell 2001).

Yeast artificial chromosomes (YACs) are made by ligating large genomic DNA fragments between the two arms of a YAC vector. The vector and DNA are then transformed into yeast for clone propagation. Each arm of the YAC vector contains a selectable marker, DNA sequences that function as telomeres, centromeric DNA segments, and an origin of replication. YACs are designed so that clones containing empty vectors are easily distinguished from clones containing the inserts. YACs have no packaging limit, so the insert size is determined mostly by the quality of genomic DNA used in the ligation. On average, YAC clones contain DNA inserts ranging from 250 kb to 400 kb (Sambrook and Russell 2001).

Bacterial artificial chromosome (BAC) vectors contain an antibiotic resistance gene; a replicon derived from the F factor of *E. coli*; an ATP-driven helicase (*repE*) for DNA replication; and the *parA*, *parB*, and *parC* loci, which ensure accurate partitioning of low-copy-number plasmids to daughter cells. The genomic DNA of interest is ligated into the BAC vector, which is then transformed into *E. coli*. BAC vectors use α-complementation to screen for clones containing inserts. Like YACs, BACs have no packaging limits, and the average insert size ranges from 120 kb to 300 kb (Sambrook and Russell 2001).

Cosmids are the oldest type of high-capacity vector in use. They are plasmids that contain one or two copies of a region of bacteriophage λ DNA, the cohesive end (*cos*) site, required for packaging viral DNA into bacteriophage λ particles. Restriction fragments containing a *cos* sequence are ligated onto each end of a DNA molecule and then cleaved to generate a linear molecule with termini that are complementary to each other. The molecules are then injected into host bacterial cells, where the termini anneal to one another, generating circular DNA molecules that also contain a colicin E1 plasmid replicon and a selectable marker. The biggest disadvantages of using cosmids is their relatively small insert size of only 43–45 kb and the more labor-intensive packaging and transformation protocols as compared to YACs and BACs (Sambrook and Russell 2001).

With the exception of sequencing the genome, physical mapping through contig assembly yields the highest resolution of physical map. Contig assembly can, however, be expensive, especially for plants with large genomes, and it is dependent on the quality of the large-insert DNA libraries used to construct the contigs. In polyploidy species or species with a high level of sequence duplication, the assignment of clones to the correct contig can be technically difficult (Cheng et al. 2001). While high-capacity vectors are routinely used in plant mapping, it is important to remember that the quality of the clones must be verified by PFGE and fingerprint analysis. Clone order in the contigs must also be eventually verified using physical mapping techniques such as FISH to confirm the chromosomal location and order of the clones in each contig.

6.3.5
Fluorescent *in situ* Hybridization

The *in situ* hybridization (ISH) technique, developed in 1969 (Gall and Pardue 1969; John et al. 1969), is used by cytogeneticists to map genes and DNA sequences directly onto chromosomes. In ISH, denatured DNA probes are hybridized to denatured chromosomes that have been prepared on a microscope slide. ISH has been used to map sequences to chromosomal regions, to determine the locations and distribution of repetitive sequences on chromosomes, to develop molecular karyotypes, and to map multicopy gene families, such as rDNA genes, on plant chromosomes. Originally, probes used for ISH were isotopic and, while they could detect single-copy DNA sequences, the technique had a limited resolution and was time-intensive. The subsequent development of non-isotopic probes, e. g., those labeled with biotinylated dUTP, has greatly improved the sensitivity and resolution of the ISH technique. Using enzymatic detection, antibodies with enzymatic reporter molecules such as horseradish peroxidase are used to visualize the signals. Enzymatic detection is advantageous in that the signals do not fade over time; however, only one probe can be visualized in a given experiment (reviewed in Jiang and Gill 1994c).

Fluorescent *in situ* hybridization (FISH), which uses fluorochromes conjugated to antibodies to detect the probes, overcomes this disadvantage. Multiple probes labeled with different conjugates can be detected simultaneously on a single slide, allowing each probe to be visualized as a different color signal using a fluorescence microscope. While the fluorochromes will fade over time, mounting agents that minimize this problem have been developed. Multicolor FISH can be used to physically order probes on chromosomes and to map probes to their locations on specific chromosomes (reviewed in Jiang and Gill 1994c). In addition, FISH has improved resolution, allowing the detection of single- and low-copy sequences in chromosome preparations (Peterson et al. 1999). FISH has been used to map 4-kb, single-copy T-DNA sequences onto metaphase chromosomes to study recombination and T-DNA integration in petunia (ten Hoopen et al. 1996). With large-insert clone libraries becoming more widely used in physical mapping, FISH has proved to be a powerful technique to map the clones onto chromosomes, confirming their positions within the assembled contigs and their relationships to markers on the genetic map (Figure 6.1; Jiang et al. 1995).

Variations on FISH are used for different physical mapping applications. Genomic *in situ* hybridization (GISH) is basically FISH that uses genomic DNA as a probe. GISH is used to estimate the amount of alien chromatin within chromosomes in interspecific hybrids and to detect the breakpoints of the translocation chromosomes. It can also be used to analyze the distribution of chromatin from different genomes in interspecific hybrids, to monitor homoeologous chromosome pairing, and to analyze chromosome structure in allopolyploids (reviewed in Jiang and Gill 1994c). Another variation of the FISH technique, chromosome painting, uses chromosome-specific probes to label chromosome regions. A widely used technique in animal studies, this has been done only recently in plants using chromosome-specific repeats as probes (Lysak et al. 2001).

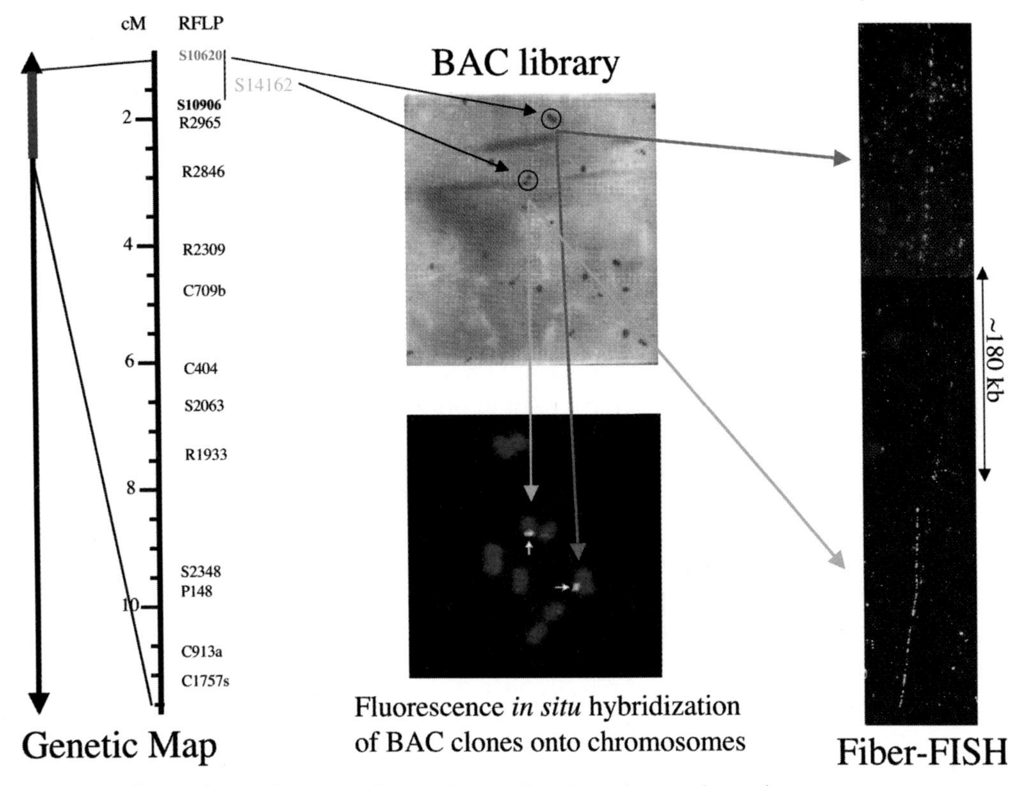

Figure 6.1. The correlation of genetic and physical maps. Genetic markers can be used to screen large-construct libraries, such as BAC libraries. The clones identified in the screen are then used as probes in FISH mapping and fiber-FISH experiments to determine the physical location of the genetic marker.

Scientists are constantly refining FISH in order to improve the resolution of the technique, which will lead ultimately toward more-accurate physical maps. One way of doing this is through the use of pachytene chromosomes rather than mitotic metaphase chromosomes. These spreads have relatively decondensed chromosomes that are about 10 times longer than metaphase spreads of the same chromosomes, and the chromosomes are more accessible to probe hybridization (Peterson et al. 1999). Cheng et al. (2001) were able to use FISH of BAC clones to pachytene chromosomes to create a physical map of rice chromosome 10 and to integrate that map with the existing genetic map. The advantages of high-resolution pachytene FISH to plant physical mapping over genetic mapping are clearly demonstrated in studies such as that of Zhong et al. (1999). In this study, FISH on tomato pachytene chromosomes located the nematode-resistance gene *Mi-1*, thought to be on the long arm of chromosome 6 due to its tight genetic linkage (1 cM) to a genetic marker on that arm, on the short arm of the chromosome, at least 40 Mb away from the marker to which it is so tightly linked on the genetic map.

The highest resolution using FISH thus far has been achieved through the development of fiber-FISH. Instead of using metaphase or pachytene chromosome preparations for FISH, interphase nuclei are lysed open and the chromatin fibers are dragged down a glass slide using a coverslip. Fibers can be stretched to approximately 3.0 kb μm^{-1}, a length close to the Watson-Crick DNA length estimate of 2.9 kb μm^{-1} (Fransz et al. 1996; Jackson et al. 1998). Fiber-FISH has been used in human studies to physically map multigene families (Nishio et al. 1996), to characterize repetitive DNA sequences (Shiels et al. 1997), and to validate the orientation, overlap, and size of BAC clones used to generate a physical map (Poulsen et al. 2001). In plants, fiber-FISH was first used by Fransz et al. (1996), who used the method to confirm the order, size, and overlap of cosmids in *Arabidopsis* and to map small repetitive sequences in tomato to a lower detection limit of 0.7 kb. Fiber-FISH has since been used in physical mapping to determine the size of gaps on molecular contig maps in *Arabidopsis* (Jackson et al. 1998).

Recently, molecular biologists have begun to embrace FISH as a tool for comparative genomic studies. Chromosome painting and FISH have been used in human studies to determine DNA sequence copy number, particularly in cancer cell lines (Kallioniemi et al. 1994) and to study intrachromosomal rearrangements, chromosomal synteny, and chromosomal evolution between humans and other mammals (Müller et al. 2000). In plants, comparative mapping has been done by sequencing orthologous regions of DNA from related species, by using PFGE followed by blotting and probe hybridization, and by doing FISH in related species using the same probes. As discussed previously, fiber-FISH provides excellent resolution for constructing physical maps and has already been used in comparative studies in *Arabidopsis* (Jackson et al. 2000), rice, maize, and soybean (Hass and Jackson, unpublished data).

6.3.6 Mapping Gene Space

We have discussed how physical mapping has aided in cloning genes and in the sequencing of genes and whole genomes. We have also seen examples of how physical mapping techniques can be used to investigate plant evolution and to compare genomes, a field called comparative genomics. Physical mapping can also be used to describe genome organization on a global level. The discovery of "gene space," a term applied to the compositional compartmentalization of genes within a narrow GC range of the genome such that genes are contained in long, gene-rich regions separated by long, gene-empty regions, has raised important questions regarding how different regions of the chromosome are evolving (Barakat et al. 1997). Examination of the gene spaces of Gramineae has shown that gene spaces in this family of plants fall within a 0.8–1.6 % GC range and represent only 12–24 % of the genome. While gene space has been described for plant species outside of the Gramineae, such as pea, it is not present in all species; gene space was not found in *Arabidopsis*, tomato, or date palm. Gene space is thought to be correlated with genome size, repeat abundance, and transposon activity within the genome (Barakat et al. 1999). The discovery of gene space can aid in focusing sequencing projects on gene-rich regions of genomes of interest, and in situ hybridization can be used to physically map gene space onto chromosomes.

6.4 Discussion

Cytogenetics has experienced a renaissance with the development of new molecular technologies and their application. In particular, FISH in its various incarnations has been useful for many modern genomic-type studies. FISH to various cytological targets has been used to integrate genetic and physical maps (Cheng et al. 2001), to associate chromosome bands with sequence data (Jiang and Gill 1994c), to gauge gaps existing in sequence maps (Jackson et al. 1998), and to map the gene space of complex genomes, which leads to more efficient gene cloning and genome sequencing (Bakarat et al. 1997).

In the future there are many exciting avenues of research where physical mapping and cytogenetics will be invaluable. The first is integrating the linear arrangement of DNA sequences with the functional organization of chromosomes and genomes. It is well known that the organization of chromatin in the eukaryotic nucleus is closely correlated with the regulated expression of genes (Lusser 2002). Researchers are just now beginning to use the tools of genomics and genome sequences to begin to investigate this association. For example, in yeast the ubiquitination of histone H2B was recently shown to regulate histone H3 methylation and gene silencing (Sun and Allis 2002). In humans, heterochromatin-mediated gene silencing has been implicated in triplet-repeat diseases such as myotonic dystrophy and Friedreich's ataxia (Saveliev et al. 2003), and chromatin modifications, particularly DNA methylation, have been linked to the regulation of tumor suppressor genes and oncogenes in cancer (reviewed in Plass 2002). Studies of *Arabidopsis* have revealed associations between gene regulation and DNA methylation, histone methylation, histone acetylation/deacetylation, and ATP-dependent chromatin remodeling via Swi2/Snf2 protein complexes (reviewed in Lusser 2002).

The second area of research where physical mapping and cytogenetics will be further employed is in the fields of comparative and evolutionary genomics. Cytogenetics has had a long history of research in chromosome and genome evolution using classical methods such as pairing studies and chromosome banding (reviewed in Knight 2003; Sengbusch 2002; Gill 1987); however, with FISH and the available genome sequences and BAC-based physical maps, researchers are now able to use new tools to ask these basic questions. One important development is the discovery that genomic clones can be used as heterologous probes in cytological studies (Jackson et al. 2000; Zwick et al. 1998), similar to the discovery that genetic mapping probes could be used in heterologous systems (Melake-Berhan et al. 1993; Devos et al. 1993). Also, integrating BAC-based physical maps from related systems is proving to be an important comparative physical mapping technique that yields information on genome evolution and DNA sequence conservation that can be used to infer gene and promoter structure (Thomas et al. 2002; http://rice.genomics.purdue.edu; http://www.omap.org). One of the challenges in the current genomics era is to provide a physical mapping toolbox to a wide variety of organisms chosen for both their economic importance and their position on the "tree of life" to represent important nodal evolutionary branches (Maddison 2001). These tools, such as BAC libraries and eventual DNA sequences, will give researchers the necessary tools to use comparative genomics to identify and clone important DNA sequences (i. e., genes) and to develop models of chromosome/genome evolution among a broad range of species.

References

Ananiev, E. V., O. Riera-Lizarazu, H. W. Rines, and R. L. Phillips (1997) "Oat-maize chromosome addition lines: A new system for mapping the maize genome." *Proc. Natl. Acad. Sci. USA* 94:3524–3529.

Ananiev, E. V., R. L. Phillips, and H. W. Rines (1998) "A knob-associated tandem repeat in maize capable of forming fold-back DNA segments: Are chromosome knobs megatransposons?" *Proc. Natl. Acad. Sci. USA* 95: 10785–10790.

Barakat, A., N. Carels, and G. Bernardi (1997) "The distribution of genes in the genomes of Gramineae." *Proc. Natl. Acad. Sci. USA* 94(13):6857–6861.

Barakat, A., D. Tran Han, A.-A. Benslimane, A. Rode, and G. Bernardi (1999) "The gene distribution in the genomes of pea, tomato and date palm." *FEBS Letters* 463: 139–142.

Carle, G. F. and M. V. Olson (1984) "Separation of chromosomal DNA molecules from yeast by orthogonal-field-alternation gel electrophoresis." *Nucleic Acids Research* 12(14):5647–5664.

Caspersson, T., M. Hulten, J. Lindsten, and L. Zech (1971) *Heredity* 67:147.

Cheng, Z., G. G. Presting, C. R. Buell, R. A. Wing, and J. Jiang (2001) "High-resolution pachytene chromosome mapping of bacterial artificial chromosomes anchored by genetic markers reveals the centromere location and the distribution of genetic recombination along chromosome 10 of rice." *Genetics* 157: 1749–1757.

Chowdhary B. P., T. Raudsepp, S. R. Kata, G. Goh, L. V. Millon, V. Allan, F. Piumi, G. Guerin, J. Swinburne, M. Binns, T. L. Lear, J. Mickelson, J. Murray, D. F. Antczak, J. E. Womack, and L. C. Skow (2003) "The first-generation whole-genome radiation hybrid map in the horse identifies conserved segments in human and mouse genomes." *Genome Research* 13(4):742–51.

Chu, G., D. Vollrath, and R. W. Davis (1986) "Separation of large DNA molecules by contour-clamped homogeneous electric fields." *Science* 234:1582–1585.

Chughtai, S. R. and D. M. Steffensen (1987) "Heterochromatic knob composition of commercial inbred lines of maize." *Maydica* 32: 171–187.

Cox, D. R., M. Burmeister, E. Roydon Price, S. Kim, and R. M. Myers (1990) "Radiation hybrid mapping: A somatic cell genetic method for constructing high-resolution maps of mammalian chromosomes." *Science* 250:245–250.

Deloukas et al. (1998) "A physical map of 30,000 human genes." *Science* 282:744–746.

Devos, K. M., M. D. Atkinson, C. N. Chinoy, R. L. Harcourt, R. M. D. Koebner, C. J. Liu, P. Masojc, D. X. Xie, and M. D. Gale (1993) *Theor. Appl. Genet.* 85:673–680.

Fransz, P. F., C. Alonso-Blanco, T. B. Liharska, A. J. M. Peeters, P. Zabel, and J. H. de Jong (1996) "High-resolution physical mapping in *Arabidopsis thaliana* and tomato by fluorescence *in situ* hybridization to extended DNA fibers." *The Plant Journal* 9(3):421–430.

Gall, J. G. and M. L. Pardue (1969) "Formation and detection of RNA-DNA hybrid molecules in cytological preparations." *Proc. Natl. Acad. Sci. USA* 63:378–383.

Gill, B. S. (1987) "Chromosome banding methods, standard chromosome band nomenclature, and applications in cytogenetic analysis." In: *Wheat and Wheat Improvement*; 2nd. ed. American Society of Agronomy, Inc., Madison, Wisconsin. pp. 243–254.

Guyon, R., T. D. Lorentzen, C. Hitte, L. Kim , E. Cadieu, H. G. Parker, P. Quignon, J. K. Lowe, C. Renier, B. Gelfenbeyn, F. Vignaux, H. B. DeFrance, S. Gloux, G. G. Mahairas, C. André, F. S. Galibert and E. A. Ostrander (2003) "A 1-Mb Resolution Radiation Hybrid Map of the Canine Genome." *Proc. Natl. Acad. Sci. USA* 100:5286–5291.

Hsu, T. C. (1973) "Longitudinal differentiation of human chromosomes." *Ann. Rev. Genet.* 7: 153–176.

Hutchinson, J. and A. G. Seal (1983) "A sequential in situ hybridization and C-banding technique." *Heredity* 51:507–509.

Jackson, S. J., M. L. Wang, H. M Goodman, and J. Jiang (1998) "Application of fiber-FISH in physical mapping of *Arabidopsis thaliana*." *Genome* 41:566–572.

Jackson, S. J., Z. Cheng, M. L. Wang, H. M. Goodman, and J. Jiang (2000) "Comparative fluorescence *in situ* hybridization mapping of a 431-kb *Arabidopsis thaliana* bacterial artificial chromosome contig reveals the role of chromosomal duplications in the expansion of the *Brassica rapa* genome." *Genetics* 156: 833–838.

Jiang, J. and B. S. Gill (1993) "Sequential chromosome banding and in situ hybridization analysis." *Genome* 36:792–795.

Jiang, J. and B. S. Gill (1994a) "Different species-specific chromosome translocations in *Triticum timopheevii* and *T. turgidum* supports diphyletic origin of polyploid wheats." *Chromosome Research* 2:59–64.

Jiang, J. and B. S. Gill (1994b) "New 18S-26S ribosomal RNA gene loci: chromosomal landmarks for the evolution of polyploidy wheat." *Chromosoma* 103:179–185.

Jiang, J. and B. S. Gill (1994c) "Nonisotopic in situ hybridization and plant genome mapping: the first 10 years." *Genome* 37:717–725.

Jiang, J., B. S. Gill, G. L. Wang, P. C. Ronald, and D. C. Ward (1995) "Metaphase and interphase fluorescence *in situ* hybridization mapping of the rice genome with bacterial artificial chromosomes." *Proc. Natl. Acad. Sci. USA* 92:4487–4491.

John, H. A., M. L. Birnstiel, and K. W. Jones (1969) "RNA-DNA hybrids at the cytological level." *Nature* (London) 223:582–587.

Joppa, L. R. (1987) "Aneuploid analysis in tetraploid wheat." In: *Wheat and Wheat Improvement*; 2nd. ed. American Society of Agronomy, Inc., Madison, Wisconsin. pp. 255–267.

Kallioniemi, A., O.-P. Kallioniemi, J. Piper, M. Tanner, T. Stokke, L. Chen, H. S. Smith, D. Pinkel, J. W. Gray, and F. M. Waldman (1994) "Detection and mapping of amplified DNA sequences in breast cancer by comparative genomic hybridization." *Proc. Natl. Acad. Sci. USA* 91:2156–2160.

Knight, Tim (2003) "Primate Cytogentics Network." Center for Mind, Brain, and Learning, University of Washington, Seattle, Washington, 98195, USA. http://staff.Washington.edu/timk/cyto/text/

Kynast, R. G., R. J. Okagaki, H. Rines, and R. L. Phillips (2002) "Maize individualized chromosome and derived radiation hybrid lines and their use in functional genomics." *Functional & Integrative Genomics* 2(1/2):60–69.

Longley, A. E. (1952) "Chromosome morphology in maize and its relatives." *The Botanical Review* 18(6):399–412.

Lusser, A. (2002) "Acetylated, methylated, remodeled: chromatin states for gene regulation." *Current Opinion in Plant Biology* 5:437–443.

Lysak, M. A., P. F. Fransz, H. B. M. Ali, and I. Schubert (2001) "Chromosome painting in *Arabidopsis thaliana*." *The Plant Journal* 28(6): 689–697.

Maddison, David R. (2001) http://tolweb.org/tree/phylogeny.html

McClintock, B. (1929) "Chromosome morphology in *Zea mays*." *Science* 69:629.

McClintock, B. (1930) "A cytological demonstration of the location of an interchange between two non-homologous chromosomes of *Zea mays*." *Proc. Natl. Acad. Sci. USA* 16: 791–796.

Melake-Berhan, A., S. H. Hulbert, L. G. Butler, and J. L. Bennetzen (1993) *Theor. Appl. Genet.* 86:598–604.

Mohammed, A. H. and E. A. Shoeib (1965) "Inheritance of quantitative characters in *Zea mays*. IV. Relationship between knob frequency and combining ability." *Can. J. Genet. Cytol.* 7:388–393.

Moll, R. H., W. D. Hanson, C. S. Livings, III, and Y. Ohta (1972) "Associations between chromosomal knobs of *Zea mays* L. and agronomic performance." *Crop Sci.* 12:585–589.

Müller, S., R. Stanyon, P. Finelli, N. Archidiacono, and J. Wienberg (2000)"Molecular cytogenetic dissection of human chromosomes 3 and 21 evolution." *Proc. Natl. Acad. Sci.USA* 97(1):206–211.

Nishio, H., M. Heiskanen, A. Palotie, L. Bélanger, and A. Dugaiczyk (1996) "Tandem arrangement of the human serum albumin multigene family in the sub-centromeric region of 4q: Evolution and chromosomal direction of transcription." *J. Mol. Biol.* 259: 113–119.

Paterson, A. H., J. W. DeVerna, B. Lanini, and S. D. Tanksley (1990) "Fine mapping of quantitative trait loci using selected overlapping recombinant chromosomes in an interspecies cross of tomato." *Genetics* 124:735–742.

Peterson, D. G., N. L. V. Lapitan, and S. M. Stack (1999) "Localization of single- and low-copy sequences on tomato synaptonemal complex spreads using fluorescence *in situ* hybridization (FISH)." *Genetics* 152:427–439.

Plass, C. (2002) "Cancer epigenomics." *Human Molecular Genetics* 11(20):2479–2488.

Poulsen, T. S., A. N. Silahtaroglu, C. G. Gisselo, E. Gaarsdal, T. Rasmussen, N. Tommerup, and H. E. Johnsen (2001) "Detection of illegitimate rearrangement within the immunoglobulin locus on 14q32.3 in B-cell malignancies using end-sequenced probes." *Genes, Chromosomes & Cancer* 32:265–274.

Qi, L. L. and B. S. Gill (2001) "High-density physical maps reveal that the dominant male-sterile gene Ms3 is located in a genomic region of low recombination in wheat and is not amenable to map-based cloning." *Theor. Appl. Genet.* 103(6/7):998–1006.

Rhoades, M. M. and E. Dempsey (1973) "Chromatin elimination induced by the B chromosome of maize. I. Mechanism of loss and the pattern of endosperm variegation." *J. Hered.* 64:13–18.

Rooney, W. L., E. N. Jellen, R. L. Phillips, R. W. Rines, and S. F. Kianian (1994) "Identification of homoeologous chromosomes in hexaploid oat (*A. byzantina* cv. Kanota) using monosomics and RFLP analysis." *Theor. Appl. Genet.* 89:329–335.

Rowe L. B., M. E. Barter, J. A. Kelmenson, and J. T. Eppig (2003) "The comprehensive mouse radiation hybrid map densely cross-referenced to the recombination map: A tool to support the sequence assemblies. *Genome Research* 13(1):122–33.

Rubin, G. M. and E. B. Lewis (2000) "A brief history of *Drosophila*'s contributions to genome research." *Science* 287:2216–2218.

Sadowski, J. and C. F. Quiros (1998) "Organization of an *Arabidopsis thaliana* gene cluster on chromosome 4 including the *RPS2* gene in the *Brassica nigra* genome." *Theor. Appl. Genet.* 96:468–474.

Sambrook and Russell (2001) *Molecular Cloning: A Laboratory Manual*; 3rd. ed. Cold Spring Harbor Laboratory Press, Clod Spring Harbor, New York.

Saveliev, A., C. Everett, T. Sharpe, Z. Webster, and R. Festenstein (2003) "DNA triplet repeats mediate heterochromatin-protein-1-sensitive variegated gene silencing." *Nature* 422(6934):909–913.

Schwartz, D. C., W. Saffran, J. Welsh, R. Haas, M. Goldenberg, and C. R. Cantor (1982) *Cold Spring Harbor Symp. Quant. Biol.* 47:198–195.

Sengbusch, P. V. (2002) "Chromosomal Banding Patterns." In: Bergfeld, A., R. Bergmann, and P. V. Sengbusch. *Botany online: The Internet Hypertextbook.* (2003) http://www.biologie.uni-hamburg.de/b-online/ereg/eindexfr.htm

Sharma, A. K. and A. Sharma (1999) *Plant Chromosomes: Analysis, Manipulation and Engineering.* Overseas Publishers Association, Amsterdam, the Netherlands, pp. 119–137.

Sharp, P. J., S. Chao, S. Desai, and M. D. Gale (1989) "The isolation, characterization and application in the Triticeae of a set of wheat RFLP probes identifying each homoeologous chromosome arm." *Theor. Appl. Genet.* 78: 342–348.

Shiels, C., C. Coutell, and C. Huxley (1997) "Analysis of ribosomal and alphoid repetitive DNA by fiber-FISH." *Cytogenet Cell Genet.* 76: 20–22.

Smith, C. L., S. K. Lawrance, G. A Gillespie, C. R. Cantor, S. M Weissman, and F. S. Collins (1987) "Strategies for mapping and cloning macroregions of mammalian genomes." In: *Methods in Enzymology*, vol. 151, Michael M. Gottesman, ed., pp. 461–489.

Sun, Z. W. and C. D. Allis (2002) "Ubiquitination of histone H2B regulates H3 methylation and gene silencing in yeast." *Nature* 417(6893):104–108.

ten Hoopen, R., T. P. Robbins, P. F. Fransz, B. M. Montijn, O. Oud, A. G. M. Gerats, and N. Nanninga (1996) "Localization of T-DNA insertions in petunia by fluorescence in situ hybridization: Physical evidence for suppression of recombination." *The Plant Cell* 8: 832–830.

Thomas, J. W., A. B. Prasad, T. J. Summers, S.-Q. Lee-Lin, V. V. B. Maduro, J. R. Idol, J. F. Ryan, P. J. Thomas, J. C. McDowell, and E. D. Green (2002) "Parallel construction of orthologous sequence-ready clone contig maps in multiple species." *Genome Research* 12:1277–1285.

Young, N. D. (2000) "Constructing a plant genetic linkage map with DNA markers." In: R. L. Phillips and J. K. Vasil, eds. *DNA-Based Markers in Plants.* Kluwer Academic Publishers, Dordrecht, the Netherlands, pp. 31–47.

Young, N. D., J. C. Miller, and S. D. Tanksley (1987) "Rapid chromosomal assignment of multiple genom.

Zhong, X.-B., J. Bodeau, P. F. Fransz, V. M. Williamson, V. van Kammen, J. H. de Jong, and P. Zabel (1999) "FISH to meiotic pachytene chromosomes of tomato locates the root-knot nematode resistance gene *Mi-1* and the acid phosphatase gene *Aps-1* near the junction of euchromatin and pericentromeric heterochromatin of chromosome arms 6S and 6L, respectively." *Theor. Appl. Genet.* 98: 365–370.

Zoller, J. F., Y. Yang, R. G. Herrmann, and U. Hohmann (2001) "Comparative genomic *in situ* hybridization (cGISH) analysis on plant chromosomes revealed by labeled *Arabidopsis* DNA." *Chromosome Research* 9: 357–375.

Zwick, M. S., M. N. Islam-Faridi, D. G. Czechin, Jr., R. A. Wing, G. E. Hart et al. (1998) "Physical mapping of the *liguleless* linkage group in *Sorghum bicolor* using rice selected RFLP Sorghum BACs." *Genetics* 148: 1983–1992.

7
Chromosome Flow Sorting and Physical Mapping

Jaroslav Doležel, Marie Kubaláková, Jan Bartoš, and Jiří Macas

Overview

Methods for chromosome analysis and sorting (flow cytogenetics) were originally developed for human and some animal species. Modification of the methodology for plants has been delayed by difficulties in preparation of suspensions of intact chromosomes suitable for flow cytometry and in discrimination of individual chromosome types. These problems have been overcome by preparing chromosome suspensions from synchronized root tips and by using cytogenetic stocks, respectively. Until now, chromosome analysis and sorting has been reported in 17 species, including legumes such as chickpea, field bean, and garden pea and cereals such as barley, maize, rye, and wheat. Chromosomes are classified by flow cytometry according to their relative DNA content, and the analysis of resulting distributions (flow karyotypes) permits quantitative detection of structural and numerical chromosome changes. Large quantities of chromosomes may be purified by flow sorting, and sorted fractions have been used for physical mapping, production of recombinant DNA libraries, targeted isolation of markers, and protein analysis. This chapter reviews the methodology and application of flow cytogenetics and presents typical results. In addition, the potential of flow cytogenetics for plant genome analysis is evaluated.

Abstract

Flow cytometry is a powerful tool for quantitative analysis and purification of mitotic chromosomes. The analysis is performed at high rates, typically 10^2–10^3 sec^{-1}, and large quantities of chromosomes stained by a DNA-binding fluorochrome are classified according to fluorescence intensity, which reflects their relative DNA content. The resulting distributions of DNA content are called flow karyotypes. Ideally, each chromosome on a flow karyotype is represented by a single well-discriminated peak. Any chromosome or a group of chromosomes that may be discriminated may also be purified by sorting. Two different chromosomes may be sorted simul-

The Handbook of Plant Genome Mapping.
Genetic and Physical Mapping. Edited by Khalid Meksem, Günter Kahl

ISBN 3-527-31116-5

taneously. Flow karyotyping has been found to be suitable for quantitative detection of specific structural and numerical chromosome changes. Large numbers of chromosomes that may be sorted onto microscopic slides facilitate high-resolution physical mapping using FISH or PRINS and discovery of rare structural changes. PCR with chromosomes sorted into reaction tubes allows physical mapping of short and single-copy targets; the use of deletion and translocation chromosomes enables subchromosomal localization of mapped sequences. DNA of sorted chromosomes is suitable for cloning and preparation of chromosome-specific DNA libraries. Short-insert libraries have been used for targeted isolation of molecular markers from specific genome regions. It is expected that the availability of chromosome- and chromosome arm–specific BAC libraries harboring large inserts will greatly simplify development of physical contig maps and gene isolation in plant species possessing large and complex genomes.

7.1 Introduction

Many crops possess genomes whose size exceeds that of human and which are characterized by a large proportion of repetitive DNA sequences. As some species are known to be ancient or recent polyploids, their genomes contain many regions that share significant sequence homologies. These features hamper genome mapping and gene isolation efforts. Fractionation of genomes into single chromosomes, which represent small and defined genome parts, appears to be an attractive way to simplify and speed up genome analysis and mapping. Flow cytometry is the only method that is capable of separating chromosomes in large quantities and high purities (Doležel et al. 1994).

To do this, flow cytometry analyzes light scatter and fluorescence properties of mitotic chromosomes in aqueous suspension. The suspension, which contains chromosomes stained by a DNA-specific/binding fluorochrome, is injected into a flow chamber and the particles are constrained to flow in single file within a fluid stream past a narrow beam of excitation light. The chromosomes cause the light to scatter and the molecules of fluorochrome(s) bound to them become excited. The chromosomes are measured individually and at high rates, typically 100–1000 sec^{-1}, and thus large populations may be evaluated in a short time and chromosomes differing in light scatter and fluorescence properties may be identified (Figure 7.1). Any chromosome that is discriminated based on the optical properties can be sorted in large quantities.

Procedures for flow cytometric chromosome analysis and sorting, later called flow cytogenetics (Carrano et al. 1983), were first developed for Chinese hamster (Gray et al. 1975; Stubblefield et al. 1975) and subsequently for human (Carrano et al. 1979) and a number of animal species (Ferguson-Smith 1997). Flow cytometry has been used for classification of chromosomes according to relative DNA content (Young 1990), AT/GC ratio (Gray and Cram 1990), and presence of antigens (Levy et al. 1991). The resulting frequency distributions were named flow karyo-

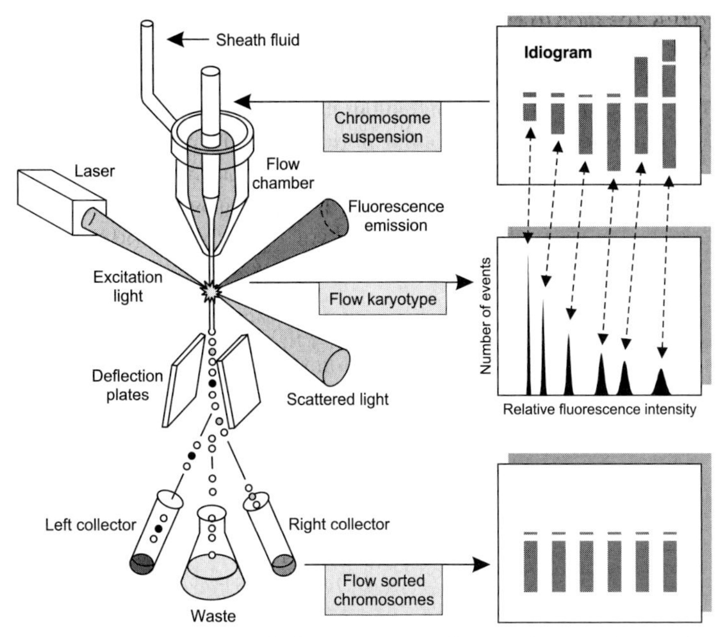

Figure 7.1. Schematic view of a flow cytometer and sorter. The example considers the analysis in a species where all chromosomes differ in DNA content and each of them is represented by a single peak on a flow karyotype.

types. While the attempts to use flow karyotyping for routine detection of structural chromosome changes in human did not meet with great success (Boschman et al. 1992), chromosome-specific DNA libraries prepared from flow-sorted chromosomes (Van Dilla et al. 1986) played an important role in mapping of the human genome. The most frequent application of sorted human and animal chromosomes has been the preparation of chromosome-specific paints (Pinkel et al. 1998), which have been utilized in phylogenetic, evolutionary, and genetic studies, as well as diagnostic tools (Tian et al. 2004; Langer et al. 2004). These successful applications stimulated the development of flow cytogenetics for plants.

7.2
Development of Flow Cytogenetics for Plants

The early efforts were hampered by difficulties in achieving a high degree of mitotic synchrony, in releasing intact chromosomes from cells with rigid walls, and in discrimination of single chromosome types (Doležel et al. 1994). The first successful report of de Laat and Blaas (1984) described discrimination and sorting of both chromosome types of *Haplopappus gracilis* ($2n=4$). The authors prepared suspensions of mitotic chromosomes after a hypotonic lysis of protoplasts obtained from synchronized suspension cultured cells. However, recent studies employ chromosome suspensions prepared from synchronized root tips (Doležel et al. 1992, 2001). Meristem cells of root tips are karyologically stable and can be synchronized to a high degree, and intact chromosomes can be easily released from them (Doležel et al. 1999a, 1999b).

In contrast to the model plant *H. gracilis*, in the majority of plants the analysis of isolated chromosomes results in flow karyotypes with composite peaks representing groups of chromosomes. Due to the similarity of relative DNA content, only one or a few chromosomes can be individually discriminated and sorted (Table 7.1). Various approaches have been tested to overcome this problem, including the analysis of AT/GC ratio and fluorescent labeling of repetitive DNA sequences (Lucretti and Doležel 1997; Lee et al. 1997; Pich et al. 1995). However, only the use of cytogenetic stocks (lines carrying chromosome deletions, translocation, and additions) was found to be useful. Lucretti et al. (1993) and Doležel and Lucretti (1995) were the first to show that alteration in chromosome size facilitates discrimination of specific chromosomes. Following their work, various cytogenetic stocks were used in a number of species to discriminate specific chromosomes and chromosome arms (Gill et al. 1999; Li et al. 2001; Kubaláková et al. 2002, 2003a, 2003b).

Table 7.1. List of plant species from which flow cytometric analysis of mitotic chromosomes has been reported.

Species	*n**	*N***	*Reference*
Avena sativa	21	0	Li et al. (2001)
Cicer arietinum	8	5	Vláčilová et al. (2002)
Haplopappus gracilis	2	2	de Laat and Blaas (1984)
Hordeum vulgare	7	1[a]	Lysák et al. (1999)
Lycopersicon esculentum	12	0	Arumuganathan et al. (1991)
Lycopersicon pennellii	12	2	Arumuganathan et al. (1991)
Melandrium album	12	2	Veuskens et al. (1995)
Nicotiana plumbaginifolia	10	0	Conia et al. (1989)
Petunia hybrida	7	1	Conia et al. (1987)
Pisum sativum	7	2[b]	Neumann et al. (1998)
Secale cereale	7	1[c]	Kubaláková et al. (2003a)
Triticum aestivum	21	1[d]	Kubaláková et al. (2002)
Triticum durum	14	0[e]	Kubaláková et al. (2003b)
Vicia faba	6	1[f]	Lucretti et al. (1993)
Zea mays	10	2[g]	Lee et al. (1996)

* Haploid chromosome number.
** Number of chromosomes that could be discriminated unambiguously in standard flow karyotype.
[a] Up to three chromosomes could be sorted from specific translocation lines.
[b] Up to four chromosomes could be sorted from specific translocation lines.
[c] Rye chromosomes 2R–7R could be discriminated from wheat-rye chromosome addition lines.
[d] Sorting of almost all chromosome arms is possible in hexaploid wheat using individual ditelosomic lines.
[e] All chromosome arms may be sorted from individual (double) ditelosomic lines.
[f] Six chromosomes may be sorted from chromosome translocation line EF.
[g] Maize chromosome 9 could be discriminated in oat-maize chromosome addition line (Li et al. 2001).

7.2.1 The Uses of Flow Karyotyping

Provided that a specific chromosome is resolved as a single peak on a flow karyotype, any change in its relative frequency and size is reflected by a change in chromosome peak area and position, respectively. This facilitated the use of flow karyotyping for quantitative detection of numerical and structural chromosome changes. Thus, Lee et al. (2000) were able to detect trisomy of chromosome 6 in barley. Alien chromosomes present in chromosome addition lines were detected in an oat-maize addition line (Li et al. 2001) and in wheat-rye addition lines (Kubaláková et al. 2003a). Flow karyotyping revealed the presence of B chromosomes in a rye cultivar, where their existence had not been reported before (Kubaláková et al. 2003a).

Following the studies of Lucretti et al. (1993) and Doležel and Lucretti (1995) on field bean, the ability of flow karyotyping to detect chromosome translocations was verified in garden pea (Neumann et al. 1998), barley (Lysák et al. 1999), rye (Kubaláková et al. 2003a), and wheat (Vrána et al. 2000; Kubaláková et al. 2002). Flow karyotyping was found to be sensitive enough to detect polymorphism in chromosome DNA content in wheat (Kubaláková et al. 2002), rye (Kubaláková et al. 2003a), maize (Lee et al. 2002), and barley (Lee et al. 2000). The "fingerprint" patterns of flow karyotypes characteristic for certain cultivars were found to be heritable. Flow karyotyping of telosomic lines of hexaploid bread wheat and tetraploid durum wheat allowed facile identification of telosomes (Gill et al. 1999; Kubaláková et al. 2002, 2003b).

7.2.2 Applications of Flow-sorted Chromosomes

Among the many uses of flow-sorted chromosomes, physical mapping using PCR with specific primers has been one of the most frequent. As sorting of hundreds of chromosomes, which are sufficient for PCR, takes only a few minutes, high-throughput mapping is feasible. Macas et al. (1993) were the first to perform PCR on sorted chromosomes and localized seed storage protein genes in field bean. Sorting of translocation chromosomes facilitated gene mapping to subchromosomal regions and integration of genetic and physical maps (Macas et al. 1993; Vaz Patto et al. 1999). Other species in which this approach was used to physically map DNA sequences and/or integrate genetic and physical maps include barley (Lysák et al. 1999), white campion (Kejnovský et al. 2001), chickpea (Vláčilová et al. 2002), and garden pea (Neumann et al. 2002).

The use of sorted chromosomes for targeted isolation of molecular markers is an attractive way towards the saturation of genetic maps in regions of interest. Thus, Arumuganathan et al. (1994) developed 11 RFLP makers from a short-insert DNA library prepared from flow-sorted tomato chromosome 2. An efficient procedure for the development of microsatellite markers from chromosome-specific DNA libraries was described by Macas et al. (1996) and Koblížková et al. (1998). Požárková et al. (2002) used the method to develop microsatellite markers from chromosome

1 of field bean. An added advantage of using flow-sorted chromosomes is that they can be used to confirm chromosome specificity of newly developed markers (Požárková et al. 2002).

Chromosomes sorted onto microscope slides are an ideal target for fluorescence in situ hybridization (FISH) and primed *in situ* DNA labeling (PRINS). They are free of cytoplasmic contamination and facilitate high-resolution cytogenetic mapping. As thousands of chromosomes can be analyzed on one slide, screening of large chromosome populations is possible. Kubaláková et al. (2002) took advantage of this and analyzed intravarietal polymorphism in genomic distribution of GAA clusters in wheat. As shown by Kubaláková et al. (2003a), this approach is attractive for detection of rare structural chromosome changes. Furthermore, flow-sorted chromosomes can be stretched linearly up to 100-fold compared with untreated chromosomes and can be used for high-resolution physical cytogenetic mapping (Valárik et al. 2004).

Chromosome-specific DNA libraries represent a valuable tool allowing for targeted genome analysis. The first chromosome-specific DNA library was produced by direct cloning of 10^5 wheat chromosomes 4A in a plasmid vector (Wang et al. 1992). In order to avoid sorting large numbers of chromosomes, several authors employed DOP-PCR amplification (Telenius et al. 1992) and its modifications and prepared chromosome-specific libraries from only hundreds or thousands of chromosomes (Arumuganathan et al. 1994; Macas et al. 1996; Požárková et al. 2002). Compared to short-insert DNA libraries, large-insert libraries represent more valuable tools for genome mapping. However, their construction was hampered by the requirement of microgram quantities of high-molecular-weight DNA. Šimková et al. (2003) developed a protocol to prepare such DNA from sorted plant chromosomes, which allowed preparation of subgenomic, chromosome-, and chromosome arm–specific BAC libraries in wheat (Šafář et al., in preparation, Janda et al., in preparation).

The range of applications of plant flow cytogenetics is similar to that of human and animals, with one great exception. While the current use of sorted chromosomes in these organisms is mainly for production of painting probes from wild-type and derivative chromosomes (Langer et al. 2004; Tian et al. 2004), attempts to produce chromosome-specific paints from sorted plant chromosomes have failed, most probably due to homogenous interchromosomal distribution of dispersed repeats (Schubert et al. 2001). Despite this limitation, plant flow cytogenetics provides a rich toolbox for plant geneticists.

7.3 Methodologies and Techniques

A general outline of the procedure for flow cytometric analysis and sorting of plant chromosomes consists of the following steps: (1) cell cycle synchronization, (2) accumulation of cells in metaphase, (3) preparation of chromosome suspensions, and (4) flow analysis and sorting. In this chapter, these steps are described and typical applications of flow karyotyping and chromosome sorting are presented.

7.3.1 Cell Cycle Synchronization and Metaphase Accumulation

Efficient cell cycle synchronization in cultured cells and root tip meristems has been achieved by using DNA synthesis inhibitors such as hydroxyurea and aphidicolin, which accumulate cycling cells at the G_1/S interface (Doležel et al. 1999a, 1999b). Removal of the block causes the cells to transit the S and G_2 phases and to enter mitosis synchronously (Doležel et al. 1999a) (Figure 7.2). Accumulation of mitotic cells in metaphase has been achieved by the action of the mitotic spindle

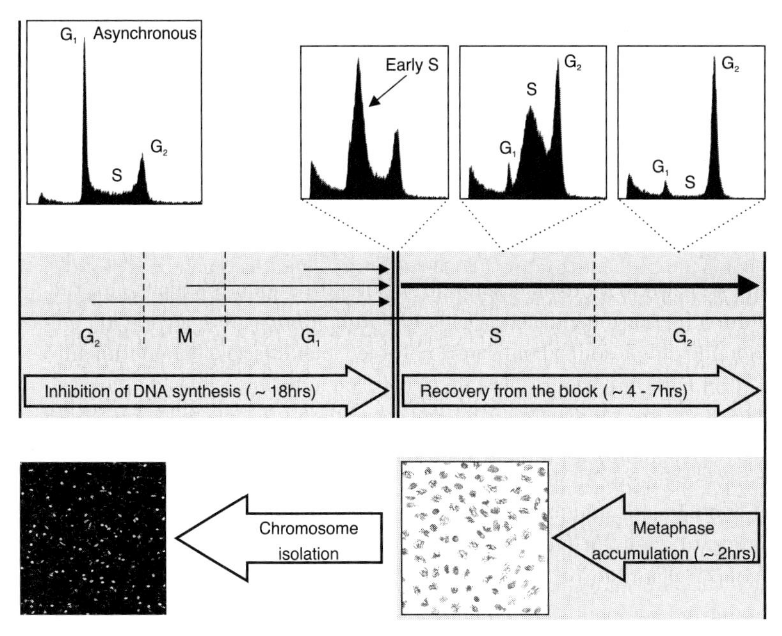

Figure 7.2. Schematic representation of a synchronization procedure based on a reversible inhibition of DNA synthesis at the G_1/S interface. After the removal from the block, the cells traverse the S and G_2 phases in a synchronous manner. Cells entering mitosis are arrested at metaphase by the action of a mitotic spindle inhibitor. Root tips with cells accumulated at metaphase are used to prepare aqueous suspension of chromosomes. Top: histograms of nuclear DNA content show cell cycle distribution in populations of meristem cells growing asynchronously (leftmost histogram) and during the procedure (x-axis: DNA content; y-axis: number of nuclei).

inhibitors, including alkaloid colchicine (de Latt and Blaas 1984; Conia et al. 1987) and the synthetic herbicides amiprophos-methyl (Doležel et al. 1992), oryzalin (Veuskens et al. 1995), and trifluralin (Lee et al. 1996). Metaphase frequencies over 50% were reached in synchronized root meristems (Lee et al. 1996, 1997; Lysák et al. 1999; Kubaláková et al. 2003a), a level considered critical for preparation of chromosome suspensions for flow cytogenetics (Doležel et al. 1999b).

7.3.2 Chromosome Isolation

Currently the only efficient approach to releasing chromosomes from synchronized plant cells involves mechanical homogenization of formaldehyde-fixed root tips using a razor blade or a mechanical homogenizer (Doležel et al. 1992; Gualberti et al. 1996). The fixation greatly increases the yield of intact chromosomes, and up to 1×10^6 chromosomes could be isolated from 30 root tips (Doležel et al. 1992). Furthermore, the fixation makes chromosomes more resistant to mechanical shearing forces during flow sorting, permitting two-step sorting to achieve high sort purities (Lucretti et al. 1993). However, some authors isolate chromosomes from non-fixed roots (Lee et al. 1996, 1997, 2000). Chemical composition of isolation buffers is critical to maintaining the structural integrity of isolated chromosomes and to protecting their DNA (Doležel et al. 1994). The most frequently used buffers are the polyamine-based buffer LB01 (Doležel et al. 1989) and magnesium sulfate–based buffers (Lee et al. 1996).

7.3.3 Chromosome Analysis

DAPI, which binds to AT-rich regions of DNA, and propidium iodide, which intercalates into double-stranded DNA, are the two most frequently used DNA fluorochromes used to stain chromosomes for flow cytometric analysis. Typically, 10,000–20,000 chromosomes are analyzed in each sample and the result is displayed as a histogram of relative florescence intensity (flow karyotype). Interpretation of chromosome analyses is facilitated by models of flow karyotypes, which help in assessing the feasibility of discriminating individual chromosome types depending on differences in relative DNA content and on the coefficient of variation of the chromosome peaks (Lysák et al. 1999; Lee et al. 2002). Spreadsheet software (Conia et al. 1989) and a dedicated computer program (Doležel 1991) have been used to model theoretical flow karyotypes.

Figure 7.3a shows a theoretical flow karyotype of hexaploid wheat ($2n=6x=42$), which predicted that most of the chromosomes formed composite peaks and that only chromosomes 1D, 4D, and 3B could be discriminated as separate peaks. However, the analysis of DAPI-stained chromosomes resulted in a flow karyotype where only chromosome 3B could be resolved (Figure 7.3b). Other chromosomes formed three composite peaks representing groups of chromosomes and could not be individually discriminated (Vrána et al. 2000; Kubaláková et al. 2002). Discrepancies

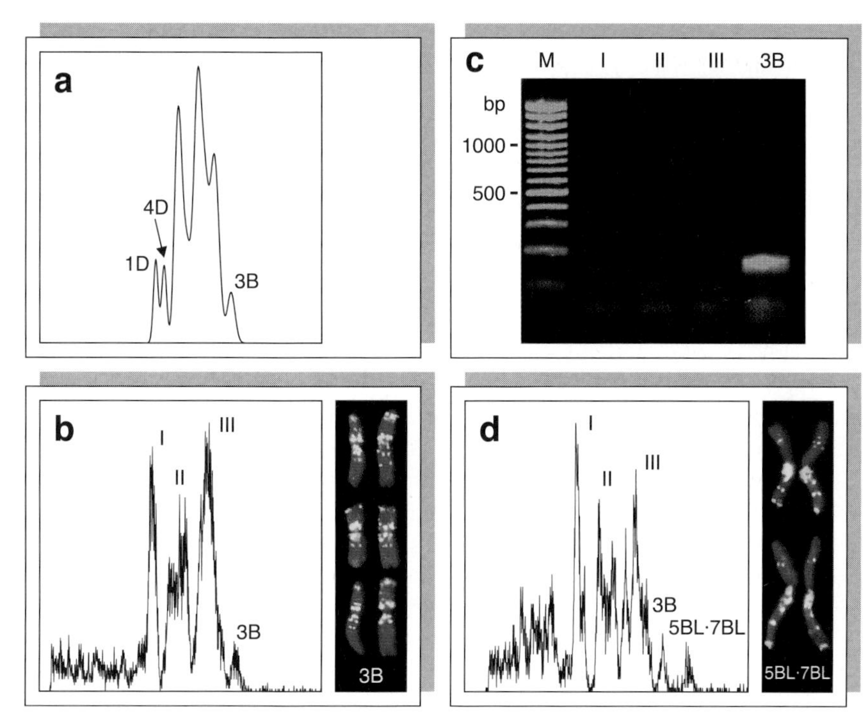

Figure 7.3. Flow cytometric chromosome analysis (flow karyotyping) in hexaploid wheat ($2n=6x=42$). (a–c) Cultivar "Chinese Spring" with a standard karyotype. (a) Theoretical flow karyotype modeled according to published chromosome lengths (Gill et al. 1991) and considering coefficient of variation of chromosome peaks 2%. The model predicts that most of the chromosomes form three composite peaks and that only chromosomes 1D, 4D, and 3B are discriminated as separate peaks. (b) Experimental flow karyotype showing three composite peaks (I–III) representing groups of chromosomes and a peak of chromosome 3B. (c) Assignment of chromosome 3B to a peak on flow karyotype using PCR with a pair of 3B chromosome-specific microsatellite primers. M: 100 bp DNA ladder; I, II, III, 3B: 500 chromosomes of each type were sorted from individual peaks on flow karyotype. (d) Translocation chromosome 5BL-7BL identified as a separate peak in the cultivar "Florida." Flow karyotypes were obtained after the analysis of about 10,000 DAPI-stained chromosomes (x-axis: relative DAPI fluorescence intensity; y-axis: number of events). Inserts show examples of sorted chromosomes, which were identified after PRINS with probes for GAA microsatellite (yellow); chromosomes were counterstained with propidium iodide (red).

between the models and the experimental karyotypes were also obtained in other species, and these could be due to the use of fluorochromes with base preference and/or differences in accessibility of chromatin to DNA fluorochromes among chromosomes. Chromosome content of individual peaks is best determined after microscopic analysis of particles sorted on a slide. Chromosomes dried on a slide are identified unambiguously using FISH or PRINS after fluorescent labeling of repetitive DNA sequences, which display characteristic genomic distribution (Kubaláková et al. 2000; Neumann et al. 2002) (Figure 7.3b, insert). Alternatively, sorted chromosomes can be identified after PCR with specific primers (Lysák et al. 1999; Vrána et al. 2000) (Figure 7.3c).

7.3.3.1 Detection of Structural Chromosome Changes

Flow cytometric analysis can be performed with a high precision, and coefficients of variation of DNA peaks as low as 1.5% can be achieved. Hence, even subtle changes in relative DNA content can be detected based on a shift of chromosome peak along the x-axis. For example, in the standard wheat cultivar "Chinese Spring," the composite peak I represents chromosomes 1D, 4D, and 6D. The peak is split in two and chromosome 4D is discriminated due to a 7% increase in DNA content in the cultivar "Mona" (Kubaláková et al. 2002). A symmetrical translocation between chromosomes 5B and 7B gives rise to the long metacentric chromosome 5BL-7BL, which is the largest in the karyotype. Using flow cytometry, Kubaláková et al. (2002) detected this chromosome in several wheat cultivars where its presence was not reported before (Figure 7.3d). A disadvantage of using flow cytometry is the need for expensive instrumentation and the inability to detect changes that do not result in measurable differences in chromosome DNA content.

7.3.3.2 Quantitative Detection of Numerical Chromosome Changes

As chromosome peak areas reflect relative frequency of chromosomes in population, flow karyotyping is suitable for quantitative analyses of numerical chromosome changes. The analysis involves comparison of peak areas of a chromosome with known frequency and of a chromosome whose frequency is not known. Kubaláková et al. (2003a) demonstrated a suitability of flow karyotyping in evaluating the frequency of alien rye chromosomes in wheat-rye chromosome addition lines. Whereas the analysis of chromosome peak areas in the 6R addition line indicated a stable (homozygous) translocation (Figure 7.4a), flow karyotyping in a line carrying rye chromosome 7R suggested that the chromosome was present in only about 30% of seeds (Figure 7.4b). The advantage of the method is that chromosomes from large numbers of individuals are analyzed and thus the frequency of a chromosome in plant population may be estimated reliably.

7.3.3.3 Chromosome Sorting

Sorting of single chromosomes in plants has been complicated by the inability to discriminate individual chromosomes on flow karyotypes. The most helpful approach has been to use chromosome addition, translocation, and deletion lines. Figure 7.5a shows a flow karyotype of a ditelosomic line of hexaploid wheat carrying a pair of telocentric chromosomes (1BS, representing the short arm of chromosome 1B). Relative DNA content of 1BS is lower compared to other wheat chromosomes, and the chromosome can be easily discriminated and sorted (Kubaláková et al. 2002). In wheat, this approach can be used to purify 40 out of 42 chromosome arms. The remaining two arms, 3BL and 5BL, can be sorted as isochromosomes

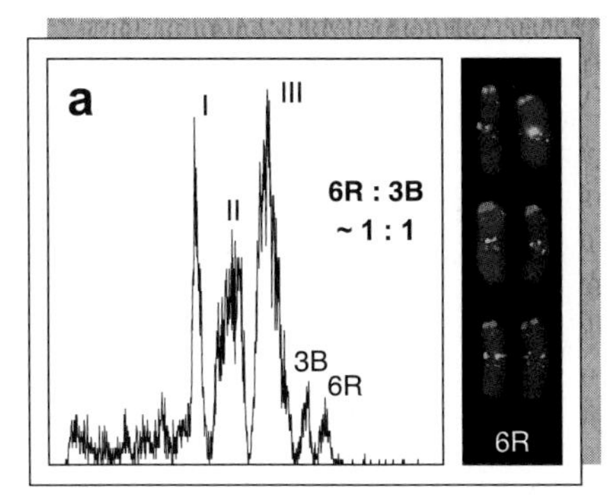

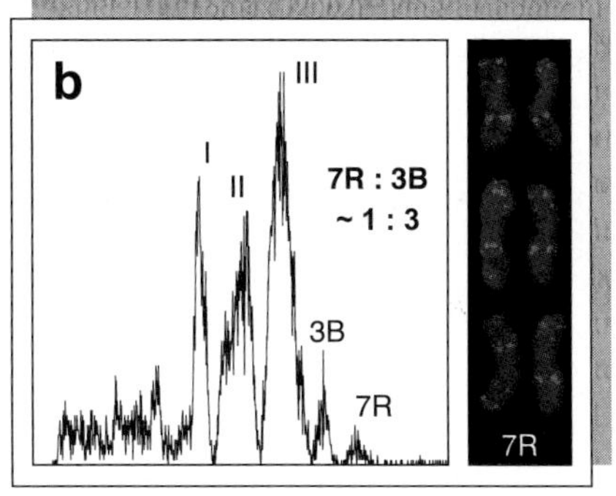

Figure 7.4. Flow karyotyping in two wheat-rye chromosome addition lines. Note that peaks representing rye chromosomes 6R (a) and 7R (b) are resolved. Comparison of chromosome peak areas (wheat chromosome 3B:rye chromosome) can be used to assess the frequency of alien chromosomes in the population. While the peak ratio 1.0 for 6R indicates its presence in every seed, a smaller ratio of peak areas indicates a lower transmission rate of chromosome 7R. Flow karyotypes were obtained after the analysis of about 10,000 DAPI-stained chromosomes. X-axis: relative DAPI fluorescence intensity; y-axis: number of events. Inserts show examples of sorted chromosomes, which were identified after double FISH with probes for pSc119.2 repeat (red) and GAA microsatellite (yellow-green); chromosomes were counterstained with DAPI (blue).

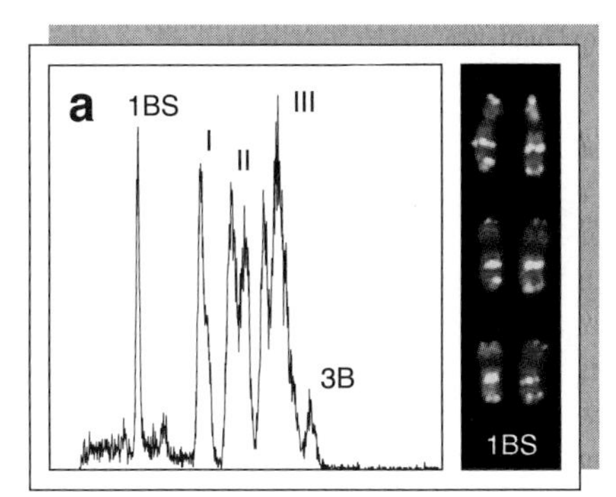

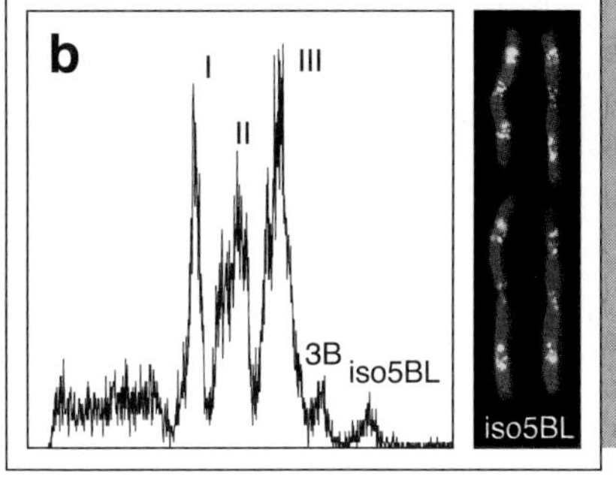

Figure 7.5. Sorting of single chromosome arms in hexaploid wheat. (a) The peak representing the short arm of chromosome 1B (1BS) is well resolved in a ditelosomic line of "Pavone." This facilitates sorting of the 1BS arm at high purity. (b) Flow karyotype of "Chinese Spring" carrying isochromosome 5BL (iso5BL), whose peak is clearly resolved. Flow karyotypes were obtained after the analysis of 10,000–20,000 DAPI-stained chromosomes (x-axis: relative DAPI fluorescence intensity; y-axis: number of events). Inserts show examples of sorted chromosomes, which were identified after PRINS with probes for GAA microsatellite (yellow); chromosomes were counterstained with propidium iodide (red).

(chromosomes with genetically identical arms that are mirror images of one another), as demonstrated in Figure 7.5b (Kubaláková et al. 2002).

The purity in sorted chromosome fractions may reach 100% (Vláčilová et al. 2002; Lysák et al. 1999). However, this is rare and sorted fractions can be contaminated to a different degree by other chromosomes. The extent of contamination depends on the presence of chromosomes, chromatids, chromosome fragments, and their clumps having the same DNA content as the chromosome selected for sorting (Doležel et al. 1994, 2001). Although their presence in a sorted fraction can be determined using PCR, microscopic analysis is preferable because the contaminating particles can be identified and their frequency quantified (Kubaláková et al. 2000). A moderate contamination is not a problem for protein localization and physical mapping using FISH or PRINS. On the other hand, physical mapping using PCR, targeted isolation of markers, and construction of chromosome-specific DNA libraries requires fractions possibly free of contamination.

7.3.3.4 Cytogenetic Mapping on Flow-sorted Chromosomes

Flow sorting offers the possibility of preparing slides with thousands of chromosomes concentrated on a small area, permitting the analysis of large numbers of chromosomes free of cell wall and cytoplasmic remnants. Kubaláková et al. (2003b) utilized flow-sorted chromosomes of durum wheat ($2n=4x=28$) to establish genomic distribution of a set of repetitive DNA sequences and developed a molecular karyotype for this crop (Figure 7.6a). Our recent results demonstrate that single-copy sequences of 1 kb can be routinely localized on flow-sorted wheat chromosomes (Figure 7.6b). In contrast to the commonly observed low frequency of hybridizing metaphase spreads, 100 % of sorted chromosomes display the FISH signals (unpublished). Apart from a need for specialized equipment, the only disadvantage of cytogenetic mapping on sorted chromosomes is that it does not permit the analysis of complete metaphases.

An attractive opportunity is the use of flow-sorted chromosomes to search for rarely occurring structural chromosome variants that require detection by FISH. This can be extremely laborious if the variant is expected to occur in frequencies of 1 % or less, calling for the screening of hundreds of metaphases and translating into many slides that must be processed and evaluated. A corresponding number of chromosomes can be easily sorted and evaluated on just one slide. Using this strategy, Kubaláková et al. (2003a) discovered two rare translocations between the A and B chromosomes of rye, occurring at frequencies of about 0.5 % (Figure 7.6c). A conventional analysis would require screening of at least 500 metaphase spreads.

Chromosomes sorted on a slide may be stretched longitudinally up to more than 100 times their original metaphase size (Valárik et al. 2004). The procedure involves proteinase K digestion of sorted chromosomes that are air-dried on a slide, followed by stretching with ethanol:acetic acid (3:1). In addition to the possibility of detecting sequences as short as 1 kb, cytogenetic mapping using super-stretched chromosomes offers the greatly improved spatial resolution of 70 kb compared to 5–10 Mb after FISH on mitotic chromosomes (Figure 7.6d). Similar sensitivity and resolution may be achieved using meiotic pachytene chromosomes. However, the use of super-stretched chromosomes for FISH is attractive in plant species with a higher number of large chromosomes, which are too long and difficult to trace individually in pachytene.

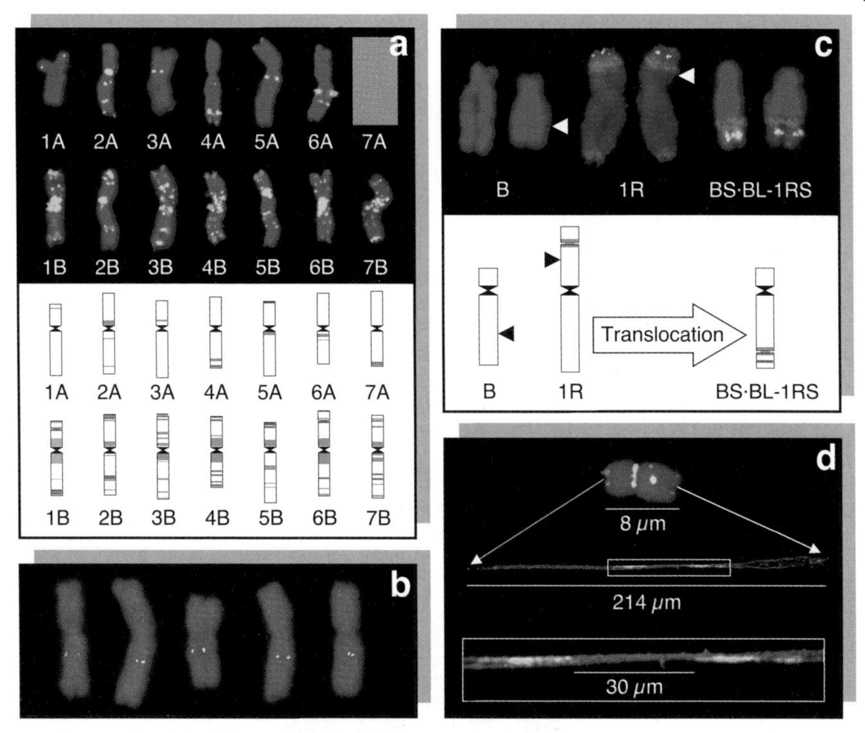

Figure 7.6. The use of flow-sorted chromosomes for physical mapping of DNA sequences using FISH. (a) Molecular karyotype developed for durum wheat ($2n=4x=28$). Genomic distribution of GAA microsatellite (yellow-green) and pSc119.2 repeat (red) was determined after double FISH and facilitated identification of every chromosome in the karyotype. (b) Localization of a single-copy, 1-kb DNA sequence on wheat chromosome 3B. Note the absence of nonspecific labeling and the presence of signals on both chromatids. (c) Double FISH with probes for 45S rDNA (red) and 5S rDNA (yellow-green) on B chromosomes of rye identified a rare translocation between a B chromosome and chromosome 1R. The presence of FISH signals on the B chromosome enabled its identification and localization of break points (arrowheads). (d) Physical mapping with increased spatial resolution on the stretched barley chromosome 1H. Double FISH was performed with probes for GAA microsatellite (yellow) and telomeric repeat (blue); the chromosomes were counterstained with propidium iodide (red). Physical distance between two clusters of the GAA microsatellite increased from 1.3 μm on the untreated chromosome to 30 μm on the chromosome stretched 27-fold.

7.3.3.5 Physical Mapping Using PCR

PCR on sorted chromosomes with sequence-specific primers is an efficient way to examine the presence of sequences of interest in specific chromosomes. This mapping approach is greatly facilitated by the ability of flow sorting to purify large amounts of chromosomes that can be sorted in aliquots and used to map large numbers of sequences in parallel. In contrast to physical mapping of sequences on chromosomes using FISH, even very short, single-copy targets are routinely detected, and the method in principle allows selective mapping of targets with minor variations in their nucleotide sequences (e. g., individual members of gene families). Sorted deletion and translocation chromosomes enable subchromosomal localization of mapped sequences (Macas et al. 1993). Compared to chromosome isolation by microdissection, flow-sorted fractions are usually contaminated by other chromosomes or their parts. This should be considered when designing and performing mapping experiments, and the sorted fractions should be inspected microscopically to assess the extent and origin of contamination (Kubaláková et al. 2000). Moreover, the conditions for PCR should be optimized to achieve efficient detection of the target but to avoid over-amplification, which could produce false-positive signals originating from contaminating particles.

A typical example of PCR mapping on sorted chromosomes is the work of Neumann et al. (2002), who linked two genetic linkage groups of garden pea ($2n=14$) to specific chromosomes. Despite the fact that pea is the organism with the longest history of genetic studies, the linkage groups IV and VII, which are supposed to correspond to either chromosome 4 or chromosome 7, remained unassigned. To solve this problem, Neumann et al. (2002) sorted three fractions from the pea line JI-148, representing translocation chromosomes 2^7 and 7^2 and all remaining wild-type chromosomes (1, 3–6). The fractions were used for PCR detection of genetic markers selected from the two linkage groups. PCR with primers for I7 and

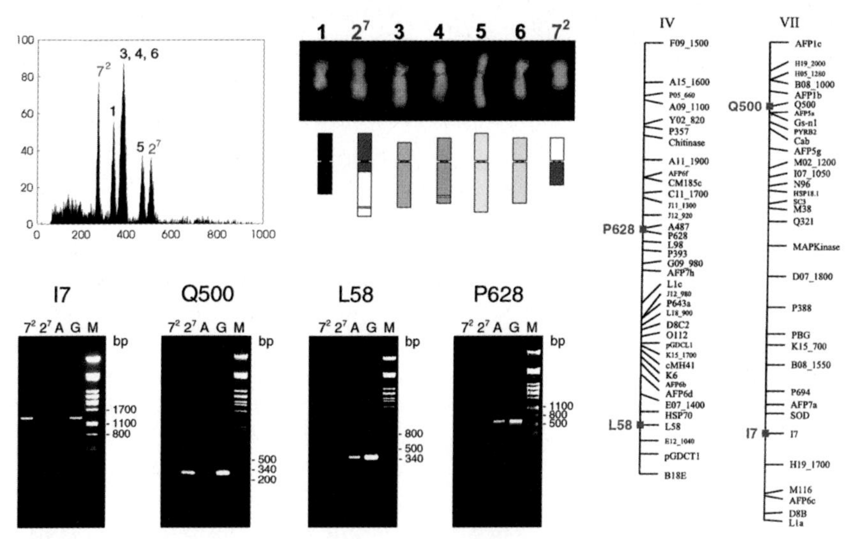

Figure 7.7. PCR-based physical mapping of genetic markers in pea (Neumann et al. 2002). The chromosomes were flow-sorted from the line JI-148 containing reciprocal translocation between chromosomes 2 and 7, which makes them distinguishable from the rest of the chromosomes on the flow karyotype (top left). Chromosomes were identified after FISH with a probe for PisTR-B repeat (red); the chromosomes were counterstained by DAPI (blue). Idiogram showing the translocation is included (top center). The presence of four markers selected from linkage groups IV and VII (right) was detected in separate fractions of chromosome 2^7, chromosome 7^2, and all remaining chromosomes (lanes marked A) using PCR and agarose gel electrophoresis of the reaction products (bottom left). Lanes marked G contain control reactions with total genomic DNA; lanes marked M contain a fragment size marker (lambda DNA digested with *Pst*I).

Q500 markers mapped to linkage group VII resulted in products of correct length in sorted fractions of chromosomes 7^2 and 2^7, respectively. This observation proved that the linkage group VII was located on chromosome 7. Furthermore, the Q500 marker was mapped to a segment of chromosome 7 involved in the translocation with chromosome 2 (Figure 7.7). The linkage group IV was assigned to chromosome 4 after detecting two molecular markers (P628 and L58) selected from this linkage group in a sorted fraction containing all wild-type chromosomes except for the chromosomes 7^2 and 2^7 (Figure 7.7).

7.3.3.6 **Chromosome and Chromosome Arm–specific DNA Libraries**

Chromosome- and chromosome arm–specific libraries facilitate targeted screening and mapping strategies, which are especially attractive for the analysis of large and complex genomes. Cloning strategies and the selection of vectors used to produce these libraries reflect their intended use and can be limited by the amounts of chromosomal DNA available for cloning. For example, the libraries designed for isolation of molecular markers can be produced by cloning of PCR-amplified fragments several kilobases to hundreds of kilobases long into plasmid vectors (short-insert libraries), which requires sorting of only hundreds of chromosomes. On the other hand, large-insert libraries containing fragments of tens to hundreds of kilobases suitable for large-scale mapping and positional cloning can be produced only by direct cloning of DNA prepared from 10^5–10^6 chromosomes.

Up to now, the only plant species in which the whole genome has been fractionated into chromosome-specific DNA libraries has been the field bean. The libraries were prepared from seven chromosome types individually purified by flow sorting from the chromosome translocation lines EF and JF (Macas et al. 1996). The chromosomes were selected so that the set of seven libraries covered the whole genome of field bean more than once. The protocol involved sorting of 250 or 1000 chromosomes, amplification of DNA according to a modified DOP-PCR protocol, and cloning into a plasmid vector. The total number of clones and the average insert size in the libraries ranged from 6×10^5 to 4.5×10^6 and from to 310 kb to 487 bp, respectively. The libraries were found to be enriched for low-copy genomic sequences, which represented about 60 % of inserts. This made them suitable for targeted isolation of molecular markers.

Large-insert DNA libraries, such as those cloned in a BAC vector, have become one of the main resources for current genomics. Until recently, preparation of large-insert libraries from sorted chromosomes has been hampered by the need for microgram quantities of high-molecular-weight DNA. Recently, Šimková et al. (2003) developed a protocol that can be used to obtain such DNA from sorted plant chromosomes. The first chromosome-specific BAC library was prepared from chromosome 3B of hexaploid wheat (Šafář et al., in preparation). The library was constructed from DNA of 2×10^6 sorted chromosomes, consists of 67,968 clones with an average insert size of 103 kb, and represents 6.2-fold coverage of the chromosome. The feasibility of constructing BAC libraries from sorted chromosomes was subsequently confirmed by Janda et al. (in preparation), who pre-

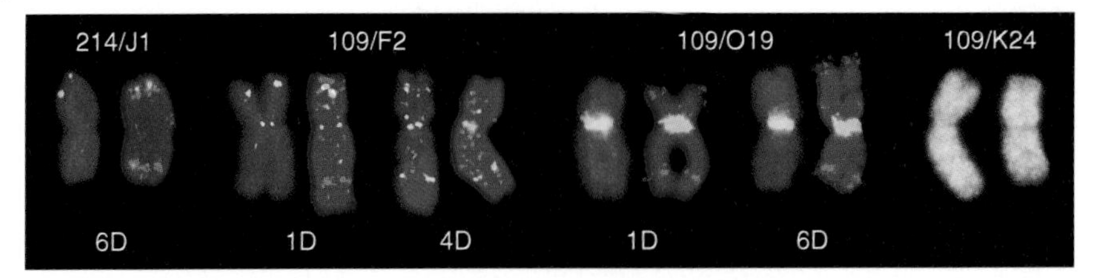

Figure 7.8. Localization of BAC clones on flow-sorted wheat D-group chromosomes using FISH. Two examples of each chromosome are shown. BAC clones 214/J1, 109/F2, and 109/O19 showed localized signals to one or a few loci. Chromosomes counterstained with propidium iodide (red) show localization of BAC clones (yellow). Chromosomes counterstained with DAPI (blue) show double FISH with probes for BAC clones (yellow-green) and *Afa* family repeat (red), which facilitated chromosome identification. BAC clone 109/K24 showed dispersed signals on all wheat chromosomes (yellow). The chromosomes were counterstained with propidium iodide (red).

pared a subgenomic BAC library from chromosomes 1D, 4D, and 6D of wheat and a chromosome arm–specific BAC library from the short arm of wheat chromosome 1B (1BS). One may envisage the use of these libraries for development of sequencing-ready global contig maps of individual chromosomes and for isolation of BAC clones from genome regions rich in genes for preferential sequencing. Figure 7.8 shows cytogenetic mapping of DNA clones isolated from a BAC library specific for chromosomes 1D, 4D, and 6D. These clones were selected for their low repetitive DNA content and were localized by FISH to provide chromosome landmarks (Janda et al., in preparation).

7.3.3.7 Targeted Isolation of Molecular Markers

High-density genetic linkage maps are invaluable for marker-assisted breeding, development of physical maps, and positional gene cloning. Traditionally, molecular markers have been isolated regardless of their position in the genome, resulting in their random distribution. As a consequence, generation of high-density maps of a particular region of interest could be achieved only by saturation of the whole genome with the new markers. However, the availability of DNA from sorted chromosomes opens a way for a targeted approach to saturating genetic maps in regions harboring traits of agronomic importance. Although the production of various types of markers might be considered, probably the most successful application of sorted chromosomes in this area involves the development of microsatellite markers. This was stimulated by the results of Macas et al. (1996), who demonstrated that short-insert DNA libraries prepared from sorted chromosomes were a useful source of microsatellite clones.

As the clones containing microsatellite motifs represent only a small fraction of cloned DNA, it is desirable to prepare microsatellite-enriched libraries, which make the screening much more efficient. Such libraries can be constructed from amplified chromosomal DNA subjected to one or more rounds of subtractive hybridization with a single-stranded probe corresponding to the sequence of interest. An improved procedure using chemically modified oligonucleotides as hybridization probes (Koblížková et al. 1998) was shown to generate amplification artifact–free libraries that are up to 100-fold enriched for a given microsatellite. Based on this protocol, Požárková et al. (2002) developed an integrated approach for targeted retrieval of microsatellite markers from selected genome regions of field bean. The procedure involves sorting of 250 copies of the target chromosome, DOP-PCR amplification of the chromosomal DNA, enrichment for specific microsatellites, cloning, isolation of polymorphic microsatellites, testing their chromosomal specific by PCR on sorted chromosomes, and finally, integration of chromosome-specific markers into a genetic map. The authors verified the procedure by developing novel microsatellite markers from a selected region of field bean chromosome 1. Although developed for field bean, the protocol is generally applicable for any species where chromosome sorting is feasible. The advantage is that only small numbers of chromosomes are sufficient to prepare enriched DNA libraries and that the chromosome specificity of candidate markers can be verified on sorted chromosomes prior to their genetic mapping.

7.4 Discussion

Flow cytogenetics has found a number of applications, and its wider use is limited mainly by the need for sophisticated equipment. However, as most of the applications require relatively small numbers of sorted chromosomes, these may be provided by specialized laboratories. The second limitation is a requirement for optimized protocols to prepare suspensions of intact chromosomes. Although the currently used procedures for chromosome isolation can be modified for other species, the optimization is laborious and time-consuming (Doležel et al. 1999b). Last but not least, strategies for discriminating and sorting individual chromosome types need to be developed for each species. Considering this, it may be expected that flow cytogenetics will be used with a limited number of species that represent economically important crops and/or are employed as models to solve important biological problems.

For the time being, the construction of chromosome- and chromosome arm–specific BAC libraries seems to be one of the most attractive uses. Genomic BAC libraries are available for most of the important crops. However, their maintenance and screening can be laborious and costly. For example, a BAC library of hexaploid wheat with a 9.3× coverage consists of 1.2×10^6 clones ordered in 3125 384-well plates (Allouis et al. 2003). In contrast, a BAC library specific for the short arm of wheat chromosome 1B with 17× coverage consists of only 65,280 clones (Janda et al., in preparation). The availability of such libraries thus helps in dissecting the complex genomes into parts that are easier to manage and analyze. The only obstacle in their construction is the need for millions of chromosomes, whose sorting may take several weeks (Šafář et al., in preparation). However, the total sorting time could be reduced by employing high-speed sorters (Ibrahim and van den Engh 2004) and, where possible, by using double ditelosomic lines to sort two different chromosome arms simultaneously.

In addition to the applications described in this chapter, other uses of sorted chromosomes appear to be promising. The first involves high-throughput physical mapping of ESTs or other low-copy DNA sequences on DNA arrays using probes prepared from sorted chromosomes. The second application involves HAPPY mapping, an in vitro approach for defining the order and spacing of DNA markers directly on a DNA molecule (Thangavelu et al. 2003). Using chromosome arm–specific DNA, this method has the potential to rapidly generate scaffolds for building physical maps. The third area focuses on the isolation of gene sequences from specific chromosomes and chromosome arms using re-association kinetics (Peterson et al. 2002).

Acknowledgments

This work was supported by research grants 521/03/0595 and 522/03/0354 from the Grant Agency of the Czech Republic, grant QC1336 from the Ministry of Agriculture of the Czech Republic, and grants ME527 and ME528 from the Ministry of Education, Youth, and Sports of the Czech Republic.

References

Allouis, S., G. Moore, A. Bellec, R. Sharp, P. Faivre, K. Mortimer, S. Pateyron, T. N. Foote, S. Griffiths, M. Caboche, and B. Chalhoub (2003) "Construction and characterisation of a hexaploid wheat (Triticum aestivum L.) BAC library from the reference germplasm 'Chinese Spring'." Cereal Res. Communications 31:331–338.

Arumuganathan, K., J. P. Slattery, S. D. Tanksley, and E. D. Earle (1991) "Preparation and flow cytometric analysis of metaphase chromosomes of tomato." Theor. Appl. Genet. 82:101–111.

Arumuganathan, K., G. B. Martin, H. Telenius, S. D. Tanksley, and E. D. Earle (1994) "Chromosome 2-specific DNA clones from flow-sorted chromosomes of tomato." Mol. Gen. Genet. 242:551–558.

Boschman, G. A., E. M. M. Manders, W. Rens, R. Slater, and J. A. Aten (1992) "Semi-automated detection of aberrant chromosomes in bivariate flow karyotypes." Cytometry 13: 469–477.

Carrano, A. V., J. W. Gray, R. G. Langlois, K. J. Burkhart-Schultz, and M. Van Dilla (1979) "Measurement and purification of human chromosomes by flow cytometry and sorting" Proc. Natl. Acad. Sci. USA 76: 1382–1384.

Carrano, A. V., J. W. Gray, R. G. Langlois, and L. C. Yu (1983) "Flow cytogenetics: Methodology and applications." In: Chromosomes and Cancer, J. D. Rowley, and J. E. Ultmann, eds., Academic Press, Inc., pp. 195–209.

Conia, J., C. Bergounioux, C. Perennes, P. Muller, S. Brown, and P. Gadal (1987) "Flow cytometric analysis and sorting of plant chromosomes from *Petunia hybrida* protoplasts." Cytometry 8:500–508.

Conia, J., P. Muller, S. Brown, C. Bergounioux, and P. Gadal (1989) "Monoparametric models of flow cytometric karyotypes with spreadsheet software." Theor. Appl. Genet. 77:295–303.

de Laat, A. M. M., and J. Blaas (1984) "Flow-cytometric characterization and sorting of plant chromosomes." Theor. Appl. Genet. 67: 463–467.

Doležel, J. (1991) "KARYOSTAR: Microcomputer program for modelling of monoparametric flow karyotypes." Biológia 46:1059–1064.

Doležel, J., and S. Lucretti (1995) "High-resolution flow karyotyping and chromosome sorting in *Vicia faba* lines with standard and reconstructed karyotypes." Theor. Appl. Genet. 90:797–802.

Doležel, J., P. Binarová, and S. Lucretti (1989) "Analysis of nuclear DNA content in plant cells by flow cytometry." Biol. Plant. 31: 113–120.

Doležel, J., J. Číhalíková, and S. Lucretti (1992) "A high-yield procedure for isolation of metaphase chromosomes from root tips of *Vicia faba* L." Planta 188:93–98.

Doležel, J., S. Lucretti, and I. Schubert (1994) "Plant chromosome analysis and sorting by flow cytometry." Crit. Rev. Plant Sci. 13: 275–309.

Doležel, J., J. Číhalíková, J. Weiserová, and S. Lucretti (1999a) "Cell cycle synchronization in plant root meristems." Meth. Cell Sci. 21: 95–107.

Doležel, J., J. Macas, and S. Lucretti (1999b) "Flow analysis and sorting of plant chromosomes." In: Current Protocols in Cytometry, J. P. Robinson, Z. Darzynkiewicz, P. N. Dean, L. G. Dressler, A. Orfao, P. S. Rabinovitch, C. C. Stewart, H. J. Tanke, and L. L. Wheeless, eds, John Wiley and Sons, Inc., pp. 5.3.1.–5.3.33.

Doležel J., M. A. Lysák, M. Kubaláková, H. Šimková, J. Macas, and S. Lucretti (2001) "Sorting of Plant Chromosomes." In: Flow Cytometry; 3rd. ed., Part B, Z. Darzynkiewicz, H. A. Crissman HA, and J. P. Robinson, eds., Academic Press, pp. 3–31.

Ferguson-Smith, M. A. (1997) "Genetic analysis by chromosome sorting and painting: phylogenetic and diagnostic applications." Eur. J. Hum. Genet. 5:253–265.

Gill B. S., B. Friebe, and T. R. Endo (1991) "Standard karyotype and nomenclature system for description of chromosome bands and structural-aberrations in wheat (Triticum aestivum)." Genome 34:830–839.

Gill, K. S., K. Arumuganathan, and J. H. Lee (1999) "Isolating individual wheat (*Triticum aestivum*) chromosome arm by flow cytometric analysis of ditelosomic lines." Theor. Appl. Genet. 98:1248–1252.

Gray, J. W., and L. S. Cram (1990) "Flow karyotyping and chromosome sorting." In: Flow Cytometry and Sorting, M. Melamed, T. Lindmo, and M. Mendelssohn, eds., Wiley-Liss, pp. 503–529.

Gray, J. W., A. V. Carrano, D. H. Moore, L. L. Steinmetz, J. Minkler, B. H. Mayall, M. L. Mendelsohn, and M. A. Van Dilla (1975) "High-speed quantitative karyotyping by flow microfluorometry." Clin. Chem. 21:1258–1262.

Gualberti, G., J. Doležel, J. Macas, and S. Lucretti (1996) "Preparation of pea (*Pisum sativum* L.) chromosome and nucleus suspensions from single root tips." Theor. Appl. Genet. 92:744–751.

Ibrahim, S. F., and G. van den Engh (2004) "High-speed chromosome sorting." Chromososme Res. 12:5–14

Kejnovský, E., J. Vrána, S. Matsunaga, P. Souček, J. Široký, J. Doležel, and B. Vyskot (2001) "Localization of male-specifically expressed *MROS* genes of *Silene latifolia* by PCR on flow-sorted sex chromosomes and autosomes." Genetics 158:1269–1277.

Koblížková, A., J. Doležel, and J. Macas (1998) "Subtraction with 3′ modified oligonucleotides eliminates amplification artifacts in DNA libraries enriched for microsatellites." BioTechniques 25:32–38.

Kubaláková, M., M. A. Lysák, J. Vrána, H. Šimková, J. Číhalíková, and J. Doležel (2000) "Rapid identification and determination of purity of flow-sorted plant chromosomes using C-PRINS." Cytometry 41:102–108.

Kubaláková, M., J. Vrána, J. Číhalíková, H. Šimková, and J. Doležel (2002) "Flow karyotyping and chromosome sorting in bread wheat (*Triticum aestivum* L.)." Theor. Appl. Genet. 104:1362–1372.

Kubaláková, M., M. Valárik, J. Bartoš, J. Vrána, J. Číhalíková, M. Molnár-Láng, and J. Doležel (2003a) "Analysis and sorting of rye (*Secale cereale* L.) chromosomes using flow cytometry." Genome 46:893–905.

Kubaláková, M., J. Bartoš, P. Kovářová, J. Číhalíková, N. Watanabe, S. F. Kianian, and J. Doležel (2003b) "Chromosome analysis and sorting in durum wheat and physical mapping of repetitive DNA sequences". In: Proceedings of the Tenth International Wheat

Genertics Symposium, Istituto Sperimentale per la Cerealicoltura, Rome, pp. 601–603.

Langer, S., J. Kraus, and M. R. Speicher (2004) "Multicolor chromosome painting in diagnostic and research applications." Chromosome Res. 12:15–23.

Lee, J. H., K. Arumuganathan, S. M. Kaeppler, H. F. Kaeppler, and C. M. Papa (1996) "Cell synchronization and isolation of metaphase chromosomes from maize (*Zea mays* L.) root tips for flow cytometric analysis and sorting." Genome 39:697–703.

Lee, J. H., K. Arumuganathan, Y. Yen, S. Kaeppler, H. Kaeppler, and P. S. Baenziger (1997) "Root tip cell cycle synchronization and metaphase-chromosome isolation suitable for flow sorting in common wheat (*Triticum aestivum* L.)." Genome 40: 633–638.

Lee, J. H., K. Arumuganathan, Y. S. Chung, K. Y. Kim, W. B. Chung, K. S. Bae, D. H. Kim, D. S. Chung, and O. C. Kwon (2000) "Flow cytometric analysis and chromosome sorting of barley (*Hordeum vulgare* L.)." Mol. Cells 10:619–625.

Lee, J. H., K. Arumuganathan, S. M. Kaeppler, S. W. Park, K. Y. Kim, Y. S. Chung, D. H. Kim, and K. Fukui (2002) "Variability of chromosomal DNA contents in maize (*Zea mays* L.) inbred and hybrid lines." Planta 215: 666–671.

Levy, H. P., R. A. Schultz, J. V. Ordónez, and M. M. Cohen (1991) "Anti-kinetochore staining for single laser, bivariate flow sorting of Indian muntjac chromosomes." Cytometry 12:695–700.

Li, L. J., K. Arumuganathan, H. W. Rines, R. L. Phillips, O. Riera-Lizarazu, D. Sandhu, Y. Zhou, and K. S. Gill (2001) "Flow cytometric sorting of maize chromosome 9 from an oat-maize chromosome addition line." Theor. Appl. Genet. 102:658–663.

Lucretti, S., and J. Doležel (1997) "Bivariate flow karyotyping in broad bean (*Vicia faba*)." Cytometry 28:236–242.

Lucretti, S., J. Doležel, I. Schubert, and J. Fuchs (1993) "Flow karyotyping and sorting of *Vicia faba* chromosomes." Theor. Appl. Genet. 85:665–672.

Lysák, M. A., J. Číhalíková, M. Kubaláková, H. Šimková, G. Künzel, and J. Doležel (1999) "Flow karyotyping and sorting of mitotic chromosomes of barley (*Hordeum vulgare* L.)." Chromosome Res. 7:431–444.

Macas, J., J. Doležel, S. Lucretti, U. Pich, A. Meister, J. Fuchs, and I. Schubert (1993) "Localization of seed protein genes on flow-sorted field bean chromosomes." Chromosome Res. 1:107–115.

Macas, J., G. Gualberti, M. Nouzová, P. Samec, S. Lucretti, and J. Doležel (1996) "Construction of chromosome-specific DNA libraries covering the whole genome of field bean (*Vicia faba* L.)." Chromosome Res. 4:531–539.

Neumann, P., M. Lysák, J. Doležel, and J. Macas (1998) "Isolation of chromosomes from *Pisum sativum* L. hairy root cultures and their analysis by flow cytometry." Plant Sci. 137:205–215.

Neumann, P., D. Požárková, J. Vrána, J. Doležel, and J. Macas (2002) "Chromosome sorting and PCR-based physical mapping in pea (*Pisum sativum* L.)." Chromosome Res. 10:63–71.

Peterson, D. G., S. R. Schulze, E. B. Sciara, S. A. Lee, J. E. Bowers, A. Nagel, N. Jiang, D. C. Tibbitts, S. R. Wessler, and A. H. Paterson (2002) "Integration of Cot analysis, DNA cloning, and high-throughput sequencing facilitates genome characterization and gene discovery." Genome Res. 12: 795–807.

Pich, U., A. Meister, J. Macas, J. Doležel, S. Lucretti, and I. Schubert (1995) "Primed *in situ* labelling facilitates flow sorting of similar sized chromosomes." Plant J. 7:1039–1044.

Pinkel, D., J. Landegent, C. Collins, J. Fuscoe, R. Seagraves, J. Lucas, and J. Gray (1998) "Fluorescence *in situ* hybridization with human chromosome-specific libraries. Detection of trisomy 21 and translocations of chromosome 4." Proc. Natl. Acad. Sci. USA 85:9138–9142.

Požárková, D., A. Koblížková, B. Román, A. M. Torres, S. Lucretti, M. Lysák, J Doležel, and J. Macas (2002) "Development and characterization of microsatellite markers from chromosome 1-specific DNA libraries of *Vicia faba*." Biol. Plant. 45:337–345.

Schubert, I., P. F. Fransz, J. Fuchs, and H. de Jong (2001) "Chromosome painting in plants." Meth. Cell Sci. 23:57–69.

Šimková, H., J. Číhalíková, J. Vrána, M. A. Lysák, and J. Doležel (2003) "Preparation of high molecular weight DNA from plant nuclei and chromosomes isolated from root tips." Biol. Plant. 46:369–373.

Stubblefield, E., S. Cram, and L. Deaven (1975) "Flow microfluorometric analysis of isolated Chinese hamster chromosomes." Exptl. Cell Res. 94:464–468.

Telenius, H., N. P. Carter, C. E. Bebb, M. Nordenskjöld, B. A. J. Ponder, and A. Tunnacliffe (1992) " Degenerate oligonucleotide-primed PCR: general amplification of target DNA by a single degenerate primer." Genomics 13:718–725.

Thangavelu, M., A. B. James, A. Bankier, G. J. Bryan, P. H. Dear, and R. Waugh (2003) "HAPPY mapping in a plant genome: reconstruction and analysis of a high-resolution physical map of a 1.9 Mbp region of Arabidopsis thaliana chromosome 4." Plant Biotechnology Journal 1:23–31.

Tian, Y., W. Nie, J. Wang, M. Ferguson-Smith, and F. Yang (2004) "Chromosome evolution in bears: reconstructing phylogenetic relationships by cross-species chromosome painting." Chromosome Res. 12:55–63.

Valárik, M., J. Bartoš, P. Kovářová, M. Kubaláková, J. H. de Jong, and J. Doležel (2004) "High-resolution FISH on super-stretched flow-sorted plant chromosomes." Plant J. 37: 940–950.

Van Dilla, M. A., L. L. Deaven, K. L. Albright, N. A. Allen, M. R. Aubuchon, M. F. Bartholdi, N. C. Brown, E. W. Campbell, A. V. Carrano, L. M. Clark, L. S. Cram, B. D. Crawford, J. C. Fuscoe, J. W. Gray, C. E. Hildebrand, P. J. Jackson, J. H. Jett, J. L. Longmire, C. R. Lozes, M. L. Luedemann, J. C. Martin, J. S. McNinch, L. J. Meincke, M. L. Mendelsohn, J. Meyne, R. K. Moyzis, A. C. Munk, J. Perlman, D. C. Peters, A. J. Silva, and B. J. Trask (1986) "Human chromosome-specific DNA libraries. Construction and availability." Biotechnology 4:537–552.

Vaz Patto, M. C., A. M. Torres, A. Koblížková, J. Macas, and J. I. Cubero (1999) "Development of a genetic composite map of *Vicia faba* using F_2 populations derived from trisomic plants." Theor. Appl. Genet. 98:736–743.

Veuskens, J., D. Marie, S. C. Brown, M. Jacobs, and I. Negrutiu (1995) "Flow sorting of the Y sex-chromosome in the dioecious plant *Melandrium album.*" Cytometry 21:363–373.

Vláčilová, K., D. Ohri, J. Vrána, J. Číhalíková, M. Kubaláková, G. Kahl, and J. Doležel (2002) "Development of flow cytogenetics and physical genome mapping in chickpea (*Cicer arietinum* L.)." Chromosome Res. 10:695–706.

Vrána, J., M. Kubaláková, H. Šimková, J. Číhalíková, M. A. Lysák, and J. Doležel (2000) "Flow-sorting of mitotic chromosomes in common wheat (*Triticum aestivum* L.)." Genetics 156:2033–2041.

Wang, M. L., A. R. Leitch, T. Schwarzacher, J. S. Heslop-Harrison, and G. Moore. (1992) "Construction of a chromosome-enriched *Hpa*II library from flow-sorted wheat chromosomes." Nucleic Acids Res. 20:1897–1901.

Young, B. D. (1990) "Chromosome analysis and sorting." In: Flow Cytometry, A Practical Approach, M. G. Ormerod, ed., Oxford University press, pp. 145–159.

8
Genomic DNA Libraries and Physical Mapping

Chengwei Ren, Zhanyou Xu, Shuku Sun, Mi-Kyung Lee, Chengcang Wu, Chantel Scheuring, and Hong-Bin Zhang

Overview

Large-insert genomic DNA libraries have become the choice of genomic resources for generation of physical maps or reconstruction of the genomes of any living organisms, especially for those of higher organisms. The development of the bacteria-based large-insert cloning system has made it possible to construct physical maps and has revolutionized genomics research. Today, hundreds of large-insert genomic DNA libraries have been constructed, and many more large-insert libraries are under construction or will be constructed for humans, plants, and animals of agronomic, economic, and environmental importance. Because of their large inserts, fewer clones are needed for a complete genomic DNA library, and all clones in the library can be arrayed and archived in microplates, allowing the library to be shared and studied collaboratively. Most importantly, the large-insert genomic DNA libraries have made many studies possible. This chapter describes the development of large-insert genomic DNA libraries and their applications in genome physical mapping. It introduces and evaluates different types of genomic DNA libraries for genome physical mapping, emphasizes the state-of-the-art technology for bacteria-based large-insert genomic DNA libraries and their uses for effective genome physical mapping, details the procedure for construction of a bacteria-based large-insert genomic DNA library, and provides some examples of the applications of bacteria-based large-insert genomic libraries for physical mapping. Finally, the further improvement and usage of genomic DNA libraries in physical mapping will be discussed.

Abstract

The bacteria-based large-insert genomic DNA cloning system has become the most widely used cloning system for genome physical mapping due to its large cloning capacity, stable maintenance of large-insert genomic DNA in *Escherichia coli*, and

The Handbook of Plant Genome Mapping.
Genetic and Physical Mapping. Edited by Khalid Meksem, Günter Kahl

ISBN 3-527-31116-5

ease of manipulation. The advent of the bacteria-based large-insert cloning system has been attributed to both the development of electroporation transformation technology and the discovery of *E. coli* host strain DH10B. Qualities of source libraries such as insert size, clone genome coverage, genome representation, and binary vectors have great influence on high-quality genome physical mapping. State-of-the-art techniques for construction of the large-insert genomic DNA libraries in bacteria have been described in detail. The bacteria-based large-insert genomic DNA libraries have widely been applied for physical mapping using a diversity of mapping methods, including fingerprint analysis FISH, optical mapping, iterative hybridization, and other methods.

8.1 Introduction

Genome physical mapping locates landmarks such as nucleotide sequences, clones, and/or other types of markers in terms of base pair distance. In increasing resolution, physical maps could include chromosome banding patterns, fluorescent *in situ* hybridization (FISH) physical maps, radiation hybrid-based physical maps, clone-based physical maps, and genome sequence maps. However, physical maps constructed from large-insert genomic DNA libraries (clone-based physical maps) are emerging as the centerpiece of genomics research. These maps are not only essential for large-scale genome sequencing and construction of genome sequence maps but also provide powerful and economical frameworks for large-scale gene mapping, gene cloning, and organizational, functional, and evolutionary analysis of genomes.

Large-insert genomic DNA libraries are essential for genome physical mapping. Clones from genomic DNA libraries provide a means to produce large amounts of easily manipulated genomic DNA of any region of a genome. A collection of clones of a whole genomic DNA library represents the entire genome. Genomic libraries are constructed in vitro by inserting the fragments of source genomic DNA into a cloning vector containing all elements required for the delivery and stable propagation of the foreign DNA fragments as a plasmid, a bacteriophage, or an artificial chromosome in the host cells.

Three cloning systems have been developed and used for construction of large-insert genomic DNA libraries and genome physical maps (see below): the bacteria-based plasmid cloning system, the yeast artificial chromosome (YAC) cloning system (Burke et al. 1987), and the bacteriophage P1-based phage cloning system (P1, Sternberg 1990). The bacteria-based plasmid cloning system includes a broad variety of vectors such as cosmids (Collins and Hohn 1978), fosmids (Kim et al. 1992), bacterial artificial chromosomes (BACs, Shizuya et al. 1992), bacteriophage P1-derived artificial chromosomes (PACs, Ioannou et al. 1994), binary BACs (Hamilton 1997), conventional plasmid-based cloning vectors (PBCs, Tao and Zhang 1998), and transformation-competent artificial chromosomes (TACs, Liu et al. 1999).

The YAC cloning system was created in 1987 (Burke et al. 1987) and revolutionized the study of large genomes due to its capacity of cloning DNA fragments greater than 1000 kb. YAC libraries were constructed for a number of species (Green et al. 1999) and used in physical mapping of human (Chumakov et al. 1995), mouse (Nusbaum et al. 1999), rice (Kurata et al. 1997; Saji et al. 2001; Wu et al. 2002), and *Arabidopsis* (Canilleri et al. 1998; Hwang et al. 1991; Matallana et al. 1992; Schmidt et al. 1995; Zachgo et al. 1996). However, they are no longer widely used as a major cloning system for genomics research and physical mapping due to several disadvantages. First, YAC libraries have a relatively high percentage of chimeric clones. Although the great majority of YAC clones contain a single YAC, the genetic selection used in YAC cloning cannot prevent the co-transformation of multiple independent YACs into the same single cell. The estimates of such multi-YAC clones in various libraries have ranged from 10% to 30%. Moreover, a chimeric insert may also occur by ligation between two or more unrelated segments of genomic DNA. Second, the instability of YACs in host cells is also a serious problem. Natural yeast chromosomes segregate with extremely high fidelity, while yeast artificial chromosomes are much less stable, especially the shorter linear YACs, and may be potentially lost from the host cells during cell division. Third, the isolation of YAC insert DNA is tedious, and the isolated DNA is easily contaminated with yeast chromosomal DNA. The DNA cloned into YACs cannot be readily purified from the host DNA by the standard chemical techniques. Instead, it has to be purified by isolation of the total intact yeast chromosomal DNA and then separation of YAC DNA from the yeast chromosomal DNA. For the purpose of physical mapping, the total yeast DNA is isolated by embedding the cells in agarose plugs or beads and enzymatically removing the cell wall, followed by cell lysis and a thorough washing. The YAC DNA is then separated by pulsed-field gel electrophoresis.

The bacteriophage P1-based phage cloning system was created by Sternberg (1990). Use of the large genome of bacteriophage P1 doubled the cloning capacity of the λ phage-based vectors, allowing cloning of DNA fragments up to 100 kb. Because of the relatively smaller sizes of its cloned DNA compared to those of the bacteria-based large-insert cloning system, the bacteriophage P1 cloning system has not been widely used for genomic DNA library construction or physical mapping of large, complex genomes.

The bacteria-based large-insert cloning system (BACs) was reported in 1992 (Shizuya et al. 1992). Due to its capacity of cloning DNA fragments of up to 300 kb and its elimination of the disadvantages of the YAC system, the BAC system has rapidly become the method of choice for large-insert genomic DNA library construction and physical mapping. The most important contributions to the development of the system were the establishment of the electroporation transformation method and the discovery of the *Escherichia coli* bacterial host strain DH10B. The electroporation transformation technology is able to introduce large-insert DNA plasmid and cosmid clones into bacterial host cells (Shizuya et al. 1992) with a much higher transformation efficiency and a much simpler technique than the previously used chemical and phage particle transformation methods. The *E. coli* strain DH10B and derivatives can stably maintain and propagate almost all existing plasmid vectors and their derivatives with large foreign DNA fragments (Tao and Zhang 1998). All bacteria-based large-insert genomic DNA libraries constructed to date are hosted in DH10B or its derivative. The key features of this strain include mutations that block recombination (*recA1*), restriction of foreign DNA by endogenous restriction endonucleases (*hsd/RMS*), and restriction of DNA containing methylated DNA (5′-methylcytosine or methyladenine residues, and 5′-hydroxymethylcytosine) (*mcrA, mcrB, mcrC,* and *mrr*).

It is obvious that the bacteria-based large-insert cloning system is superior to the YAC and bacteriophage P1 cloning systems in many respects. Due to the limited applications of YAC and bacteriophage P1 libraries in current genomics research, only bacteria-based large-insert genomic DNA libraries will be described and discussed below in detail.

8.2 Methodologies and Techniques

8.2.1 Bacteria-based Large-insert DNA Clones

The bacteria-based large-insert cloning system represents the most advanced technology for large-insert genomic DNA library construction. The large-insert clones in bacterial vectors are probably best known as BACs, PACs, BIBACs, and TACs. Nevertheless, the vectors used in cloning of these "artificial chromosomes" are modified plasmids and their clones are maintained as large-insert plasmids in the bacterial hosts; no components of the vectors were derived from the bacterial chromosomes. Tao and Zhang (1998) discovered that conventional plasmid-based cloning (PBC) vectors previously developed for different research purposes, including plasmids and cosmids, have the same capacity as the "artificial chromosomes" when the electroporation technology was used along with the bacterial strain DH10B. Therefore, BACs, PACs, BIBACs, PBCs, and TACs are collectively referred as to bacteria-based large-insert clones here. Figure 8.1 shows the typical vectors of BACs, PACs, BIBACs, PBCs, and TACs, and Figure 8.2 shows an example of bacteria-based large-insert clones cloned in the BAC and BIBAC vectors and analyzed on pulsed-field gels.

Figure 8.1. Vectors used for construction of bacteria-based large-insert genomic libraries. The restriction enzymes in bold are cloning sites. The BAC and PAC can be used only to construct genomic libraries, whereas BIBAC, binary PBC, and TAC can be used not only for library construction but also for plant transformation via *Agrobacterium*. ▶

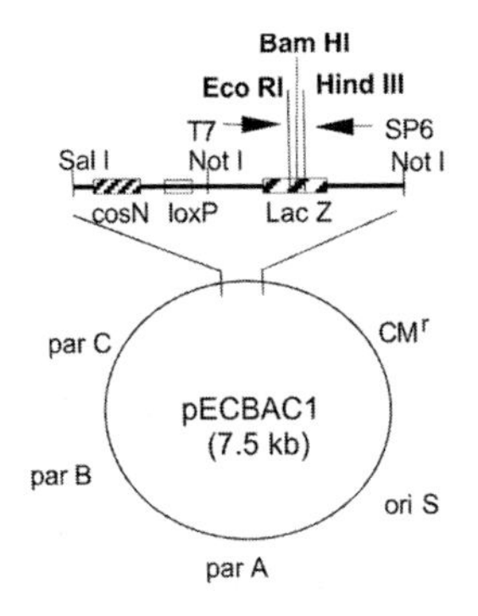

Bacterial artificial chromosome (BAC)

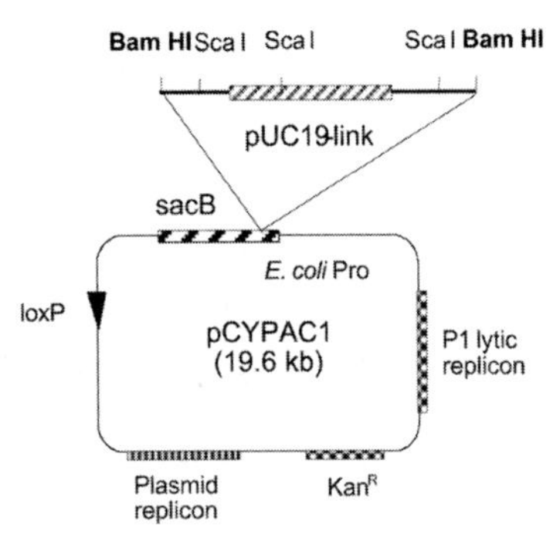

Bacteriophage P1-derived artificial chromosome (PAC)

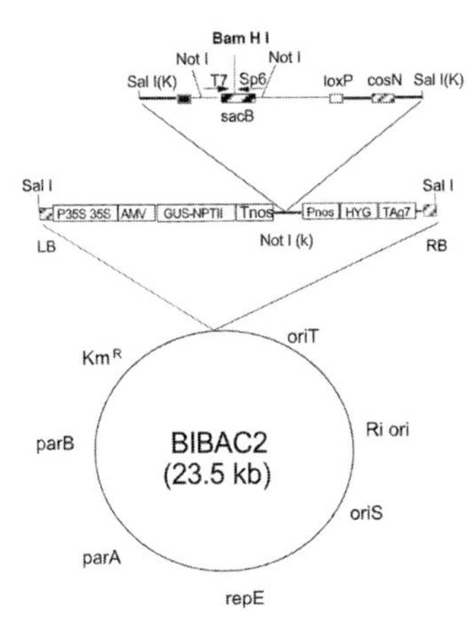

Binary BAC (BIBAC)

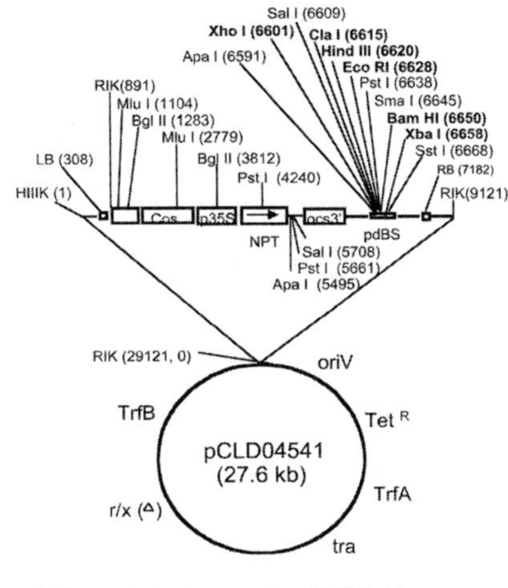

Plasmid-based cloning vector (PBC): binary cosmid

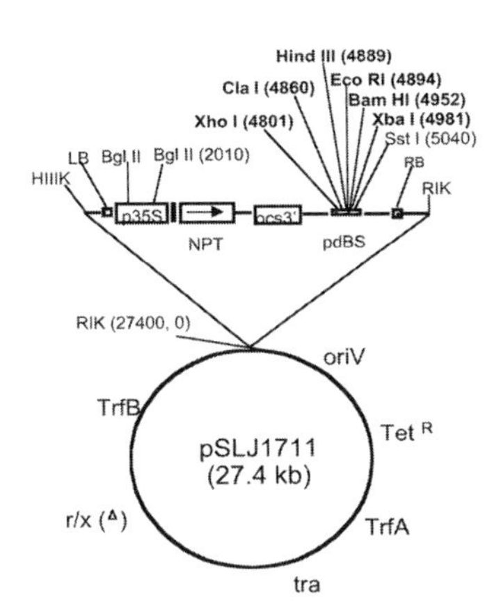

Plasmid-based cloning vector (PBC): binary plasmid

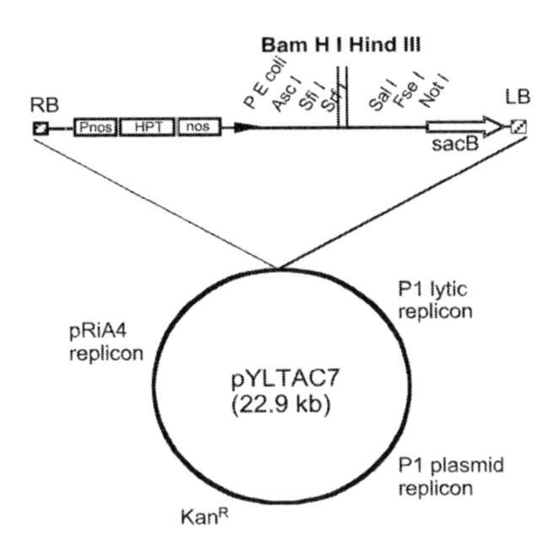

Transformation-competent artificial chromosome (TAC)

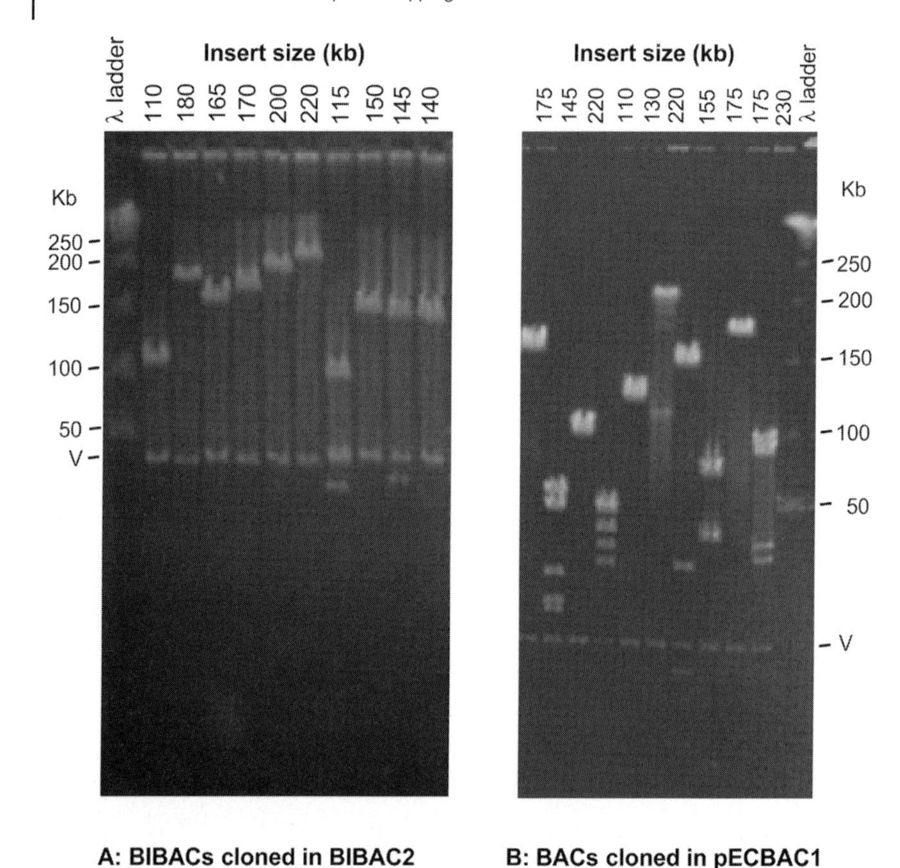

Figure 8.2. Plant-transformation-competent *Arabidopsis* BIBACs (from Chang et al. 2003) and wheat BACs. DNA was isolated, digested with *Not* I to release the inserts from the cloning vector (V), run on pulsed-field gels, stained in EB, and photographed. The insert size of each was estimated by adding up all insert fragments of the clone.

8.2.1.1 BAC

BAC vectors were developed based on the *E. coli* fertility (F-factor) plasmid (Shizuya et al. 1992). The F-factor plasmid naturally occurs as a 100-kb circular DNA molecule. Since the replication of the plasmid in *E. coli* is strictly controlled, it is usually maintained in low copy number (one or two copies per cell). The first BAC cloning plasmid vector, pBAC108L, reported by Shizuya et al. (1992), was constructed by deleting some unnecessary fragments and adding marker genes to the natural F-factor plasmid. pBAC108L is capable of cloning and stably maintaining DNA fragments larger than 300 kb in *E. coli*, but it lacks a selectable marker for recombinant clones. The recombinant clones in pBAC108L must be identified by colony hybridization, which is not suitable for library construction. Therefore, the *lacZ* gene was introduced into pBAC108L for recombinant selection, producing the BAC vector pBeloBAC11 (Kim et al. 1996). pBeloBAC11 has two cloning sites, *Hin-*

dIII and *Bam*HI. To facilitate cloning of large DNA fragments generated with *Eco*RI, pBeloBAC11 was further modified into pECBAC1 by destroying the *Eco*RI site in the CM^r gene of pBeloBAC11 (Frijters et al. 1997). Thus, pECBAC1 is capable of cloning DNA fragments generated with *Eco*RI, in addition to those generated with *Hind*III and *Bam*HI. Both pBeloBAC11 and pECBAC1 have been widely used for BAC library construction.

The molecular structure of pECBAC1 is shown in Figure 8.1. It contains the minimal sequences needed for the autonomous replication and copy-number control of the F-factor plasmid (*oriS*, *parA*, and *parB*). The *oriS* gene mediates the unidirectional replication of the F-factor while *parA*, *parB*, and *par C* maintain copy number at a level of one or two per *E. coli* cell. pECBAC1 contains three restriction sites, *Hind*III, *Bam*HI, and *Eco*RI, for cloning within the *lacZ* gene. When grown on a medium containing the histochemical substrate 5-bromo-4-chloro-3-indolyl-β-D-galactoside (X-gal) and the inducer isopropyl β-D-thiogalactoside (IPTG), the recombinant BACs in which DNA inserts have disrupted the *lacZ* gene are identified as white colonies, while non-recombinant, insert-empty clones are blue. The plasmid also contains a gene for chloramphenicol resistance (CM^r) as a selectable marker for transformants. Two *Not*I sites flanking the multiple cloning sites allow excision of the insert from the vector. The bacteriophage T7 and the bacteriophage SP6 promoter sequences flanking the cloning sites allow preparation of probes from the ends of cloned sequences by *in vitro* transcription of RNA or by PCR methods. Other functional structures included in this plasmid are the bacteriophage λ*cosN* site and the bacteriophage P1 *loxP* site.

8.2.1.2 PAC

As discussed above, the first BAC vector, pBAC108L, lacks selectable markers for recombinants and is not convenient for construction of genomic DNA libraries on a large scale. To overcome this disadvantage, Ioannou et al. (1994) combined the features of the bacteriophage P1 and F-factor plasmid and constructed a new vector pCYPAC-1. This vector was named PAC. It was derived from the P1 vector pAd10SacBII by deleting the adenovirus stuffer fragment and inserting a modified pUC19 plasmid into the *Bam*HI cloning site.

The PAC has most of the features of the BAC and is largely interchangeable with BACs in library construction, provided that attention is paid to the antibiotic selection (kanamycin is used for PACs, whereas chloramphenicol is used for BACs) and cloning site (e. g., *Bam*HI for the PAC vector pCYPAC1). However, the molecular structures of PAC differ from those of BAC in several aspects (Figure 8.1). First, the commonly used BAC vectors, with the exception of pBAC108L, contain the *LacZ* gene for selection of recombinants, while the PAC vectors have the *SacB* gene from bacteriophage P1 for recombinant selection. The *SacB* gene is lethal to *E. coli*, and *Agrobacterium tumefaciens* when 5 % sucrose is present in agar medium. Thus, the host cell containing recombinant vector (with a foreign DNA fragment inserted in the cloning site *Bam*HI) can survive and grow on the 5 % sucrose medium, forming a white colony, whereas the non-recombinant cells cannot survive on the 5 % sucrose. Second, like bacteriophage P1, the PACs are usually maintained in a cell as a single copy due to their P1 replicon and partition system, but they also contain an IPTG-inducible high-copy-number origin of replication (P1 lytic replicon), which can be used to amplify and reproduce the PAC DNA in the host cell in high yield (Ioannou et al. 1994). In addition, PACs contain another high-copy-number origin of DNA replication in the "pUC19-link," which is used for convenient vector propagation and is later removed by *Bam*HI digestion during vector preparation for library construction.

8.2.1.3 **BIBACs**

In the plant kingdom, many traits are controlled by clusters of physically closely linked genes (e.g., the genes for plant disease resistances). Transformation of such gene clusters in plants requires a vector that is capable of cloning large DNA fragments and of transforming the cloned DNA into plants via *Agrobacterium.* Such a vector will also streamline positional cloning and genetic engineering of genes and quantitative trait loci (QTL) in plants. Therefore, binary BAC (BIBAC) vectors and related large DNA fragment transforming systems were designed to meet these requirements (Hamilton et al. 1996; Hamilton 1997).

The BIBAC vector is a combination of BAC plasmid pBAC108L and *Agrobacterium* Ri plasmid and thus replicates as a single-copy plasmid in both *E. coli* and *A. tumefaciens.* Figure 8.1 shows the molecular structure of the vector BIBAC2 (Hamilton 1997). Like BAC vectors, it contains the minimal sequences needed for the autonomous replication and copy-number control of the F-factor plasmid (*oriS, repE, parA,* and *parB*). The λ*cosN* and P1 *loxP* sites from pBAC108L and the T7 and Sp6 promoters from bacteriophage are also incorporated into the BIBAC. The recombinant selection marker of the BIBAC vector is the *SacB* gene from the bacteriophage P1 vector pAd10sacBII, in which the insert cloning site *Bam*HI is situated. The antibiotic selectable marker in BIBAC is the Km^R gene, which confers resistance to the antibiotic kanamycin. The plant-selectable marker GUS-NPTII was introduced to the left border of the insert DNA to facilitate the detection of efficient transformation of T-DNA in plants. GUS-NPTII is a bifunctional fusion peptide of β-glucuronidase (GUS) and neomycin phosphotransferase II (NPTII). NPTII confers resistance to kanamycin, and GUS can be easily assayed to determine the level of protein expression. BIBAC2 has an additional plant selectable marker, HYG, which encodes hygromycin phosphotransferase and provides resistance to hygromycin. The BIBAC2 has an origin of replication from the Ri plasmid, which maintains a single copy of the plasmid in the host cell of *A. tumefaciens.* The BIBAC2 also has an origin of conjugal transfer (*ori T*). When all other transfer functions are provided in *trans* by a helper plasmid, the *oriT* allows for the conjugal transfer of any covalently linked DNA. Therefore, BIBAC clones can be introduced into *A. tumefaciens* by electroporation or triparental mating. Other structures in BIBAC2 include P35S 35S (tandem 35S promoter from CaMV), Tnos (nopaline synthase terminator), Pnos (nopaline synthase promoter), TAg7 (*A. tumefaciens* gene 7 terminator), and AMV (alfalfa mosaic virus enhancer).

The BIBAC system takes advantage of the natural ability of *A. tumefaciens* to transform plants with the aid of the helper plasmid. To improve the efficiency of transfer of large T-DNA molecules into plant chromosomes, the helper plasmids pCH30 and pCH32 have been developed by increasing the gene dosage of *virG* and *virE*, the transcriptional activator of the *Agrobacterium* virulence (*vir*) genes. For more information on the helper plasmid, see Hamilton (1997).

8.2.1.4 **PBC**

The vectors introduced above were designed particularly for cloning large-insert genomic DNA. However, Tao and Zhang (1998) found that electroporation can introduce plasmid-based conventional vectors containing large DNA fragments into the host cells, and these conventional vectors are capable of cloning and stably maintaining DNA fragments as large as those of BACs, PACs, and BIBACs. These results suggest that many, if not all, of the existing plasmid-based vectors, including plant and animal transformation and expression binary vectors, could be directly used for cloning of very large DNA fragments with the electroporation technology. Experiments using a conventional binary plasmid vector pSLJ1711 and a conventional binary cosmid vector pCLD04541showed that both vectors can clone foreign DNA inserts over 300 kb, and at least four to five copies of the clones can be stably maintained in the host *E. coli* strain DH10B. Tao and Zhang (1998) designated the conventional plasmid-based large-insert clones as PBCs.

The native P1 plasmid (which is different from bacteriophage P1) RK2 is 56 kb in size, has a broad host range (including *E. coli* and *A. tumefaciens*), and maintains five to eight copies in a host cell (Figurski and Helinski 1979). The pRK290 plasmid was derived by deleting some unnecessary sequences from RK2, resulting in a vector of 20 kb (Ditta et al. 1980). The binary plasmid vector pSLJ1711 was constructed by insertion of a Bluescript polylinker into pRK290 (Jones et al. 1992), while the cosmid vector pCLD04541 was further modified by inserting a cos site into pSLJ1711 (Jones et al. 1992). Figure 8.1 gives an overview of the structures of vectors pSLJ1711 and pCLD04541. Both vectors use the tetracycline-resistant gene *Tet*^{R} for transformant selection, the *LacZ* gene for recombinant selection, the P35S promoter-driven NPT gene (coding neomycin phosphotransferase and conferring kanamycin resistance) for plant selection, and a dark Bluescript (dBS) polylinker between the T-DNA borders (inside *LacZ*) as cloning sites. The regions *oriV, TrfA*, and *TrfB* function in DNA replication, and *rlx* and *tra* function in conjugal transformation. ocs3′ indicates the octopine synthase 3′ end.

The binary vectors pCLD04541 and pSLJ1711 have a few additional advantages for large-insert genomic DNA library development. First, they have five to eight copies per cell, so that the DNA of their clones is higher in yield and much easier to purify. Second, both of these cosmid and plasmid vectors have cloning sites for *Xho*I, *Cla*I, *Hin*dIII, *Eco*RI, and *Bam*HI, which give additional restriction enzymes for large-insert genomic DNA library construction than do BAC, BIBAC2, PAC, and TAC. Third, pCLD04541 and pSLJ1711 have the *LacZ* gene as the selectable marker for recombinant clones, whereas the BIBAC vector has the *SacB* gene as the selectable marker for recombinants. In our minds, the *SacB* gene is not as reliable as the *LacZ* gene for recombinant selection because the clones selected with the *SacB* gene have a relatively high (~10 %) false recombinants frequency (Hamilton et al. 1999; Chang et al. 2003). Although BACs have the *lacZ* gene as the recombinant selection marker, the colony color of their vector clones is not as intense as that of the pCLD04541 and pSLJ1711 clones. Additionally, BAC can be used only for large-insert DNA library development, whereas pCLD04541 and pSLJ1711 are binary vectors that can also be directly used in plant transformation of large-insert clones through *Agrobacterium.*

8.2.1.5 TAC

The bacteriophage P1-derived binary plasmid vector pYLTAC7 was developed and designated by Liu et al. (1999) as a transformation-competent artificial chromosome (TAC). This vector is suitable for stable maintenance of large genomic DNA fragments in both *E. coli* and *A. tumefaciens* and is competent for transfer of large cloning insert DNA into plant genomes by *Agrobacterium*-mediated transformation. The components of the TAC vector are derived from various plasmids in the minimum sizes possible (Figure 8.1). It carries the bacteriophage P1 replicon and the plasmid Ri replicon for stable maintenance of single copies in both *E. coli* and *A. tumefaciens*, the P1 lytic replicon for high yield of cloned DNA inducible by IPTG, the *SacB* gene for recombinant selection in host cells, the *HPT* (hygromycin phosphotransferase) gene for selection in plants resistant to hygromycin in the media, and the kanamycin-resistant marker gene Km^R for transformant selection in *E. coli.* The unique cloning sites of the TAC vector are the *Hin*dIII and *Bam*HI sites located within the *SacB* gene. The recombinant TACs are transformed by electroporation into *E. coli* or *A. tumefaciens*, and the transformed *E. coli* cells are grown on the media containing 5 % sucrose. It should be noted that there is only one *Not*I site flanking the cloning sites. Thus, when checking the cloned insert size with *Not*I digestion, the vector molecule is not excised from the insert DNA, and the insert size must be calculated by subtracting the vector size (23 kb) from the total size of the bands.

8.2.2 Genomic DNA Library Quality and Genome Physical Mapping

High-quality large-insert genomic DNA libraries are crucial to genome physical mapping. In comparison to YAC libraries, it has been demonstrated that the foreign DNA inserts in bacteria-based large-insert libraries are stable for many generations with few chimeric DNA fragments, even with the PBC vectors, where five to eight copies of recombinant clones exist in a single cell (Tao and Zhang 1998; Zhang et al. 1996). In addition, a high-quality genomic DNA library to be used for genome physical mapping should have large insert sizes and desirable genome coverage and be representative of the entire genome. When preparing a library, the long-term uses and versatility of different research projects, such as direct plant transformation, should be considered. Therefore, we will focus our discussion on the influences of insert size, genome coverage, genome representation, and plant transformation competence of bacteria-based large-insert genomic DNA libraries on genome-wide physical mapping.

8.2.2.1 **Insert Sizes**

The cloned insert sizes of genomic DNA libraries depend to a great degree on the transformation methods. Before the advent of electroporation transformation technology, plasmid constructs were transformed into bacterial host cells by DNA uptake through $CaCl_2$-treated competent cells, and cosmid and bacteriophage P1 constructs were transformed by in vitro bacteriophage particle head packaging, both of which have their limitations in transferring large (<100 kb) DNA fragments into bacterial cells. The application of electroporation in transformation has significantly increased the capacity of transforming large DNA fragments into bacterial cells not only for the specially developed large DNA fragment cloning vectors such as BAC, PAC, BIBAC, and TAC, but also for many conventional plasmid-based vectors (PBC) that were previously developed for different biological research purposes. It has been proved that that all of these vectors, whether "artificial chromosome" or conventional plasmid-based vectors, can clone and stably maintain fragments of up to more than 300 kb of foreign DNA in bacteria.

The average insert size of a genomic DNA library is one of the most important criteria for genome physical mapping. It is apparent that for a genome of a given size, the larger the clone inserts of the library, the fewer clones are needed to represent the entire genome sequence and the fewer contigs will be required to span the genome in the physical map. In a given genome coverage or representation, physical maps with fewer contigs result in fewer gaps present and a higher quality. However, it is not well known how significantly the source clone insert sizes affect the number or size of contigs spanning the genome.

Soderlund et al. (2000) studied the influence of library insert size on the construction of BAC contigs essential for genome physical mapping using BAC libraries with different insert sizes (Figure 8.3). The BAC libraries represented an equivalent of 20× coverage of a 110-Mb genome, and contigs were constructed using the computer program FingerPrinted Contig (FPC). When the average insert size of the library was increased from 100 kb to 165 kb, the number of contigs was reduced and the average size of the contigs was increased by 5.8-fold (212/38). However, when the average insert size of the library was increased further from 165 kb to 200 kb, the number of contigs and the average size of the contigs was improved by only 1.4-fold (38/26). Therefore, a genomic DNA library with an average insert size of greater than 150~160 kb is desirable for efficient development of an optimal physical map.

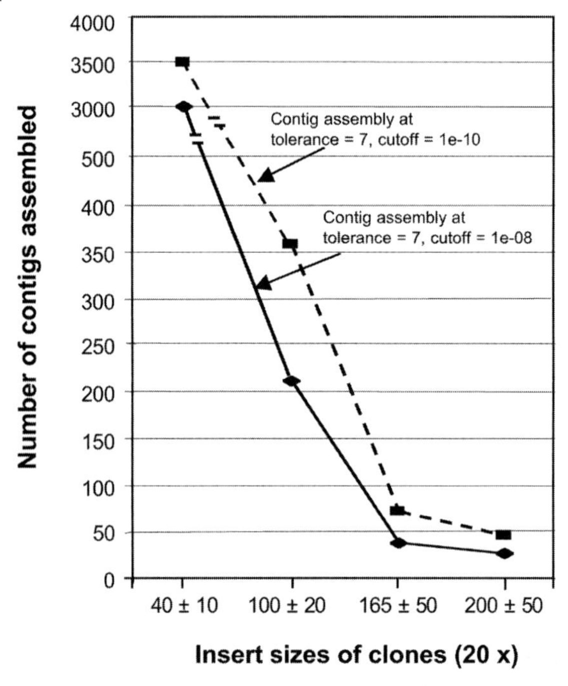

Figure 8.3. Influence of clone insert sizes on the number or size of fingerprint physical map contigs. The contigs were assembled from the fingerprints of 20× genome coverage clones having different insert sizes for 100-Mb genomic sequence using the program FingerPrinted Contig (FPC) V4.7 (data used are from Soderlund et al. 1997).

8.2.2.2 **Clone Genome Coverage**

The coverage of the entire genome by the clones in a genomic DNA library is also of significance for physical mapping. Theoretically, the probability (P) that a given genomic sequence is present in a library can be estimated by the following formula:

$$P = 1 - e^{-n}$$

where n is the number of genome equivalents of the library, which is the product of the number of total clones and the average insert size divided by the haploid genome size.

BAC libraries equivalent to genome coverage ranging from 7.0× to 20.0× were used in the development of whole-genome BAC-based integrated physical and genetic maps of *Arabidopsis thaliana* (Marra et al. 1999; Chang et al. 2001), indica rice (Tao et al. 2001), human (IHGMC 2001), japonica rice (Chen et al. 2002; Li et al. 2003), and mouse (Gregory et al. 2002). The minimum genome coverage required to efficiently and economically construct a high-quality genome-wide physical map has been debated. To answer this question, computer simulation studies were conducted by mimicking different fingerprinting methods to assemble contigs from known sequences of *Arabidopsis* chromosomes 2 and 4, human chromosome 22, and the three-chromosome combination (Xu et al. 2003, 2004). The results (Figure 8.4) showed that the number of contigs spanning the chromosomes rapidly decreased as the genome coverage was increased from 5× to 10×. However, when

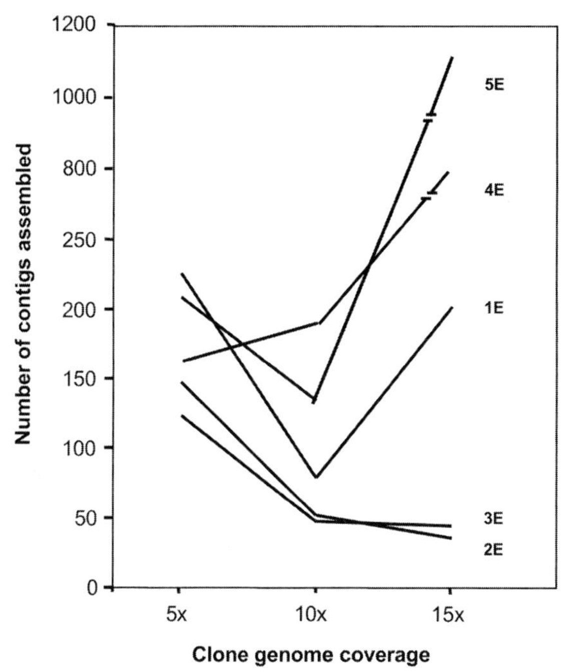

Figure 8.4. Influence of clone genome coverage on the number or size of fingerprint physical map contigs assembled, based on computer simulation. The 71.9-Mb genome sequences of *Arabidopsis* chromosomes 2 and 4 and human chromosome 22 were generated into BACs having an insert size range from 120 kb to 180 kb (150±30 kb) and fingerprinted using the agarose gel-based, one-enzyme method (1E) (Marra et al. 1997), the manual sequencing gel-based, two-enzyme method (2E) (Zhang and Wing 1997), the capillary sequencing-based, three-enzyme method (3E) (Xu et al. 2003), the automatic sequencing gel-based, four-enzyme method (4E) (Ding et al. 1999), and the capillary sequencing-based, five-enzyme (SnapShot) method (5E) (Xu et al. 2003), respectively. The contigs were assembled using FPC V6.0 (Soderlund et al. 2000) by lowering cutoff values until no questionable clones were produced (from Xu et al. 2004).

the genome coverage was increased from 10× to 15×, the number of contigs spanning the chromosomes decreased at a lower rate or even actually increased (Figure 8.4), implying that the number of contigs is not always inversely related to the genome coverage used for genome physical mapping. The results indicated that the number of clones equivalent to approximately 10× genome coverage is the most efficient for whole-genome physical mapping with bacteria-based large-insert clones.

8.2.2.3 **Representation for Genome**

Studies show that the theoretical probability (calculated by the formula in Section 8.2.2.2) or genome representation of a large-insert genomic DNA library constructed with a single restriction enzyme is about 15 % lower than the genome representation estimated by library screening (Tao et al. 2001). This implies that when the library is screened with random DNA markers, only about 85 % of the markers are represented by one or more positive clones, whereas the remaining 15 % are not represented in the library. The studies also show that the genome coverage of the library cannot be increased significantly by simply increasing the number of clones in the library. This observation is due to at least two reasons. First, the restriction sites of the enzyme used for the library construction are not uniformly distributed along the genome. Consequently, the genomic regions having too many or too few restriction sites of that particular enzyme are less likely to be cloned because the small (<50 kb) and large (>400 kb) DNA fragments generated by partial digestion are removed during the size selection of large-insert DNA library construction. Therefore, it is recommended that at least two libraries prepared with single restriction enzymes that have different nucleotide contents (AT-rich and GC-rich) in their restriction sites be used for physical mapping applications. Physical mapping practice has demonstrated that this approach would minimize the bias of clone distribution along the genome, thus achieving truly high genome representation of libraries complementary to each other, similar to mechanically sheared shotgun libraries. Two or more complementary large-insert genomic DNA libraries constructed with different restriction enzymes have been successfully used for whole-genome physical mapping of several species such as *Arabidopsis* (Chang et al. 2001), indica rice (Tao et al. 2001), human (IHGMC 2001), mouse (Gregory et al. 2002), japonica rice (Chen et al. 2002; Li et al. 2003), soybean (Wu et al. 2004), and chicken (Ren et al. 2003; Lee et al. 2003).

Several complementary genomic DNA libraries are especially needed for species with huge genome sizes, e. g., common wheat (15,600 Mbp/1C). Obtaining a higher genome representation by simply increasing the library clone coverage using a single restriction enzyme for such a large genome size would increase the cost and efforts for library production, storage, and manipulation. Therefore, efficient genome representation by complementary DNA libraries would result in more economical physical maps for genome research.

Although a DNA library with 4.7× genome coverage is equivalent to a 99 % library according to the above formula, a total number of clones with an average insert size of 150 kb or larger constructed with two or more restriction enzymes and representing approximately 10× haploid genomes is sufficient for most genome research purposes, including global genome physical mapping and genome sequencing (Xu et al. 2004). This is further supported by the whole-genome shotgun sequencing of *Haemophilus influenzae*, *Mycoplasma genitalium*, *Anopheles gambiae*, and the euchromatic genome portion of *D. melanogaster*, in which the numbers of clones equivalent to 6–10× haploid genomes were analyzed.

The second reason for the discrepancy between theoretical and actual representation of a genome by a genomic DNA library is be that some genomic regions may be lethal, less likely to be cloned, or poorly maintained and/or propagated in particular host system. To clone these DNA sequences, a different vector and/or host system is needed. For example, several genomic libraries may be constructed using vectors such as F plasmid-based (BAC or BIBAC), P1 plasmid-based (PBC-pCLD04541), and P1 bacteriophage-based (PAC or TAC) vectors. The YAC cloning system may also be used as a complementary cloning system to the bacteria-based large-insert libraries because the clones are hosted in the single-cell eukaryotic organism yeast, which differs from the bacteria-based large-insert cloning system. For example, the physical map of human was constructed from BAC and PAC libraries generated with *Hin*dIII, *Eco*RI, and *Bam*HI restriction enzymes, but some contig gaps in the map were closed using YACs (IHGMC 2001).

8.2.2.4 The Importance of Binary Vectors in Plant Physical Mapping

Unlike small-insert clones, the entire inserts of large-insert clones are difficult to subclone into a transformation-competent binary vector for functional analysis and genetic engineering of the inserts by genetic transformation. It is difficult to find an enzyme with restriction sites flanking the cloning site of the vector that is also absent in the large clone insert. Although the restriction enzyme *Not* I, a rare 8-bp cutter, has been used for subcloning, it is present at an average frequency of every 65,536 bp in a genome, which is smaller than the 150-kb average insert size of a large-insert DNA library. Studies also show that the frequency of the *Not* I sites is much higher in the genomes of monocotyledonous plant species than in those of dicotyledonous plant species. Although it is not difficult to subclone a part of a large-insert clone, it is time-consuming and tedious to subclone the entire insert for a number of large-insert clones. Therefore, it is desirable to construct large-insert DNA libraries in binary vectors that are capable of stably cloning and maintaining large DNA inserts and also are able to directly transform their insert DNA fragments into plants. If the large-insert binary libraries are used for whole-genome physical mapping and sequencing, the ability of the binary clones to directly transform plants would significantly streamline the functional analysis of genomic sequences and use of the physical mapping and sequencing results in plant breeding by genetic transformation. Therefore, it is recommended that for plant physical mapping projects (even for other genomic research), at least one large-insert DNA library be constructed using a binary vector.

The BIBAC2 vector has been demonstrated to transform tobacco and tomato with a 150-kb human DNA fragment (Hamilton et al. 1996, 1999), and pYLTAC7 has been demonstrated to transform *Arabidopsis* with about 100-kb DNA fragments (Liu et al. 1999). However, the binary vectors have one (BIBAC2) or two (pYLTAC7) cloning sites and the *sacB* gene for recombinant selection. In some studies, ~10% of the clones selected were found to be false recombinants (Hamilton et al. 1999; Chang et al. 2003; T. Uhm and H.-B. Zhang, unpublished data). These limitations have influenced the utility of the BIBAC2 and pYLTAC7 vectors for large-insert binary clone library construction. Alternatively, the PBC binary vectors pCLD04541 and pSLJ1711 or other binary vectors may be the vectors of choice for this purpose since they have multiple cloning sites, higher DNA yields, and the *LacZ* gene for recombinant selection. In each of the physical mapping projects of *Arabidopsis* (Chang et al. 2001), soybean (Wu et al. 2004), and japonica rice (Li et al. 2003), one of the libraries used was constructed using the binary vector pCLD04541.

These physical maps have been used in functional analysis and genetic engineering of the insert DNA by genetic transformation.

8.2.3 Construction of Bacteria-based Large-insert Genomic DNA Libraries

The same procedure has been used for construction of bacteria-based large-insert libraries using different vectors, except for some modifications required by different vectors in choosing restriction enzymes, antibiotics for transformant selection, and cell culture media for recombinant selection. The procedure described below has been used successfully in the construction of bacteria-based large-insert genomic DNA libraries for many species of plants, animals, insects, and microbes using the vectors BAC, BIBAC, PAC, TAC, and PBC (Zhang 2000). The procedure can also be downloaded from the website of GENE*finder* Genomic Resources (http://hbz.tamu.edu – Analysis Tools – BAC Tools).

8.2.3.1 Megebase-size Nuclear DNA Preparation

DNA of a high molecular weight (HMW) (>600 kb) is essential for large-insert genomic DNA library construction. Unlike conventional-sized (<150 kb) DNA preparation, HMW DNA must be protected from physical shearing during preparation. Therefore, protoplasts (plants and fungi), entire cells (animals and microbes), or nuclei (plants, animals, and insects) are isolated and then embedded in low-melting-point (LMP) agarose in the form of plugs or microbeads. The cell lysis and DNA purification and manipulation are then conducted in the plugs or microbeads. Because the preparation of HMW DNA from nuclei is straightforward, results in low contamination with cytoplast DNA, and is economical and applicable for preparation of HMW DNA from a wide variety of plant and animal species (Zhang et al. 1995), the nuclei method has rapidly become the method of choice and is widely used for preparation of HMW DNA in plants. The development of the nuclei method has made it possible to prepare high-quality HMW DNA and to construct bacteria-based large-insert genomic DNA libraries for different plant species. Here we introduce the procedure of Zhang et al. (1995) with the latest updates for preparation of HMW DNA from plants.

Plant Materials

Young plant leaves or whole plant seedlings of divergent species, including grasses, legumes, vegetables, and trees, can be used as materials for preparation of HMW DNA by this method (Zhang et al. 1995). The tissues can either be frozen in liquid nitrogen and stored at –80 °C or kept fresh on ice before use.

Reagents

- 10 × homogenization buffer (HB) stock: 0.1 M Trizma base, 0.8 M KCl, 0.1 M EDTA, 10 mM spermidine, and 10 mM spermine. The pH of the solution is adjusted to 9.4–9.5 with NaOH. The stock is stored at 4 °C before use.

- 1 × HB: A suitable amount of sucrose is mixed with a suitable volume of 10 × HB stock. The final concentration of sucrose is 0.5 M and HB stock is 1 ×. The resultant 1 × HB is stored at 4 °C.
- Nuclei extraction and wash buffer (1 × HB plus 0.5 % Triton X-100 and 0.15 % β-mercaptoethanol): The buffer is prepared by mixing 1 × HB with 20 % Triton X-100 in 1 × HB for a final concentration of 0.5 % of Triton X-100. β-mercaptoethanol is added just prior to use for a final concentration of 0.15 %.
- Lysis buffer: 0.5 M EDTA, pH 9.0–9.3, 1% sodium lauryl sarcosine, and 0.2–1.0 mg mL^{-1} proteinase K, depending on plant species. The proteinase K powder is added just before use.

Preparation of Intact Nuclei

1. Grind about 50 g of the tissue into a very fine powder in a large amount of liquid nitrogen with a large mortar and pestle (about 1 h), and immediately transfer the powder into an ice-cold 1000-mL beaker containing 800–1000 mL ice-cold nuclei extraction and wash buffer.
2. Gently swirl the contents with a magnetic stir bar for 10 min on ice and filter into six ice-cold 250-mL centrifuge bottles through two layers of cheesecloth and one layer of Miracloth by squeezing with gloved hands.
3. Pellet the homogenate by centrifugation with a fixed-angle rotor at 3000 g at 4 °C for 20 min.
4. Discard the supernatant fluid and add approximately 1 mL of ice-cold nuclei extraction and wash buffer to each bottle.
5. Gently resuspend the pellet with the assistance of a small paintbrush soaked in ice-cold nuclei extraction and wash buffer, combine the resuspended nuclei from all bottles into a 40-mL centrifuge tube, and fill the tube with ice-cold nuclei extraction and wash buffer. If particulate matter remains in the suspension, filter the resuspended nuclei into the 40-mL centrifuge tube through two layers of Miracloth by gravity.
6. Pellet the nuclei by centrifugation at 3000 g at 4 °C for 15 min in a swinging bucket centrifuge.
7. Wash the pellet one to three additional times (until the supernatant is no longer green) by resuspension in nuclei extraction and wash buffer using a paintbrush followed by centrifugation at 3000 g at 4 °C for 15 min. Note that this step is necessary to minimize the contamination of cytoplast and mitochondrial organelles in the nuclei.
8. After the final wash, resuspend the pelleted nuclei in a small amount (about 1 mL) of 1× HB, count the nuclei, if possible, under the contrast phase of a microscope, bring to approximately 5×10^7 nuclei per milliliter (for a species with a genome size of about 1000 Mb/1C) with addition of 1× HB, and store on ice. The concentration of nuclei can also be estimated empirically; a concentration of nuclei that is just transparent under light is estimated to be $5–10 \times 10^7$ nuclei per milliliter. The concentration of nuclei varies, depending on the genome sizes of different species. In general, 5–10 µg DNA per 100-µL plug is suited for large-insert genomic DNA library construction.

Embedding the Nuclei in LMP Agarose Plugs

The nuclei isolated can be embedded into LMP agarose plugs or microbeads according to Zhang et al. (1995) and Zhang (2000). Since plugs are relatively easier to prepare and handle than microbeads, and since the digestion of DNA embedded in plugs is no longer problematic using the procedure described below, LMP agarose plugs have more popularly been used for preparation of HMW DNA. Therefore, here we describe only the procedure of preparing HMW DNA in LMP agarose plugs.

1. Prepare 10 mL of 1 % LMP agarose in 1 × HB, cool down to 45 °C, and maintain in a 45 °C water bath before use.
2. Prewarm the nuclei to 45 °C in a 45 °C water bath (about 5 min), and mix gently but thoroughly with an equal volume of the prewarmed 1 % LMP agarose in 1× HB using a cut off pipette tip.
3. Aliquot the mixture into ice-cold plug molds (Bio-Rad) on ice with the same pipette tip, 100 μL per plug. When the agarose is completely solidified, transfer the plugs into 5–10 volumes of lysis buffer.
4. Incubate the agarose plugs in the lysis buffer for 24–48 h at 50 °C with gentle shaking.
5. Wash the plugs once in 0.05 M EDTA, pH 8.0, for 1 h on ice and store in 0.05 M EDTA, pH 8.0, at 4 °C. The DNA at this step can be stored at 4 °C for one year without significant degradation. If the DNA is used immediately, the plugs can

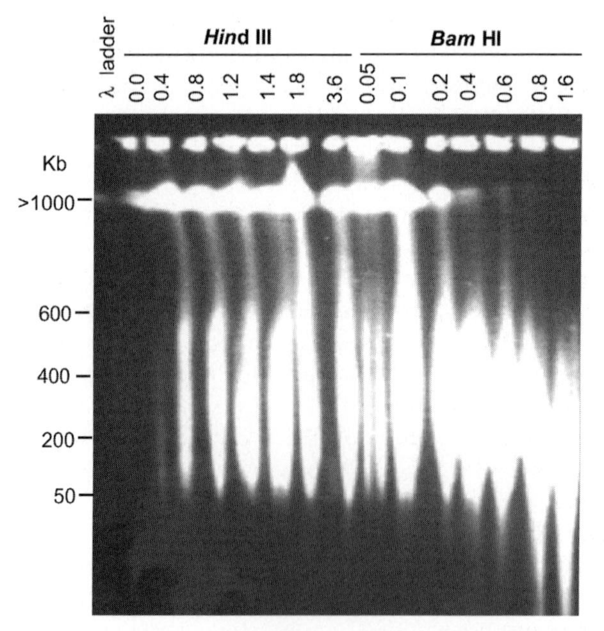

Figure 8.5. Megabase-size DNA isolated from wheat nuclei according to Zhang et al. (1995). DNA was partially digested with a series of amounts of *Hind* III and *Bam* HI, respectively, at 37 °C for 8 min, stopped, and analyzed on a pulsed-field gel.

directly be transferred into ice-cold TE (10 mM Tris-HCl, 1 mM EDTA, pH 8.0) from the lysis buffer. After two washes in TE, the plugs can then be washed in ice-cold TE plus phenylmethyl sulfonyl fluoride (PMSF) (see below).

Figure 8.5 shows MHW genomic DNA isolated from wheat nuclei using this method. The fragments of uncut DNA (the lane 0.0) are larger than 1000 kb in size and are readily digested with restriction enzyme, suggesting that the DNA isolated is suitable for large-insert genomic DNA library construction.

8.2.3.2 Vector Preparation

Preparation of DNA cloning vectors is one of the most critical steps for large DNA fragment cloning. The vector DNA to be used must be very pure, completely digested (but not over-digested), and completely dephosphorylated (>95 % recombinant clones in a ligation test). Presented here is the procedure that has been widely used in our laboratory for the vectors pBeloBAC11, pECBAC1, pCLD04541, pSLJ1711, BIBAC2, and pYLTAC7 (Zhang 2000; http://hbz.tamu.edu – Analysis Tools – BAC Tools). The vectors prepared have successfully been used to construct over 100 bacteria-based large-insert genomic DNA libraries.

Vector DNA Isolation and Purification

1. Streak the stock cells containing a cloning vector on a Luria broth (LB) agar plate plus appropriate antibiotics and grow at 37 °C overnight to obtain single colonies. For the vector cells with blue and white (*LacZ*) selection such as pBeloBAC11, pECBAC1, pCLD04541, and pSLJ1171, use the LB agar plate containing appropriate antibiotics, X-gal, and IPTG. For the vector cells with *SacB* selection such as BIBAC2 and pYLTAC7, use the LB agar plates containing appropriate antibiotics with or without 5 % sucrose.
2. Select a single blue colony (*LacZ*) or a colony (*SacB*) that grows on the medium without sucrose, but not on the medium with 5 % sucrose, and inoculate into 50–100 mL LB or Terrific broth (TB) liquid medium containing appropriate antibiotics in a 250-mL glass flask. Alternatively, the culture medium can directly be inoculated from the frozen stock of vector cells derived from a single colony at 1 μL per liter of medium. Grow the cells overnight at 37 °C and 250 rpm.
3. Take 10 mL of the overnight culture, inoculate into 1 L of LB or TB medium with appropriate antibiotics, and then grow at 37 °C and 250 rpm for 16–20 h. To obtain a sufficient amount of vector DNA, grow 0.5–1.0 L of culture for multiple-copy vectors such as pCLD04541 and pSLJ1711, and 1.5–2.0 L of culture for single-copy vectors such as pBeloBAC11, pECBAC1, BIBAC2, and pYLTAC7.
4. Harvest the culture cells in 250-mL centrifuge bottles and centrifuge at 6000 rpm and 4 °C for 15 min.
5. Discard the supernatant and resuspend the bacterial cell pellet in each bottle (from 200 mL culture) in 10 mL of solution I (50 mM glucose, 10 mM

EDTA, 25 mM Tris-HCL, pH 8.0), add 1 mL of a freshly prepared lysozyme solution at 10 mg mL^{-1} in 10 mM Tris-HCl, pH 8.0, and incubate at room temperature for 5–10 min to digest the cell walls. Note that lysozyme will not work efficiently if the pH of the solution is <8.0.

6. Add 20 mL of freshly prepared solution II (0.2 N NaOH, 1% SDS) per bottle and mix well. The solution should become clear and viscous immediately. Incubate on ice for 5–10 min.
7. Add 15 mL of ice-cold solution III (5 M KOAc, pH 4.8–5.3). Gently invert and swirl to mix the contents. A white precipitate should form immediately. Incubate on ice for 5–10 min. The precipitate that forms consists of chromosomal DNA, high-molecular-weight RNA, and potassium/SDS/protein/membrane complexes.
8. Centrifuge the bacterial lysate at 5000 rpm and 4 °C for 10 min and filter the supernatants in all bottles through four layers of cheesecloth into a clean flask to collect any remaining precipitate.
9. Transfer the supernatant into fresh centrifuge bottles, add 0.6 volume of isopropanol, mix well, and store at room temperature for 10 min.
10. Recover the DNA by centrifugation at 10,000 rpm at room temperature for 10 min.
11. Decant the supernatant carefully and invert the open bottle to allow the last drops of supernatant to drain away. Rinse the pellet and the walls of the bottle with 70% ethanol at room temperature. Drain off the ethanol and place the inverted, open bottle on paper towels for 15–20 min at room temperature to allow all traces of ethanol to evaporate.
12. Dissolve the DNA pellet in 8 mL of TE per 1-L culture.
13. Transfer the DNA solution into a graduated cylinder of known weight and measure the volume of the DNA solution. For every milliliter of DNA solution, add exactly 1 g of solid CsCl. Mix the solution gently until the CsCl salt is completely dissolved.
14. Add 10 mg mL^{-1} ethidium bromide (EB) stock to the DNA/CsCl solution at a rate of 0.8 mL EB per 10 mL DNA solution. Mix well, weigh the DNA/CsCl/EB solution in the cylinder, and adjust the density of the solution to 1.50–1.60 g mL^{-1}.
15. As needed, centrifuge the solution at 8000 rpm at room temperature for 5 min. The furry scum that floats to the top consists of complexes formed between the EB and bacterial proteins.
16. Transfer the clear, red solution to a suitable tube for subsequent ultracentrifugation using a Pasteur pipette, balance the tubes, and seal them according to the manufacturer's instructions.
17. Centrifuge the density gradients at 45,000 rpm for 16 h (VTi 65 rotor), 45,000 rpm for 48 h (Ti50), 60,000 rpm for 24 h (Ti65), or 60,000 rpm for 24 h (Ti70.1) at 20 °C. Two bands of DNA, located in the center of the gradient, should be visible in ordinary light. The upper band, which usually contains less material, consists of linear bacterial chromosomal DNA and nicked circular plasmid DNA, while the lower band consists of closed circular plasmid DNA.

The deep-red pellet on the bottom of the tube consists of ER/RNA complexes. If no bands are observed, further dilute the DNA with TE and prepare and rerun the CsCl gradient as above.

18. Insert a 21-gauge needle into the top of the tube to allow air to enter and collect the lower circular plasmid DNA band with an 18-gauge needle. If necessary, repeat steps 16–18.
19. Extract the plasmid DNA solution with an equal volume of H_2O-saturated isoamyl alcohol four to five times to completely remove the EB in the solution (until no color remains in the sample).
20. Dilute the DNA/CsCl solution with 3–4 volumes of water and precipitate the DNA by adding one volume of isopropanol or two volumes of ethanol and incubating at 4 °C for 15 min, followed by centrifugation at 10,000 rpm and 4 °C for 15 min
21. Discard the supernatant, rinse the DNA pellet with 70 % ethanol twice, air-dry, dissolve the DNA in 0.5 mL TE, and measure the concentration of the DNA.

Vector DNA Digestion

1. Set up the digestion as below and incubate at 37 °C for 2 h.

H_2O	238 µL
Vector DNA	100 µL (5–10 µg)
10× reaction buffer	40 µL
40 mM Spermidine (Spd)	20 µL
10 U μL^{-1} *Bam*HI, *Hind*III, or *Eco*RI	2 µL
Total volume	400 µL

2. Add another 1 µl (10 units) of the restriction enzyme to the digestion and incubate at 37 °C for an additional hour.
3. Transfer the reaction onto ice and check the digestion on a 1 % agarose gel to ensure that the digestion is complete (Figure 8.6).
4. Extract the digest with an equal volume of saturated phenol/chloroform (1:1), and spin at 10,000 rpm at room temperature for 5 min. Use only freshly opened phenol (<1 week) to purify the vector DNA.
5. Precipitate the DNA by adding 1/10 volume of 3 M NaAC, pH 5.2, and 1 volume of isopropanol and incubating at –80 °C for 10 min, followed by centrifugation at 10,000 rpm for 15 min
6. Discard the supernatant, wash the pellet carefully with 70 % ethanol, air-dry, dissolve in 100 µl H_2O, and measure the concentration of the DNA using a series of known concentrations of lambda DNA on a 1 % agarose gel.

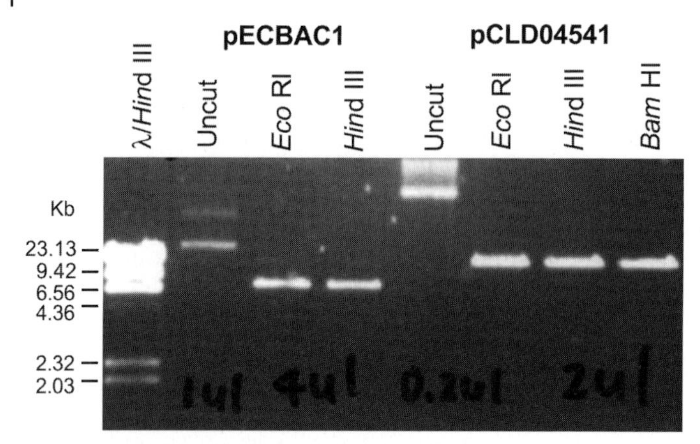

Figure 8.6. Vector preparation. CsCl-gradient-purified vector DNA was digested completely, run on a 1% agarose gel, stained, and photo-documented.

Dephosphorylation of Linearized Vector DNA

1. Set up the dephosphorylation reaction as below and incubate at 37 °C for 30 min.

H_2O	256 µL
Digested vector DNA	100 µL (10 µg)
10× CIAP Reaction buffer	40 µL
1 U μL^{-1} CIAP (BRL Gibco)*	3.89 µL
Total volume	400 µL

(*Add 3.89 U CIAP/10 µg DNA for pBeloBAC11 or pECBAC1, 1.23 U CIAP/10 µg DNA for BIBAC2 and pYLTAC7, and 1.0 U/10 µg DNA for pCLD04541 or pSLJ1711).

2. Stop the reaction immediately by adding 4 µL 0.5 M EDTA, pH 8.0, 20 µL 10% SDS, and 40 µL 1 mg mL^{-1} proteinase K in cold TE to the tube. Incubate at 56 °C for 30 min.
3. Cool down to room temperature and extract once with an equal volume of saturated phenol (freshly opened) and then once with an equal volume of saturated phenol/chloroform/iso-amyl alcohol (25:24:1).
4. Precipitate the DNA by adding 1/10 volume of 3 M NaAC, pH 7.0, and 1 volume of isopropanol and incubating at –80 °C for 10 min followed by centrifugation at 10,000 rpm for 15 min.
5. Discard the supernatant, wash the pellet carefully with 70% ethanol, air-dry, dissolve in 200 µL H_2O, and measure the concentration of the DNA by comparing to lambda DNA of known concentrations on a 1% agarose gel.
6. Adjust the concentration of the DNA to 10 ng μL^{-1} for pBeloBAC11 or pECBAC1 and to 40 ng μL^{-1} for BIBAC2, pYLTAC7, pCLD04541, or pSLJ1711; ali-

quot the DNA into 1.5-mL microtubes; and store in a –20 °C freezer. The vector DNA stored in the –20 °C freezer is good for cloning for at least six months.

7. Optional: Check the dephosphorylation of the vector using the conventional ligation test. When the vector DNA is ligated to digested lambda DNA at a molar rate of 3 vector:1 digested lambda DNA and transformed into *E. coli* DH10 cells

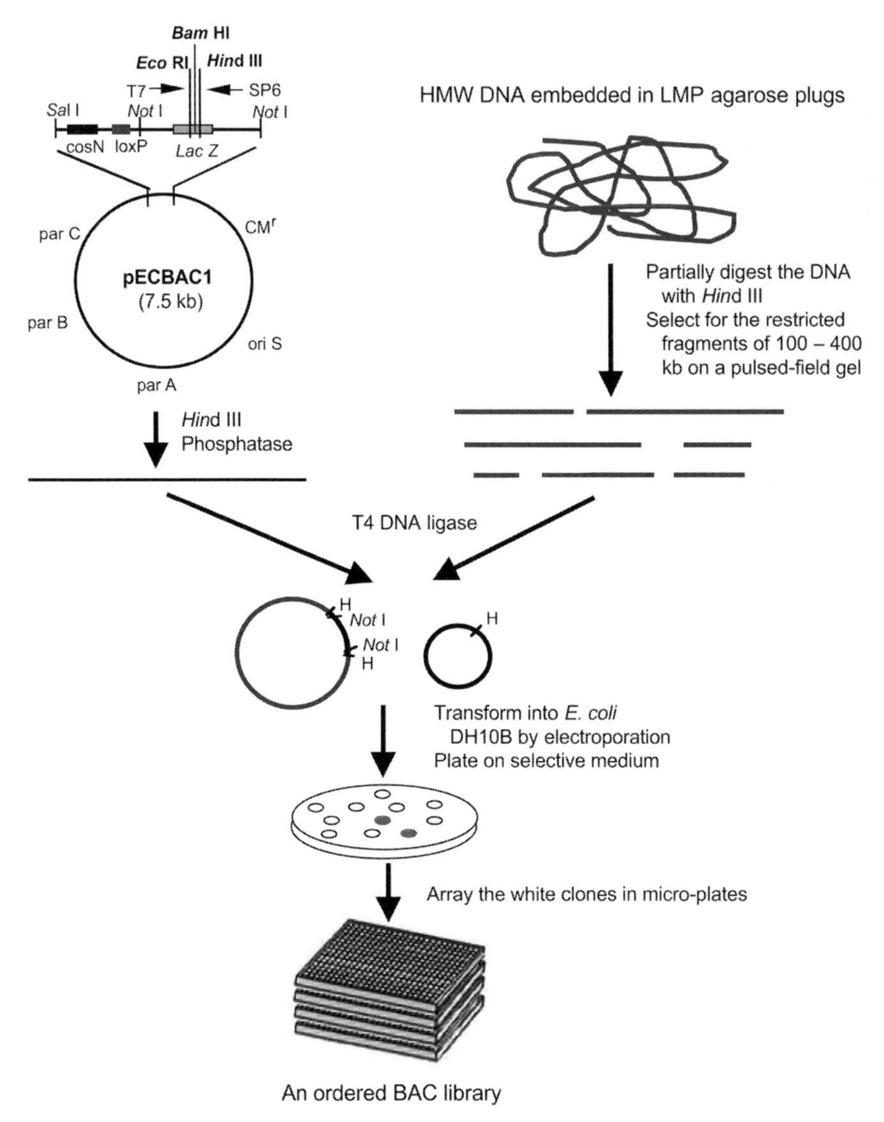

Figure 8.7. A general procedure for construction of bacteria-based large-insert arrayed genomic DNA libraries. Although the construction of a large-insert arrayed BAC library was used for demonstration, the procedure is applicable to construction of all bacteria-based large-insert arrayed genomic libraries, including BAC, PAC, BIBAC, PBC, and TAC libraries.

by electroporation, the percentage of recombinant (white) clones should be greater than 95 %.

8.2.3.3 **Library Construction**

The procedure presented below for bacteria-based large-insert genomic DNA library construction is based on that of Zhang (2000) (http://hbz.tamu.edu – Analysis Tools – BAC Tools) with the latest updates according to Chang et al. (2003) and Tao et al. (2002) (Figure 8.7).

Partial Digestion of HMW DNA Embedded in LMP Agarose Plugs

Generation of clonable and overlapping large DNA fragments is an essential step for large-insert genomic DNA library construction. Such large DNA fragments are usually generated by partial digestion of megabase DNA with a restriction enzyme that cuts relatively frequently within a genome. To achieve the desired partially digested genomic DNA for library construction, at least four methods have been used: (1) varying the concentration of the restriction enzyme, (2) varying the time of digestion, (3) varying the concentration of a restriction enzyme cofactor (e. g., Mg^{2+}), and (4) varying the ratio of the restriction enzyme to the corresponding methylase. In this procedure, we will determine the optimal enzyme concentration for the generation of clonable and overlapping DNA fragments desired for large-insert genomic DNA library construction by varying the concentration of the restriction enzyme.

Pre-purification of the Megabase DNA Embedded in Agarose Plugs

1. Wash the agarose plugs containing HMW DNA stored in 0.05 M EDTA, pH 8.0, or directly from the lysis buffer twice in 10–20 volumes of ice-cold TE, 30 min per wash, and three times in 10–20 volumes of ice-cold TE plus 0.1 mM PMSF, one hour each wash.
2. Further wash the plugs three times for 1 h each in 10–20 volumes of TE on ice.
3. Store the plugs in TE at 4 °C before use. At this stage the plugs can be stored for several months without significant degradation.

Partial Digestion Test

A series of small pilot digestions is commonly carried out to determine the optimal concentration range of the restriction enzyme required to generate DNA fragments desired for library construction before large-scale partial digestion.

1. Cut two to three 100-µL plugs each into nine equal slices using a glass coverslip. Use three slices for each reaction.

2. Equilibrate the DNA slices in the following equilibration buffer in a 15-mL Falcon tube (for six reactions) for 1 h with one change of the buffer after 30 min:

H_2O	867×6 = 5202 μL
Megabase DNA	30 (3/9 plug) × 6 = 180 μL (18 slices)
10× enzyme buffer	100×6 = 600 μL
1 M spermidine	2×6 = 12 μL
1 M DTT	1×6 = 6 μL
Total volume	1000×6 = 6000 uL

3. Set up the digestion buffer as follows in each 1.5-mL Eppendorf tube:

H_2O	137 μL
10× enzyme buffer	17 μL
1 M spermidine	0.34 μL
1 M DTT	0.17 μL
10 mg mL^{-1} BSA	10 μL
Total volume	164.5 μL

4. Transfer three plug slices into each tube and add the restriction enzyme with a concentration series ranging from 0 to 10 units per tube (such as 0, 0.3, 0.6, 1.2, 2.4, 4.8 units per tube) and a volume of 2–10 μL of each. Incubate on ice for 60–120 min to allow the enzyme to access the DNA in the agarose, depending on the porosity of the HMW DNA plugs. The longer incubation on ice would reduce enzyme activity but enhance the partial digestion results.
5. Transfer the digestion reactions into a 37 °C water bath and incubate for exactly 8 min. Immediately transfer the reactions onto ice, stop them by adding 1/10 volume of 0.5 M EDTA, pH 8.0, and then keep on ice.
6. Check the partial digestion results by subjecting the plug slices to pulsed-field gel electrophoresis (PFGE) under the following conditions: 1% agarose in 0.5× TBE, 12.5 °C (cooler setting), 80 (pump setting), 120° angle, 6 V cm^{-1}, initial pulse time of 50 s, and final pulse time of 50 s for 18 h.
7. Stain the gel and photograph (Figure 8.5). The concentration of the enzyme resulting in the majority of the partially restricted fragments ranging in size from 100 kb to 400 kb on the gel is selected for large-scale partial digestion.

Large-scale Partial Digestion

Large-scale partial digestion should be conducted under conditions identical to those used in the above small-scale partial digestion test. The number of 100-μL plugs to be used for the large-scale partial digestion is dependent on the concentration of the DNA per plug. In general, 6–10 100-μL plugs should be sufficient to construct a bacteria-based large-insert genomic DNA library for most species.

1. Make 2×18,000 μL equilibration buffer for 18 reactions according to the ratio listed above.
2. Take six 100-μL HMW DNA plugs, cut each into nine slices, and incubate in 18,000 μL equilibration buffer for 1 h, with one change of the buffer after 30 min.
3. Make 20×170 μL = 3400 μL digestion buffer as above and dilute the restriction enzyme used in the partial digestion test, using the digestion buffer so that 2–10 μL of the dilution containing the optimal units of the enzyme selected in the partial digestion test is added to each microtube (see step 5).
4. Transfer the plug slices into 18 1.5-mL Eppendorf tube, three slices per tube, and then add 164.5 μL of the digestion buffer per tube.
5. Add the diluted restriction enzyme to each tube at 2–10 μL per tube containing the optimal amount of enzyme determined by the partial digestion test above.
6. Incubate on ice exactly as long as in the pilot experiments and then transfer the tubes of digestion reactions into a 37 °C water bath and incubate for exactly 8 min. Immediately transfer the reactions onto ice, stop them by adding 1/10 volume of 0.5 M EDTA, pH 8.0, and keep the reactions on ice.

DNA Fragment Size Selection

After partial digestion, the DNA fragments must be selected to obtain the fragments that have a desirable size range for library construction and to remove the smaller fragments that can compete more effectively for vector ends. Size selection is performed one or two times by PFGE (Figure 8.8).

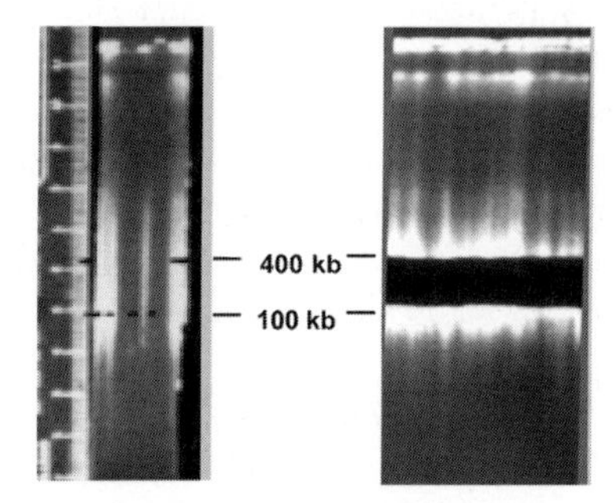

A. After the first size selection

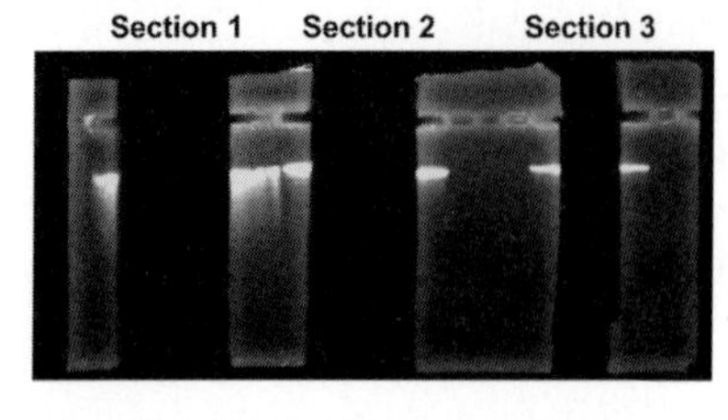

B. After the second size selection

Figure 8.8. Size selection for optimal partially restricted DNA fragments on pulsed-field gels for construction of large-insert clone libraries. Section 1 was from the 100–200 kb zone, section 2 from the 200–300 kb zone, and section 3 from the 300–400 kb zone. Frequently, section 2 gives clones having desirable insert sizes and generates targeted libraries.

First Size Selection

1. Prepare a 1% agarose PFGE gel in 0.5× TBE (45 mM Tris, 45 mM boric acid, 1 mM EDTA, pH 8.3) with a sample trough at the top portion of the gel and 2 L of 0.5× TBE running buffer, and precool the buffer to 11 °C in the electrophoresis chamber.
2. Load the plug slices containing the partially digested DNA into the sample trough, and size markers (lambda concatemers and/or yeast chromosomes) in the wells on both sides of the trough. Seal the sample trough and marker wells with 1% molten agarose gel kept in a 65 °C water bath. Run the PFGE using the CHEF DRIII system at 6.0 V cm^{-1}, 11 °C, 120° angle, and initial and final pulse times of 90 s in 0.5× TBE for 16 h.
3. Cut both the sides of the gel containing the size markers and the 0.5-cm edge of the DNA sample with a clean razor blade and keep the central portion of the gel on the gel plate on ice.
4. Stain and photograph the gel pieces containing the samples adjacent to the size markers along with a ruler. Do not stain the central portion of the gel containing the majority of the sample.
5. Using the photograph as a reference, locate the gel zone containing the DNA fragments ranging from 100 kbp to 400 kbp in the central portion of the gel kept on the ice, excise it, and cut it into three sections horizontally, 100–200 kb, 200–300 kb, and 300–400 kb, roughly 0.5–0.6 cm in width per section (Figure 8.8A). Note that the width of the gel section can be used to adjust the volume of the electroeluted DNA solution. If the section is too wide, it will be difficult to insert the gel section into the dialysis tubing, whereas if it is too small, the volume of the buffer added to the dialysis tubing will be large. As a result, the concentration of the DNA will be low.
6. Insert each gel section containing the desirable size DNA fragments into a piece of dialysis tubing (1/4 inch in diameter and with a molecular weight exclusion limit of 12,000–14,000 Da) that was prerinsed with 0.5× TBE, close one end of the tubing with a tubing closure, fill the tubing with 0.5× TBE, remove all bubbles in the tubing, and close the other end of the tubing with the second tubing closure. Recover the size-selected DNA by electroelution in 0.5× TBE using the CHEF DRIII at 6 V cm^{-1}, 11 °C, and 35-s switch time for 4 h, followed by turning the dialysis tubing 180° in the CHEF apparatus and continuing the electrophoresis for 60 s in 0.5× TBE to release the DNA from the tubing wall. The eluted DNA fragments can be used for ligation and library construction after dialysis in 0.5× TE (see below) or can be carefully transferred from the dialysis tubing to a microfuge tube with a wide-bore pipette tip for the second size selection.

Second Size Selection

1. Prepare a 1% agarose PFGE gel in 0.5× TBE with a sample trough at the top portion of the gel and 2 L of 0.5× TBE running buffer, and precool the buffer to 11 °C in the electrophoresis chamber.

2. Load λ ladder size markers on both sides of the sample trough, seal the marker wells with 1% molten agarose gel as above, and place them in the electrophoresis chamber containing 0.5× TBE buffer.
3. Mix the three sections of the eluted DNA fragments from the first size selection with the gel-loading dye by inverting the tube gently, and load it into the sample trough. Run the gel at 4 V cm^{-1}, 11 °C, and 5-s switch time in 0.5× TBE for 6–8 h.
4. Cut the gel portion containing the size markers and 0.5-cm of the DNA sample, stain it in EB solution, and mark the gel zone containing the compressed DNA fragments (Figure 8.8B) on the stained portion of the gel. During this process, the gel portion containing the majority of the DNA samples should be kept on ice as above.
5. Locate the DNA sample compressed on the gel using the marks made on the stained portion of the gel, excise the gel section containing >100-kb DNA fragments, and electroelute the DNA from the gel as above.
6. Dialyze the DNA fragments electroeluted from the gel in the same dialysis tubing against 1 L of 0.5× TE for at least 3 h at 4 °C with one change of 0.5× TE every hour.
7. Carefully collect the DNA from the dialysis tubing to a 1.5-mL microtube with a wide-bore pipette tip. The immediate ligation of the DNA into a vector following dialysis often gives a much better result for library construction

Ligation of the Large DNA Fragments Into the Cloning Vector

1. Estimate the concentration of the insert DNA on a 1% conventional agarose mini-gel using the known concentrations of λ DNA as standards.
2. Set up the ligation mixture using the following criteria: molar ratio of vector:-insert = 4–10:1; final concentration of insert DNA in the ligation mixture = 1~2 ng μL^{-1}; and T4 DNA ligase = 1 unit per 50 µL ligation. To control the ratio between recombinant and non-recombinant clones in the transformants, the molar ratio of vector:insert can be adjusted, with a smaller ratio for a higher recombinant/non-recombinant clone ratio. However, this may lead to a smaller total number of recombinant clones.
3. Incubate the ligation at 16 °C for 6–10 h. The ligation can immediately be used for transformation or be aliquoted and stored in a frost-free –20 °C freezer. The ligation can be stored for several months without significant problems. Although it was observed that storage of the ligated DNA might slightly reduce the transformation efficiency, it did not significantly reduce the insert sizes of the clones.

Transformation of the Recombinant DNA Constructs (Ligation) into *E. coli* Host Cells

The electroporation method and *E. coli* strain DH10B (Hanahan et al. 1991) are used for transformation, as discussed in the Introduction.

1. Mix 1–2 µL of the ligation mixture with 20 µL of *E. coli* DH10B electroporation-competent cells.

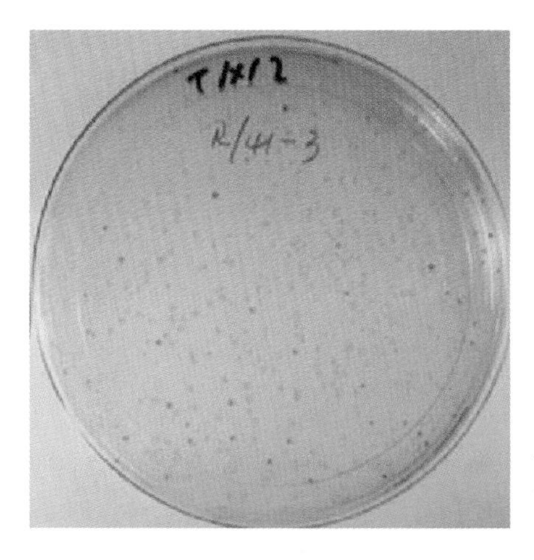

Figure 8.9. Plant-transformation-competent BIBACs cloned in the PBC vector pCLD04541. Blue colonies indicate clones with no inserts (non-recombinants) and the colorless colonies indicate clones with inserts (recombinants).

2. Transform into the cells using the cell porator and voltage booster electroporation system (now available from Labrepco) at the following settings:

Cell porator		Voltage booster	
Voltage	375 V	Resistance	4 kΩ
Capacitance	330 μF		
Impedance	Low ohms		
Charge rate	Fast		

3. Immediately transfer the electroporated cells into 1 mL of SOC medium and incubate at 37 °C and 250 rpm for 1 h to allow the antibiotics resistance gene to be fully expressed.
4. Plate the transformed cells onto LB agar medium containing appropriate antibiotics and/or amendments for transformant and recombinant selection.
5. For vectors such as pBeloBAC 11, pECBAC1, and pCLD04541, which use the *Lac*Z gene for recombinant selection, incubate at 37 °C for 36 h to allow the colony color (blue) to fully develop; for vectors such as BIBAC2 and pYLTAC7, which use the *Sac*B gene for recombinant selection, incubate at 37 °C for 20 h. Only the colorless (white) colonies constructed with the *Lac* Z gene-containing vectors are expected to contain inserts (recombinants) (Figure 8.9), whereas *all* of the clones of the *Sac*B gene-containing vectors growing on the selective medium are expected to contain inserts.

Characterization of Bacteria-based Large-insert Genomic DNA Libraries

Analysis of the clones resulting from a particular ligation is essential before large-scale transformation and library assembly. The parameters usually used for this purpose include average insert size, insert size distribution, and the percentage of empty clones. In general, a large-insert genomic DNA library desirable for genome research should have a large average insert size (≥150 kb), a low percentage of insert-empty clones (≤5 %), and a reasonable distribution of clone insert size (providing better genome coverage). Therefore, at least 100 randomly selected clones obtained from each ligation should be analyzed. Figure 8.2 shows a picture of PFGE to check the insert sizes of wheat libraries.

1. Inoculate the recombinant clones in 5 mL of LB medium with appropriate antibiotics and incubate at 37 °C with shaking at 250 rpm overnight.
2. Isolate DNA using the alkaline lysis method used for conventional plasmid DNA isolation.
3. Digest with *Not* I, a rare cutting enzyme.
4. Run on a 1% pulsed-field gel in 0.5× TBE at 6 V cm^{-1}, 11 °C, 5-s initial switch time, and 15-s final switch time, with a linear ramp for 16 h. the lambda DNA ladder is often used as a size marker.
5. Stain the gel with EB, photograph it, and estimate the insert size of each clone by adding the sizes of all DNA bands derived from the clone insert. Count the number of clones with no inserts.

Storage of Bacteria-based Large-insert Genomic DNA Libraries

1. Select the ligation that gives the largest insert sizes, fewest insert-empty clones, and highest transformation efficiency for large-scale transformation and library assembly.
2. Array the recombinant clones into 384-well microtiter plates containing 50 μL of freezing medium (LB broth plus 1× freezing buffer [36 mM K_2HPO_4, 13.2 mM KH_2PO4, 1.7 mM citrate, 0.4 mM $MgSO_4$, 6.8 mM $(NH_4)_2SO_4$, 4.4% (v/v) glycerol]) per well with appropriate antibiotics, grow at 37 °C overnight, and store at –80 °C for long-term maintenance. The library can be stored in the buffer at –80 °C for at least five years without significant problems.
3. Usually, the library is duplicated into three copies and stored in separate –80 °C freezers to guard against unforeseen disasters. The three copies are designated the working copy, the master copy, and the backup copy.

8.2.4
Applications of Bacteria-based Large-insert Genomic DNA Libraries in Genome Physical Mapping

Genomic libraries, especially bacteria-based large-insert genomic DNA libraries, have been developed for many of the economically important plant, animal and insect species. Most of these libraries have been made available to the public and can be accessed through the following websites:

- Texas A&M University GENE*finder* Genomic Resources: http://hbz.tamu.edu
- Caltech Genome Research Laboratory: http://www.tree.caltech.edu/
- Children's Hospital Oakland Research Institute: http://bacpac.chori.org
- Invitrogen Life Technologies: http://mp.invitrogen.com/

Genomic DNA libraries, especially the bacteria-based large-insert cloning libraries, are essential for physical mapping. In addition, they are also crucial tools for genomic, genetic, and molecular biological research in many aspects, including gene discovery and cloning, comparative and functional genomics research, gene transformation and expression analysis, etc. This chapter will only provide examples of applications of large-insert genomic DNA libraries for genome physical mapping. It is worthwhile to note that a combination of different types of libraries and different techniques are usually applied to a physical mapping project to obtain a high-quality map. For example, in the physical mapping projects of *Arabidopsis* (Marra et al. 1999; Chang et al. 2001), rice (Tao et al. 2001; Chen et al. 2002; Li et al. 2003), soybean (Wu et al. 2004), human (IHGMC 2001), mouse (Gregory et al. 2002), and chicken (Ren et al. 2003), in addition to random clone fingerprinting and contig assembly, a large number of genetic markers have been used to verify the map contig accuracy and to anchor the physical map contigs onto genetic linkage maps, allowing the physical map contigs to merge into larger contigs. The clone end sequences have also been used to merge map contigs by sequence alignment along related genome sequences (IHGMC 2001; Chen et al. 2002; Gregory et al. 2002; Li et al. 2003). In addition, YAC clones are sometimes used to bridge the gaps of map contigs generated by BAC and/or PAC clones (IHGMC 2001).

8.2.4.1 **Physical Mapping by Fingerprint Analysis**

The clone-based physical map is a physical map consisting of continuously overlapping genomic DNA clones that are ordered to their chromosomes of origin on the basis of overlapping of adjacent clones. Several approaches have been developed to construct genome-wide physical maps from large-insert genomic DNA libraries. The most common technique is fingerprint analysis.

A fingerprint of a genomic DNA clone is the unique restriction profile (bands) of the clone. It is created by digesting the clone DNA with one or more restriction enzymes and fractionating the restricted fragments on gel matrices or through capillaries. Since clones derived from the same regions of a genome share similar restriction patterns, analysis of clone fingerprints with computer programs will allow overlapping clones in a genomic region to be assembled into contigs or

Table 8.1. Summary of whole-genome physical maps of higher organisms generated by fingerprint analysis of bacteria-based large-insert clones.

Species	*Genome size (Mb)*	*Fingerprinted clone coverage*	*Reference*
Arabidopsis	130	18×	Marra et al. 1999
Arabidopsis	130	7.2×	Chang et al. 2001
Chicken	1140	7.9×	Ren et al. 2003
Human	3200	17×	IHGMC 2001
Indica rice	430	6.9×	Tao et al. 2001
Japonica rice	430	25.6×	Chen et al. 2002
Japonica rice	430	7.2×	Li et al. 2003
Mouse	2800	NA×	Gregory et al. 2002
Soybean	1100	9.6×	Wu et al. 2004

clone groups according to the similarity of the clone fingerprints. The advent of bacterial-based large-insert genomic DNA libraries has facilitated application of the clone fingerprint analysis strategy in the development of physical maps of large, complex genomes. To facilitate genome physical mapping with bacteria-based large-insert genomic DNA libraries, several fingerprinting methods have been developed; these methods are discussed in Chapter 6. Genome-wide physical maps have been developed for several plant and animal species by fingerprint analysis of bacteria-based large-insert clones (Table 8.1).

8.2.4.2 Physical Mapping by Fluorescent *in situ* Hybridization

(This topic is discussed in more detail in Chapter 6.) The technology of *in situ* hybridization of specific nucleotide sequences to chromosomes was developed over 30 years ago (Pardue and Gall 1969; John et al. 1969) and evolved to fluorescence *in situ* hybridization (FISH) in the early 1980s (Bauman et al. 1980). A wide range of probes has been employed for FISH experiments, including species-specific total genomic DNA and chromosome-specific DNA, tandem repeats (α-satellite, β-satellite, and telomere probes), interspersed repetitive sequences, library clones, cDNA, and RNA probes. Probes smaller than several kilobase pairs are difficult to detect by FISH, while probes of large DNA fragments from genomic DNA clones can obtain excellent results. Due to instability and difficulty in DNA isolation, YACs have limitations as probes for FISH. However, bacterial-based large-insert DNA clones have been widely used as probes in FISH for various research projects.

BAC clones have been used to link cytogenetic markers to DNA sequences. Some human genetic diseases are caused by chromosome rearrangements (abnormalities) and have been cytogenetically marked on chromosomes. The sequences in the marker areas are particularly important to further research of the diseases. The BAC Resource Consortium (2001) of the human genome sequencing project is working on a project to connect the cytogenetic marker maps to sequence maps via BAC clones using the BAC-FISH approach. The BAC probes used for in situ hybridization contain one or more unique sequence tags that allow each BAC clone to be positioned on the human genome sequence maps. In 2001, more than 8000 BAC clones were mapped to 23 of the 24 human chromosomes, resulting in an average of at least one clone per megabase of DNA (BAC Resource Consortium 2001; Korenberg et al. 1999). This integrated BAC-FISH sequence map can now be used to rapidly identify mutated genes affected by chromosomal rearrangements seen in genetic disorders and cancers.

8.2.4.3 Physical Mapping by Optical Mapping

The restriction sites of random clones can also be examined directly by looking at the restricted DNA molecules under a microscope, a technique called optical mapping. Since the development of optical mapping by Schwartz et al. (1993), the techniques have been improved and automated (Jing et al. 1998). Cloned large-insert

DNA molecules are isolated and mounted to a chemically treated glass slide surface, where the DNA molecules stretch out and become linear fibers by fluid fixation. The fixed DNA fibers are digested with a certain restriction enzyme, followed by staining with fluorescent dye. The restricted sites produce nicks along the linear DNA and can be observed using a fluorescent light microscope. The images are then collected using a cooled charge-coupled device digital camera. The restriction sites and the produced fragment lengths produced are unique for each clone. By analysis of restricted fragment lengths of each clone, significant overlapping between clones implies neighboring clones, and hence overlapping clone contigs are constructed.

One major advantage of optical mapping as a tool for physical mapping over fingerprinting methods is that optical mapping can provide information for all of the fragment lengths of a clone digested by a restriction enzyme as well as their orders along the linear DNA molecule. However, the application of this technology is not as common as restriction fingerprinting methods for physical mapping, especially not for whole-genome physical mapping of higher organisms. The major application of optical mapping is to construct physical maps as a scaffold for accurately aligning sequence contigs so that the sequence gaps can be bridged or the sequence maps can be verified. For instance, physical maps prepared by optical mapping were constructed based on the BAC libraries for human chromosomes 11, 22 (Cai et al. 1998), and Y (Giacalone et al. 2000) and used for the human genome sequencing project. For microbes with small genome sizes, optical mapping can be conducted directly on shotgun sheared-DNA fragments from living cells, without the need for constructing genomic DNA libraries. Such physical maps have been constructed for a human pathogenic bacterium (Zhou et al. 2002), *E. coli* (Lim et al. 2001), and other microbes and have played roles in their genome sequencing.

8.2.4.4 **Physical Mapping by Iterative Hybridization**

Iterative hybridization was first developed by Hoheisel et al. (1993) and applied to generate high-resolution cosmid clone– and P1 clone–based physical maps of the fission yeast *Schizosaccharomyces pombe*. Further application was reported in *Arabidopsis* physical mapping by Mozo et al. (1998, 1999). By this approach, a batch of randomly selected BAC clones (e. g., 100 clones) is used as a probe to hybridize the whole BAC library spotted on membranes, and the positive clones are localized. A second batch of clones is randomly selected from the clones that are not detected by the first round of hybridization and used as the probe to conduct the second round of hybridization to the whole genomic DNA BAC library. Then the subsequent rounds of hybridization are conducted continually in the same way until 99.5 % of the clones of the whole library have been hit. The clone contigs are then assembled by a probe-probe comparison based on the hybridization frequencies using a computer program. However, the success of this approach is significantly influenced by an abundance of repetitive sequences in the genome (Marra et al. 1999), implying that it is difficult to generate a genome-wide physical map of species having large, complex genomes using this approach (Chumakov et al. 1995).

8.2.4.5 Physical Mapping by Other Techniques

Chromosome Walking

Chromosome walking was the first method devised for construction of the clone contigs of physical maps. A genomic DNA library is screened with a DNA probe of known location or by PCR using a primer pair designed from a particular sequence, and a set of positive clones is isolated. This is the first step of chromosome walking. The insert ends of the outermost clones of the first set are isolated and used as probes to screen the library again. The second set of clones is then isolated. This process continues until clones spanning the entire genomic region are isolated and ordered. Chromosome walking has been used mainly to construct overlapping clone contigs from a DNA marker to a gene of interest and to isolate the gene by positional cloning. A major problem with chromosome walking is that if a probe contains a repeat sequence, it will hybridize to the clones located in different genomic regions. To solve this problem, only the DNA fragments that are single-copy are used as probes in library screening for chromosome walking. Therefore, chromosome walking is laborious, and it is rarely possible to construct contigs for an extensive region of a large, complex genome.

DNA Marker-based Chromosome Landing

DNA marker-based chromosome landing is a hybridization- and/or PCR-based strategy to land clones or contigs on a genetic linkage map or to integrate a physical map with a genetic linkage map. DNA markers are selected from a molecular genetic linkage map and used as probes or PCR primers to screen a large-insert genomic DNA library. Overlapping clones are assembled into contigs, and the contigs are extended and anchored to the genetic linkage map according to the position of the DNA markers on the genetic map. This physical mapping strategy was first applied to develop large-insert YAC-based physical maps of human (Chumakov et al. 1995) and has been widely used in the development of YAC- or BAC-based physical maps of *Arabidopsis*, rice, mouse, chicken, and the euchromatic portion of the *Drosophila* genome.

To rapidly screen the libraries more efficiently with a large number of probes, the pooling strategy may be adopted (Chang et al. 2001; Tao et al. 2001). If PCR screening is used with PCR-based markers, such as sequence-tagged sites (STSs) and simple sequence repeats (SSRs), the clones of the genomic DNA library are fabricated into clone pools, including multiple-plate superpools, single-plate pools, column pools, and row pools (three-dimension pools). Alternatively, if hybridization screening is used with the probes of PCR products (e. g., STSs and SSRs) or with the probes of restriction fragment length polymorphisms (RFLPs) or small-insert DNA clones, the marker probes can be pooled to screen the library clones gridded onto high-density colony filters (Chang et al. 2001; Tao et al. 2001).

The probes used for hybridization screening can also be made from overlapping oligos (overgos) designed according to the known sequences of the probes obtained by bioinformatics tools. Each overgo consists of a pair of oligos of 22~24 nucleotides, with a complementary 3′ end to each other of 8 bases. Thus, the total length of each pair of oligonucleotides spans 36~40 bases. The pair of oligonucleotides is radioactively labeled by a Klenow fill-in reaction, producing 36~40 double-stranded DNA fragments that are used as probes to screen the library clones. Overgo screening has many advantages over clone probe screening. The most important is that the pair of oligonucleotide probes is a single-copy sequence of the genome because it is designed based on marker sequence information, which minimizes the influence from repetitive sequences. Secondly, overgo probes have high hybridization specificities due to their single-copy and perfect hybridization features, hence increasing the accuracy. Again, overgo hybridization shows strong signals and low background because the overgo probes are labeled by a Klenow fill-in reaction. Finally, these short probes (36~40 bases) are easily washed off the library clone filters; thus the membrane may be reused several times without increasing the background. Nevertheless, the cost for synthesis of the overgos should be considered.

A significant advantage of DNA marker-based chromosome landing for physical mapping is the simultaneous development and integration of the physical map with linkage genetic maps. The integration of a physical map with a genetic map associates the large-insert clones of the physical map with useful genes or traits mapped on the genetic map, which will greatly expedite isolation of the genes by positional cloning. However, it is prerequisite to have a high-density, evenly distributed DNA marker genetic map. Although high-density DNA marker maps have been developed for the genomes of human and several model species, they are not available for many of the plant and animal species of economic importance.

Facilitating Genome Sequencing

Genome sequencing is a powerful tool for large-scale gene discovery and decoding. The advent of bacteria-based large-insert clones has accelerated by several-fold the process of large-scale sequencing of large, complex genomes such as *Arabidopsis*, *Drosophila*, human, rice, and other plant and animal species of economic importance. Two sequencing strategies are being used: the clone-by-clone shotgun and the whole-genome shotgun. It seems evident that bacteria-based large-insert clones are indispensable for both of these methods (Zhang and Wu 2001).

Using the clone-by-clone shotgun strategy, a clone-based physical map is first constructed. Minimally overlapping (also called tiling path) clones that completely cover the genome are selected from the physical map and used as substrates for shotgun sequencing, one clone at a time. The sequencing data of shotgun clones are assembled into the sequences of each large-insert clone, which is then used to assemble the sequences of the whole genome. A typical example in plants is from the genome sequencing project of *Arabidopsis*, which started in 1996 and was completed at the end of 2000 (Arabidopsis Genome Initiative 2000). In the project, BAC-, cosmid-, TAC-, and P1 clone–based physical maps were constructed for the five chromosomes of the genome, respectively, by the random clone fingerprinting method with the assistance of DNA marker landing. The clones in tiling paths were selected as sequencing substrates. After assembly, a sequence map of 115,409,949 bp was obtained with 99.99% accuracy, which covers 92% of the whole genome. Using the same approach, the human genome was also sequenced (IHGMC 2001).

In whole-genome shotgun sequencing, high-quality genomic DNA is purified, physically sheared, or partially digested and then cloned into plasmid vectors with different insert sizes, including bacteria-based large-insert genomic DNA clones. The shotgun library clones with small inserts (1.0–2.0 kb) are used for draft sequence production and sequence contig assembly. The further extension of the sequence contigs and linkage of the contigs into scaffolds require large-insert clones as the connectors. Usually, both ends of each selected large-insert clone are sequenced and the clone end-sequence data are aligned with sequence contigs by sequence match. In this way, adjacent sequence contigs can be merged into larger contigs, whereas uncontinuous contigs can be made into scaffolds. Gaps are thus bridged. In 2002, the draft sequences of both cultivated rice subspecies, indica and japonica, were finished using the whole-genome shotgun strategy, and a large number of sequence contigs and scaffolds have been generated. Further accomplishment of the sequence maps requires filling in the gaps between contigs.

8.3 Discussion

Bacteria-based large-insert genomic DNA libraries have been indispensable resources and tools for many aspects of genomics, genetics, and molecular biology and are expected to play more significant roles in future research. For whole-genome physical mapping, it has been proved that large-insert genomic DNA libraries constructed with two or more restriction enzymes having different G/C or A/T contents in recognition sites will facilitate construction of high-quality contig physical maps. This is because such libraries represent the whole genome more evenly. Binary vectors such as BIBAC2, pCLD04541, and pYLTAC7 not only are capable of cloning large DNA fragments, as do the BAC and PAC vectors, but also are competent for direct transformation in plants. The latter feature will facilitate functional analysis of the genome sequence in the post-genome sequencing era and its utility in plant molecular breeding. Therefore, incorporation of clones constructed in binary vectors into the construction of physical maps will enhance the utility of the resultant physical maps in gene cloning and functional analysis and engineering of genome sequences.

Genomic DNA libraries with larger insert sizes are much more efficient tools for genome physical mapping and sequencing. The current bacteria-based large-insert cloning system has allowed generation of genomic DNA libraries with average insert sizes ranging from 100 kb to 200 kb, although it has been demonstrated that this cloning system is capable of cloning and stably maintaining DNA fragments of up to at least 350 kb. The construction of libraries with average insert sizes greater than 150 kb is still restricted to a few laboratories worldwide. Therefore, further improvement in the techniques of cloning large DNA fragments in bacteria will be of significance for comprehensive genome research. It is the electroporation transformation technology that has increased the cloning capacity of the conventional plasmid-based cloning vectors from ca. 20 kb to 350 kb. Therefore, if the existing electroporation transformation technique could be further improved or if a novel method is developed, the insert sizes of genomic DNA libraries may be further increased significantly by the bacteria-based large-insert cloning system. Consequently, development of whole-genome physical maps of high quality will be enhanced greatly.

Based on many earlier publications on large DNA fragment cloning in bacteria (e. g., Kim et al. 1992; Shizuya et al. 1992; Ioannou et al. 1994; Hamilton 1997; Liu et al. 1999), it seems that the single-copy status of large-insert clones in a host cell was required for their stable maintenance in the host cell. Nevertheless, Tao and Zhang (1998) reported that the large-insert clones of multiple-copy (5–8) conventional binary plasmid (pSLJ1711) and binary cosmid (pCLD04541) were stable in the bacterial strain DH10B as the single-copy BACs, PACs, and BIBACs. Ioannou et al. (1994) also observed that large-insert PACs were stable in the bacterial strain DH10B after they were induced into multiple copies (5–8) per cell. Furthermore, when the BIBACs (Hamilton 1997; Song et al. 2003) and TACs (Liu et al. 1999; Song et al. 2002) were transformed into different *Agrobacterium* strains for genetic transformation (although they were strictly controlled to a single-copy in the *Agrobacterium* cells), they were stable in some of the strains but not in others. Similar results were obtained when the multiple-copy large-insert pCLD04541 clones were transformed into different *Agrobacterium* strains (Wu et al. 2000; Men et al. 2001). These results disagreed with the concept that the stability of large-insert clones is dependent on their single-copy status in the bacterial host cells. Therefore, further studies are needed to determine the key factors that influence the stability of large-insert clones in host cells. In comparison, vectors with a few copies per host cell have advantages over single-copy vectors for cloned DNA isolation and thus will greatly enhance large-scale analysis of cloned DNA for whole-genome physical mapping and genome sequencing.

Acknowledgments

This article was supported in part by the USDA/CSREES National Research Initiative (award numbers 99-35205-8566 and 2001-52100-11225), the National Science Foundation Plant Genome Program (award numbers 9872635 and 0077766), the USDA/BARD (award number US-3034-98R), and the Texas Agricultural Experiment Station (Acc. #8536-203104).

References

Arabidopsis Genome Initiative (2000) Analysis of the genome sequence of the flowering plant *Arabidopsis thaliana*. *Nature* 408: 796–815.

BAC Resource Consortium (2001) Integration of cytogenetic landmarks into the draft sequence of the human genome. *Nature*. 409: 953–958.

Bauman, J. G. J., J. Wiegant, P. Borst and P. van Duijn (1980) A new method for fluorescence microscopical localization of specific DNA sequences by *in situ* hybridization of fluorochrome labeled RNA. *Exp. Cell Res.* 138:485–490.

Burke, D. T., G. F. Carle and M. V. Olson (1987) Cloning of large segments of exogenous DNA into yeast by means of artificial chromosome vectors. *Science* 236:806–812.

Cai, W., J. Jing, B. Irvin, L. Ohler, E. Rose et al. (1998) High-resolution restriction maps of bacterial artificial chromosomes constructed by optical mapping. *Proc. Natl. Acad. Sci. USA* 95:3390–3395.

Canilleri, C., J. Lafleuriel, C. Macadre, F. Varoquaux, Y. Parmentier, G. Picard, M. Caboche and D. Bouchez (1998) A YAC contig map of *Arabidopsis thaliana* chromosome 3. *Plant J.* 14:633–42.

Chang, Y. L., Q. Tao, C. Scheuring, K. Ding, K. Meksem and H.-B. Zhang (2001) An integrated map of *Arabidopsis thaliana* for functional analysis of its genome sequence. *Genetics* 159:1231–1242.

Chang, Y. L., X. Henriquez, D. Preuss, G. P. Copenhaver and H.-B. Zhang (2003) A plant transformation-competent binary BAC (BIBAC) library from the *Arabidopsis thaliana* Landsberg ecotype for functional and comparative genomics. *Theor. Appl. Genet.* 106: 269–276.

Chen, M., G. Presting, W. G. Barbazuk, J. L. Goicoechea, B. Blackmon et al. (2002) An integrated physical and genetic map of the rice genome. *Plant Cell* 14:537–545.

Chumakov, I. M., P. Rigault, I. Le Gall, C. Bellanne-Chantelot, A. Billault et al. (1995) *Nature* 377:157–298.

Collins, J. and B. Hohn (1978) Cosmids: a type of plasmid gene-cloning vector that is packageable *in vitro* in bacteriophage λ heads. *Proc. Natl. Acad. Sci.* 75:4242–4246.

Ding, Y., M. D. Johnson, R. Colayco, Y. J. Chen, J. Melnyk, H. Schmitt and H. Shizuya (1999) Contig assembly of bacterial artificial chromosome clones through multiplexed fluorescence-labeled fingerprinting. *Genomics* 56:237–246.

Ditta, G., S. Stanfield, D. Corbin and D. R. Helinski (1980) Broad host range DNA cloning system for Gram-negative bacteria: Construction of a gene bank of *Rhizobium meliloti*. *Proc. Natl. Acad. Sci. USA* 77:7347–7351.

Figurski, D. H. and D. R. Helinski (1979) Replication of an origin-containing derivative of plasmid RK2 dependent on a plasmid function provided *in trans*. *Proc. Natl. Acad. Sci. USA* 76:1648–1652.

Frijters, A. C.J., Z. Zhang, M. van Damme, G. L. Wang, P. C. Ronald and R. W. Michelmore (1997) Construction of a bacterial artificial chromosome library containing large *Eco*RI and *Hind* III genomic fragments of lettuce. *Theor. Appl. Genet.* 94:390–399.

Giacalone, J., S. Delobette, V. Gibaja, L. Ni, Y. Skiadas et al. (2000) Optical mapping of BAC clones from the human Y chromosome DAZ locus. *Genome Res.* 10:1421–1429.

Green, E. D., P. Hieter and F. A. Spencer (1999) Yeast artificial chromosomes. *In: Genome Analysis–A Laboratory Manual (eds Birren, B.,E. D. Green, S. Klapholz, R. M. Myers, H. Riethman and J. Roskams), Cold Spring Harbor Laboratory Press. Vol. 3*, pp. 297–565.

Gregory, S. G., M. Sekhon, J. Schein, S. Zhao, K. Osoegawa et al. (2002) A physical map of the mouse genome. *Nature* 418:743–750.

Hamilton, C. M. (1997) A binary-BAC system for plant transformation with high-molecular-weight DNA. *Gene* 200:107–116.

Hamilton, C. M., A. Frary, C. Lewis and S. D. Tanksley (1996) Stable transfer of intact high molecular weight DNA into plant chromosomes. *Proc. Natl. Acad. Sci. USA* 93:9975–9979.

Hamilton, C. M., A. Frary, Y. Xu, S. D. Tanksley and H.-B. Zhang (1999) Construction of tomato genomic DNA libraries in a binary-BAC (BIBAC) vector. *Plant J.* 18:223–229.

Hanahan, D., J. Jersee and F. R. Bloom (1991) Plasmid transformation of *Escherichia coli* and other bacteria. *Methods Enzymol.* 204:53–113.

Hoheisel, J. D., E. Maier, R. Mott, L. McCarthy, A. V. Grigoriev, L. C. Schalkwyk, D. Nizetic, F. Francis and H. Lehrach (1993) High resolution cosmid and P1 maps spanning the 14 Mb genome of the fission yeast *S. pombe*. *Cell* 73:109–120

Hwang, I., T. Kohchi, B. M. Hauge and J. M. Goodman (1991) Identification and map position of YAC clones comprising one-third of the *Arabidopsis* genome. *Plant J.* 1:367–374.

(IHGMC) International Human Genome Mapping Consortium (2001) A physical map of the human genome. *Nature* 409:934–941.

Ioannou, P. A., C. T. Amemiya, J. Garnes, P. M. Kroisel, H. Shizuya, C. Chen, M. A. Batzer and P. J. de Jong (1994) A new bacteriophage P1-derived vector for the propagation of large human DNA fragments. *Nature Genet.* 6:84–89.

Jing, J., J. Reed, J. Huang, X. Hu, V. Clarke et al. (1998) High resolution optical mapping using arrayed, fluid-fixed DNA molecules. *Proc. Natl. Acad. Sci. USA* 95:8046–8051.

John, H., M. Birnstiel and K. Jones (1969) RNA-DNA hybrids at the cytological level. *Nature* 223:582–587.

Jones, J. D.G., L. Shlumukov, F. Carland, J. English, S. R. Scofield, G. J. Bishop and K. Harrison (1992) Effective vectors for transformation, expression of heterologous genes, and assaying transposon excision in transgenic plants. *Transgenic Research* 1:285–297.

Kim, U. J., H. Shizuya, P. J. de Jong, B. Birren and M. I. Simon (1992) *Nucleic Acids Res.* 20:1083–1085.

Kim, U. J., B. Birren, T. Slepak, V. Mancino, C. Boysen, H. L. Kang, M. I. Simon and H. Shizuya (1996) Construction and characterization of a human bacterial artificial chromosome library. *Genomics* 34:213–218.

Korenberg, J. R., X. N. Chen, Z. Sun, Z. Y. Shi, S. Ma et al. (1999) Human genome anatomy: BACs integrating the genetic and cytogenetic maps for bridging genome and biomedicine. *Genome Res.* 10:994–1001.

Kurata, N., Y. Umehara, H. Tanoue and T. Sasaki (1997) Physical mapping of the rice genome with YAC clones.
Plant Mol. Biol. 35:101–113.

Lee, M.-K., C. Ren, B. Yan, K. Ding, B. Cox, H.-B. Zhang, M. N. Romanov, F. G. Sizemore, S. P. Suchyta, E. Peters and J. B. Dodgson (2003) Construction and characterization of three complementary BAC libraries for analysis of the chicken genome. *Animal Genet.* 34:146–160.

Li, Y., T. Uhm, T. S. Santos, C. Ren, M.-K. Lee, B. Yan, F. Santos, A. Zhang, D. Liu and H.-B. Zhang (2003) A plant-transformation-complement BIBAC-based integrated physical map of japonica rice for functional analysis of the rice genome sequence. *In: International Plant, Animal & Microbe Genomes 11 Conference (PAG-XI), San Diego, CA, USA. P342.*

Lim, A., E. T. Dimalanta, K. D. Potamousis, G. Yen, J. Apodoca et al. (2001) Shotgun optical maps of the whole *Escherichia coli* O157:H7 genome. *Genome Res.* 11:1584–1593.

Liu, Y. G., Y. Shirano, H. Fukaki, Y. Yanai, M. Tasaka, S. Tabata, and D. Shibata (1999) Complementation of plant mutants with large genomic DNA fragments by a transformation-competent artificial chromosome vector accelerates positional cloning. *Proc. Natl. Acad. Sci. USA* 96:6535–6540.

Marra, M., T. Kucaba, N. L. Dietrich, E. D. Green, B. Brownstein et al. (1997) High throughput fingerprint analysis of large-insert clones. *Genome Res.* 7:1072–1084.

Marra, M., T. Kucaba, M. Sekhon, L. D. Hillier, R. Martienssen et al. (1999) A map for sequence analysis of the *Arabidopsis thaliana* genome. *Nature Genetics* 22:265–270.

Matallana, E., C. J. Bell, P. J. Dunn, M. Lu and J. R. Ecker (1992) Genetic and physical linkage of the *Arabidopsis* genome: methods for anchoring yeast artificial chromosomes. *In: Methods in Arabidopsis Research (eds Koncz, C., N. H. Chua and J. Schell), World Scientific Publishing Co., Singapore*, pp. 144–169.

Men, A., K. Meksem, M. A. Kassem, D. Lohar, J. Stiller, D. Lightfoot and P. M. Gresshoff (2001) A bacterial artificial chromosome library of *Lotus japonicus* constructed in an *Agrobacterium tumefaciens*- transformable vector. *Mol. Plant Microbe Interact.* 14:422–425.

Mozo, T., S. Fischer, S. Meier-Ewert, H. Lehrach and T. Altmann (1998) Use of the IGF BAC library for physical mapping of the *Arabidopsis thaliana* genome. *Plant J.* 16:377–384.

Mozo, T., K. Dewar, P. Dunn, J. R. Ecker, S. Fischer et al. (1999) A complete BAC-based physical map of the *Arabidopsis thaliana* genome. *Nature Genet.* 22:271–275.

Nusbaum, C., D. K. Slonim, K. L. Harris, B. W. Birren, R. G. Steen et al. (1999) A YAC-based physical map of the mouse genome. *Nature Genet.* 22:388–393.

Pardue, M. L. and J. G. Gall (1969) Molecular hybridization of radioactive DNA to the DNA of cytological preparations. *Proc. Natl. Acad. Sci. USA* 64:600–604.

Ren, C., M.-K. Lee, B. Yan, K. Ding, B. Cox, M. N. Romanov, J. A. Price, J. B. Dodgson and H.-B. Zhang (2003) BAC-based physical mapping of chicken genome. *Genome Res.* 13:2754–2758.

Saji, S., Y. Umehara, B. A. Antonio, H. Yamane, H. Tanoue et al. (2001) A physical map with yeast artificial chromosome (YAC) clones covering 63 % of the 12 rice chromosomes. *Genome* 44:32–37.

Schmidt, R., J. West, K. Love, Z. Lenehan, C. Lister, H. Y. Thompson, D. Bouchez and C. Dean (1995) Physical map and organization of *Arabidopsis thaliana* chromosome 4. *Science* 270:480–483.

Schwartz, D. C., X. Li, L. I. Hernandez, S. P. Ramnarain, E. J. Huff and Y. K. Wang (1993) Ordered restriction maps of *Saccharomyces cerevisiae* chromosomes constructed by optical mapping. *Science* 262:110–114.

Shizuya, H., B. Birren, U.-J. Kim, V. Mancino, T. Slepak, Y. Tachiri and M. Simon (1992) Cloning and stable maintenance of 300-kilobase-pair fragments of human DNA in *Echerichia coli* using an F-factor-based vector. *Proc. Natl. Acad. Sci. USA* 89:8794–8797

Soderlund, C., L. Longden and R. Mott (1997) FPC: a system for building contigs from restriction fingerprinted clones. *CABIOS* 13:523–535.

Soderlund, C., S. Humphray, A. Dunham and L. French (2000) Contigs built with fingerprints, markers, and FPC V4.7. *Genome Res.* 10:1772–1787.

Song, J., J. M. Bradeen, S. K. Naess, J. P. Helgeson and J. Jiang (2002) BIBAC and TAC clones containing potato genomic DNA fragments larger than 100kb are not stable in Agrobacterium. *Theor. Appl. Genet.* 107:958–964.

Sternberg, N. (1990) Bacteriophage P1 cloning system for the isolation, amplification, and recovery of DNA fragments as large as 100 kilobase pairs. *Proc. Natl. Acad. Sci. USA* 87:103–107.

Tao, Q. and H.-B. Zhang (1998) Cloning and stable maintenance of DNA fragments over 300 kb in *Escherichia coli* with conventional plasmid-based vectors. *Nucleic Acids Res. 1998*, 26:4901–4909.

Tao, Q., Y. L. Chang, J. Wang, H. Chen, M. N. Islam-Faridi, C. Scheuring, B. Wang, D. M. Stelly and H.-B. Zhang (2001) Bacterial artificial chromosome-based physical map of the rice genome constructed by restriction fingerprint analysis. *Genetics* 158:1711–1724.

Tao, Q., A. Wang and H.-B. Zhang (2002) One large-insert plant-transformation-competent BIBAC library and three BAC libraries of Japonica rice for genome research in rice and other grasses, *Theor. Appl. Genet.* 105:1058–1066.

Wu, C., S. Sun, P. Nimmakayala, F. A. Santos, K. Meksem, R. Springman, D. Ding, D. A. Lightfoot and H.-B. Zhang (2004) A BAC- and BIBAC-based physical map of the soybean genome. *Genome Res.* 14:319–26.

Wu, J., T. Maehara, T. Shimokawa, S. Yamamoto, C. Harada et al. (2002) A comprehensive rice transcript map containing 6591 expressed sequence tag sites. *Plant Cell* 14:525–535.

Wu, Y., L. Tulsieram, Q. Tao, H.-B. Zhang and S. J. Rothstein (2000) A binary vector-based large insert library for *Brassica napus* and identification of clones linked to a fertility restorer locus for *Ogura* cytoplasmic male sterility (CMS). *Genome* 43:102–109.

Xu, Z., Y. L. Chang, K. Ding, L. Covaleda, S. Sun, C. Wu and H.-B. Zhang (2003) Automation of the procedure for whole-genome physical mapping from large-insert random BACs and BIBACs. *In: International Plant, Animal & Microbe Genomes 11 Conference (PAG-XI), San Diego, CA, USA. P123.*

Xu, Z., S. Sun, L. Covaleda, K. Ding, A. Zhang, C. Wu, C. Scheuring and H.-B. Zhang (2004) Genome physical mapping with large-insert bacterial clones by fingerprint analysis: methodologies, source clone genome coverage and contig map quality (submitted).

Zachgo, E. A., M. L. Wang, J. Dewdney, D. Bouchez, C. Camilleri, S. Belmonte, L. Huang, M. Dolan and H. M. Goodman (1996) A physical map of chromosome 2 of *Arabidopsis thaliana. Genome Res.* 6:19–25.

Zhang, H.-B. (2000) *Manual: Construction and Manipulation of Large-Insert Bacterial Clone Libraries, Texas A & M University, Texas, USA.*

Zhang, H.-B. and R. A. Wing (1997) Physical mapping of the rice genome with BACs. *Plant Mol. Biol.* 35:115–127.

Zhang, H.-B. and C. Wu (2001) BAC as tool for genome sequencing. *Plant Physiol. Biochem.* 39:195–209.

Zhang, H.-B., X. P. Zhao, X. L. Ding, A. H. Paterson and R. A. Wing (1995) Preparation of megabase-size DNA from plant nuclei. *Plant J.* 7:175–184.

Zhang, H.-B., S. S. Woo and R. A. Wing (1996) BAC, YAC and cosmid library construction. *In: Plant Gene Isolation: Principles and Practice (eds Foster, GD andD. Twell), John Wiley & Sons Ltd, New York,* pp. 75–99.

Zhou, S., W. Deng, T. S. Anantharaman, A. Lim, E. T. Dimalanta et al. (2002) A whole-genome shotgun optical map of *Yersinia pestis* strain KIM. *Appl. Environ. Microbiol.* 68:6321–6331.

9
Integration of Physical and Genetic Maps

Khalid Meksem, Hirofumi Ishihara, and Tacco Jesse

9.1
Introduction

The demand for integrated physical-genetic maps has increased steadily in the past decade. These maps have been used intensively for gene cloning and lately have been very useful in supporting several genome sequencing initiatives, both whole-genome and region-specific DNA sequencing. Whole-genome integrated physical-genetic maps have been or are being generated for several plant genomes, including the model plant *Arabidopsis thaliana* (Marra et al. 1999), sorghum (Klein et al. 2000), corn (Coe et al. 2002), and soybean (Wu et al. 2004).

To develop an integrated genetic and physical map resource for plant genomes, a comprehensive approach that includes several components is needed: a high-resolution DNA marker map that provides essential genetic anchor points for ordering the physical map and for utilizing comparative information from other genomes; a large-insert DNA library, with ordered clones, consisting of contigs sometimes assembled into a physical map; and a set of bioinformatics tools that allow the integration of several databases to store, search, analyze, and display the integrated physical-genetic map.

Anchoring large-insert DNA clones (such as BAC) and contigs with genetically mapped probes and DNA marker sequences can be achieved by using two strategies: (1) PCR-based integration of large-insert DNA pools with DNA markers and (2) hybridization of DNA probes to large-insert DNA filters. For PCR-based screening, the DNA is usually isolated from large-insert DNA libraries and pooled using a six-dimensional scheme (Klein et al. 2000) to facilitate the screen. While constructing an integrated physical-genetic map, it is necessary to keep track of each large-insert DNA clone, the contig and its DNA anchor, and the DNA marker sequence information associated with each anchor point. This task is extremely difficult to do manually, as it would require many people and a lot of time. Therefore, automatic tracking via a system that can be cheaper, quicker, and more accurate is indeed necessary and highly in demand. Few computer tools that facilitate viewing of the maps and the underlying integrated physical-genetic data attached to it, join the

The Handbook of Plant Genome Mapping.
Genetic and Physical Mapping. Edited by Khalid Meksem, Günter Kahl

ISBN 3-527-31116-5

sequences into their proper genome location, and present the results on the Internet for everyone to see are available.

9.2 Colony Hybridization Techniques for the Integration of DNA Sequences Into a Physical Map

The clones of a large-insert library are usually double-griddled onto Hybond N^+ membranes in a 3×3 format using an automated arraying robot. Once ready, the membranes are probed with the clones representing the targeted sequence as described by Kanazin et al. (1996). The inserts of the selected probe clones are purified, labeled by random-priming, and hybridized to the high-density clone membranes. Once identified, the positive clones of each probe are re-arrayed in 384-well plates.

High-density colony Hybond N^+ nylon filters (Amersham Pharmacia Biotech Inc., Piscataway, MD) are prepared using an automated arraying robot, where the clones are spotted onto the filters in two copies from each colony of the 384-well microtiter plates. One filter of 20×20 cm may carry up to 50,000 clones. The filters containing colonies are placed onto LB (Luria-Bertani) agar mix petri plates (Fisher Scientific, Pittsburgh, PA), which contain the corresponding antibiotic, and incubated at 37 °C. When the colonies grow to 1–2 mm in diameter, the filters are prepared for hybridization (Meksem et al. 2000).

Membranes with cell colonies are placed on sheets of filter paper and saturated with the following solutions in flat trays:

Trays	Solutions used to saturate the filter paper	Incubation time on the paper
1	10% (w/v) SDS	10 min
2	Denaturing solution	10 min
3	Neutralizing solution	10 min
4	2× SSC, 0.1% (w/v) SDS	10 min
5	0.4 N NaOH	20 min
6	5× SSC, 0.1% (w/v) SDS	20 min
7	2× SSC, 0.1% (w/v) SDS	20 min

Neutralizing solution	Denaturing solution
500 mL 1 M Tris-HCl, pH 8.0	125 mL 400 mM NaOH
300 mL 5 M NaCl	300 mL 5 M NaCl
2 mL 0.5 M EDTA, pH 8.0	575 mL dd. H_2O
198 mL dd. H_2O	
1000 mL	1000 mL

Membranes are dried in an oven at 80 °C for 2 h to remove all excess liquid. The filters are then placed in hybridization buffer (5× SSC [20× SSC stock solution: 3.0 M sodium chloride, 0.3 M sodium citrate, pH 8.0, 0.5 M EDTA], 5× Denhardt's reagent [50× Denhardt's reagent: 1% (w/v) Ficoll 400, 1% (w/v) polyvinylpyrrolidone, 1% (w/v) bovine serum albumin], 0.5% (w/v) SDS) at 65 °C for 2 h. Twenty microliters of the DNA probe (50–100 ng) and 25 μL water are mixed and placed on the heating block at 95 °C for 5 min. The mixture is cooled to room temperature, and 5 μL of 3000 Ci/mM α-^{32}P dCTP is added to the mixture, which is then transferred to a Ready-To-Go DNA labeling beads tube (Amersham Pharmacia Biotech Inc., Piscataway, MD). The probe mixture is incubated at room temperature for 30 min and then heated for 5 min at 95 °C. Filtering of the mixture through a MicroSpin S-200 HR column (Amersham Pharmacia Biotech Inc., Piscataway, MD, U. S.A.) is performed by centrifugation at 400 g for 2 min. Flow-through radioactive probe is transferred to the tube containing the pre-hybridization buffer and the filter. After 16 h incubation at 65 °C, the filters are washed twice with 150 mL of 2× (w/v) SSC with 0.1% (w/v) SDS, 10 min each time, and then washed twice with 150 mL 0.1× (w/v) SSC with 150 mL 0.1% (w/v) SDS for 10 min each time. Plastic is wrapped around the filter to cover it, and the filters are subsequently exposed to BioMax-MR X-ray film (Kodak, Rochester, NY) for 16 h. The film is then developed and analyzed (Figure 9.1).

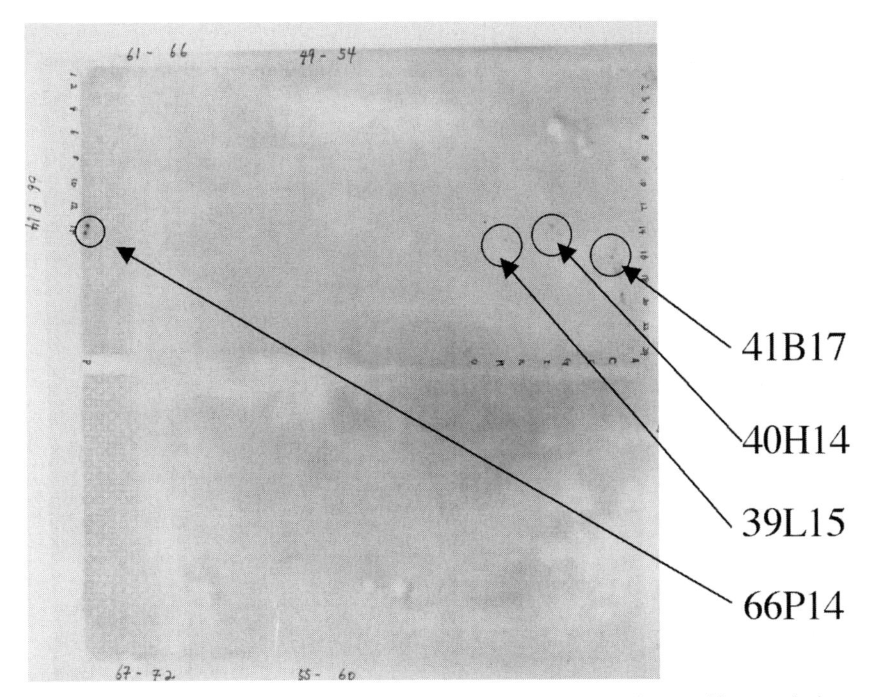

Figure 9.1. Colony hybridization screening of the soybean Forrest *Hind*III BAC library with the soybean *Hs1(pro1)* gene.

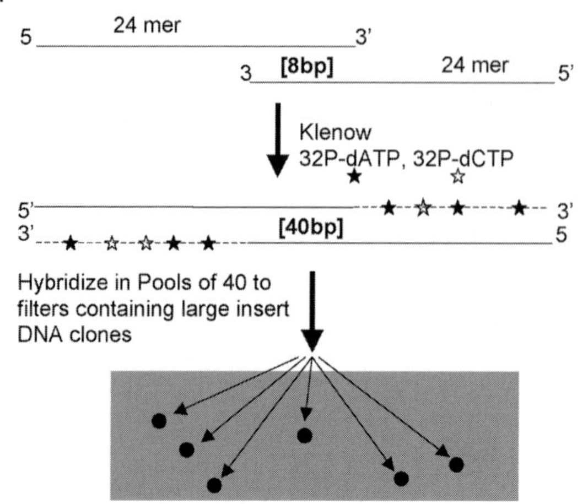

Figure 9.2. Overgos: overlapping oligonucleotide probes scheme.

9.3 Integrating Gene-rich Sequences into the Physical Map Via Overgo Hybridization

Several mapping projects found the use of the overgo hybridization strategy very successful for mapping EST sequences to a physical map. This information, when incorporated into FPC software (FingerprintContig), led to a significant reduction in the overall number of contigs in physical maps. Overgo hybridization is a procedure by which distinct 40-bp sequences are used with high specificity to integrate genetic and physical maps (See Figure 9.2).

After a set of sequences (genes or EST sequences) targeted for physical map anchoring are identified. From a fasta input file, the OvergoMaker 40 overgo designer – software developed by John D. McPherson, Washington University School of Medicine, St. Louis, MO – is used to design specific overlapping oligonucleotide probes. The overgos are selected to have as close to a 50 % distributed GC content as possible. OvergoMaker 40 designs two 24mers with an eight-base overlap on their 3′ ends. After annealing, these will form a 40-base overgo with two 16-base overhangs that can be labeled by filling in with the Klenow fragment of DNA polymerase I. DNA from positive controls (known sequences of plasmids) is spotted in a unique pattern and used to align the filters. Overgo modules of 576 probes, assembled in a 24×24 array, can be hybridized to high-density large-insert DNA filters using the procedures outlined below.

9.3.1 Overlapping Oligo Labeling

The stock solution (10 pmol/μL) of mixed oligos is heated at 80 °C for 5 min and then incubated at 37 °C for 10 min and stored on ice.

Labeling reaction (10 µL):

Oligonucleotides	10 pmol each
BSA (2 mg mL^{-1})	0.5 µL
OLB (-A,-C,-N6)	2.0 µL
32P-dATP*	0.5 µL
32P-dCTP*	0.5 µL
H_2O	to 10 µL
Klenow fragment	1 µL (2 U μL^{-1})

The labeling reaction tube is incubated at room temperature. After 1 h, the unincorporated nucleotides are removed using Sephadex G50 columns and the probe is denatured before hybridization.

OLB(-A,-C,-N6) = A:B:C (1:2.5:1.5), stored at –20 °C.

Solution O	Solution A	Solution B
1.25 M Tris-HCL, pH 8.0	1 mL solution O	2 M HEPES-NaOH, pH 6.6
125 mM $MgCl_2$	18 µL 2-mercaptoethanol	Solution C
	5 µL 0.1 M dTTP	3 mM Tris-HCl, pH 7.4
	5 µL 0.1 M dGTP	0.2 mM EDTA

9.3.2 Hybridization

The hybridization is usually performed at 58 °C for overgos with GC content between 40 % and 60 %. Using overgos that are AT-rich may require lowering the hybridization temperature. Fifty milliliters of warmed hybridization solution is added to a 30×4 cm hybridization bottle. Filters are pre-wet with warmed 2× SSC and pre-hybridized for 4 h. After pre-hybridization, 10 mL of the hybridization buffer is removed from each bottle and combined, and the labeled oligos are denatured at 90 °C for 10 min. They are then added and mixed by pipetting. Ten milliliters is then added to each bottle to transfer the probe. Probes are allowed to hybridize overnight typically; however, 2-day hybridizations may be allowed to give stronger signals.

Hybridization solution:

Composition (1000 mL)	
1 % BSA (Fraction V, Sigma)	10 g
1 mM EDTA	2 mL 0.5 M EDTA (pH 8.0)
7 % SDS (use 99.9 % pure SDS)	70 g
0.5 M sodium phosphate	500 mL 1 M sodium phosphate

(1 M sodium phosphate:: 134 g $Na_2HPO_4 \cdot 7H_2O$, add 4 mL 85 % H_3PO_4, water is added to 1000 mL)

9.3.3 Washing

The hybridization solution is removed and the filters are washed with 2× SSC, 0.1% SDS at room temperature and then transferred to larger tubes on a rotary platform and washed as follows: 2 L 1.5× SSC, 0.1% SDS at 58 °C for 30 min; 2 L 0.5× SSC, 0.1% SDS at 58 °C for 30 min.

9.3.4 Autoradiography

The filters are then sealed in plastic bags and exposed to an XAR5 Kodak film at −80 °C overnight; the time can be increased for aged filters. The film is then developed and the positive signals on the film are oriented with overlay template and membrane/filter information for the specific library. The probes that were used for that specific hybridization are noted on the film along with date and time of exposure.

9.4 Pooling Strategy for the Integration of DNA Sequences Into a Physical Map

9.4.1 DNA Pooling

Large-insert libraries are usually organized into pools to allow an efficient integration of DNA sequences into the physical map and to reduce the effort and cost needed to identify clones from a library for a targeted sequence. A procedure described by Klein et al. (2000) is used as a community standard to organize large-insert clones into pools, where a BAC library with 81,408 clones can be pooled into 287 pools.

A large-insert library with 81,408 clones is organized and stored in 212 individual 384-well plates. The library will be arranged into a stack consisting of 53 layers with each layer containing four plates. The four plates in a layer will be arranged in a 2×2 plate pattern. Because each 384-well plate is an array of 16 rows and 24 columns of wells, this 2×2 plate pattern results in each plate layer containing wells in 32-row by 48-column arrays (1536 wells/layer). Every well in a stack has a unique address by its x, y, and z coordinates relative to the axes of the stack. The stack of plates will be sampled in six distinct ways to generate the large-insert DNA pools (Figure 9.1). Each pool represents the intersection of a plane with the stack (with the plane passing through the wells). Planes intersecting the stack parallel to three mutually perpendicular surfaces define three of the pool types. Layer pools (LPs) will be prepared from each layer of the stack. Row pools (RPs) will be prepared from all wells in the same row of all layers. Column pools (CPs) will be prepared from all wells in the same column of all layers.

Table 9.1. Summary statistics for pooling of a large insert library for each type of pools generated from a stack with dimensions 53×32×48.

Pool type	*No. of pools*	*Pool size*	*No. of clones*	*Total no. of clones*
Layer pool	53	32×48	1536	81,408
Row pool	32	48×53	2544	81,408
Column pool	48	32×53	1696	81,408
Layer diagonal pool	48	32×53	1696	81,408
Row diagonal pool	53	32×48	1536	81,408
Column diagonal pool	53	32×48	1536	81,408
Total	287			

The three types of diagonal pools – layer diagonal pools (LDPs), row diagonal pools (RDPs), and column diagonal pools (CDPs) – are then generated as illustrated in Figure 9.1. To keep the number of wells constant in the same type of diagonal pool, wrapping is done as illustrated in Figure 9.2. A total of the 287 pools will be generated. Table 9.1 summarizes the number of pools and the number of large-insert DNA clones in each pool for the six types of pools.

Two hundred and twelve 384-well plates containing 81,408 large-insert DNA clones are organized into a three-dimensional stack consisting of 53 layers, with each layer containing four plates of 32 rows by 48 columns. The stack will be

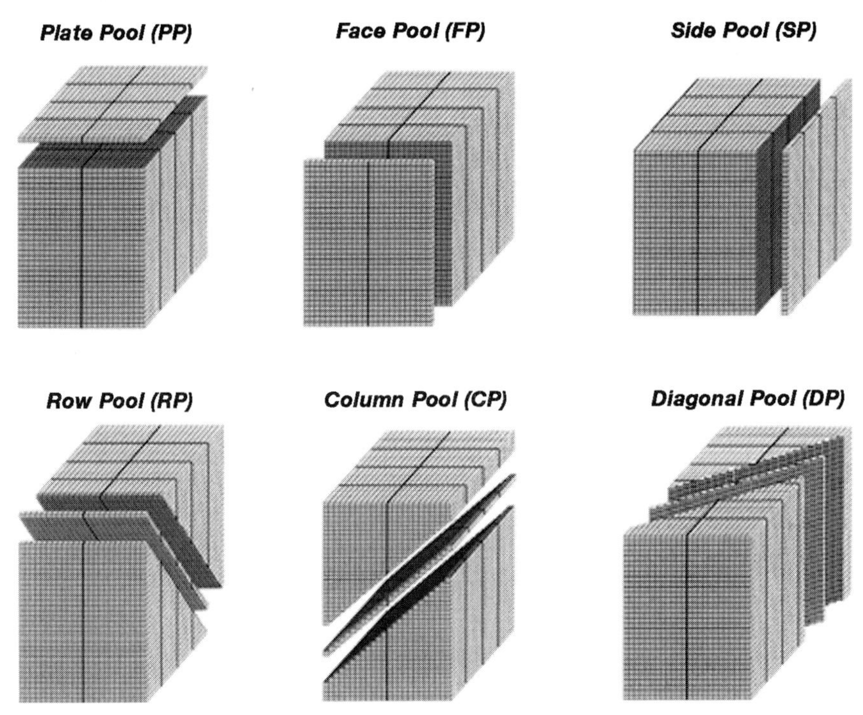

Figure 9.3. Six-fold large-insert DNA pooling strategy.

pooled along the six unique depicted coordinates to generate a total of 287 DNA pools.

9.4.2
PCR Multiplexing

Multiplexing PCR techniques, in which up to 10 PCR primer pairs can be tested in a single PCR reaction, can be used for screening the library. Specific primer pairs to the target DNA sequence are tested on total genomic DNA from the source genome to determine the size of the amplified fragment. The primer pairs producing clearly distinguishable size differences of their amplified fragments will be selected for multiplexing. Computer programs (Klein et al. 2000) can be used to resolve candidate clones that contain the target DNA sequence in the pools. All candidate clones are confirmed by testing individual clones with the specific primer pairs.

9.4.3
DNA Isolation from Pooled Large-insert Clones

DNA from large-insert DNA clone pools is isolated as described by Klein et al. (2000), with minor modifications. Each of the 384-well plates containing the DNA clones for a library will be replicated in deep, 384-well cell culture plates containing 100 μL LB medium with the corresponding antibiotics. A 384-pin tool is used to inoculate the deep, 384-well cell culture plates with clones from corresponding 384-well stock plates. The inoculated plates are incubated at 37 °C for 20–24 h without shaking. Then the plates are arranged into the stack design and sampled in one of the six distinct ways described above to generate a particular pool type. Ninety microliters of the bacterial culture is removed from each well. The pooled culture is then transferred to a 250-mL centrifuge bottle and incubated on ice for 15 min. The cells are collected by centrifugation at 4400 g at 4 °C for 15 min. The supernatant is discarded and cell pellets are resuspended in 2.4 mL of a solution containing 50 mM glucose, 10 mM EDTA, 25 mM Tris-HCl (pH 8.0) by vortexing. The resuspended cells are transferred to a 30-mL centrifuge tube containing 4.8 mL of freshly prepared lyses solution containing 0.2 M NaOH and 1% SDS. The samples are mixed gently by inversion 10–12 times and incubated at room temperature for 10 min. After incubation, 3.6 mL of 3 M potassium acetate (pH 5.2) is added and the samples are mixed by inversion 12 times. Then the samples are incubated on ice for 20 min and centrifuged at 31,000 g at 20 °C for 20 min. The supernatant is transferred to a new tube containing 10 mL of isopropanol; the tube is inverted several times and incubated at –20 °C overnight. The samples are centrifuged at 13,800 g at 4 °C for 30 min, and the supernatant is discarded. The precipitated DNA is washed twice with 80% ethanol, air-dried, resuspended in TE buffer, and quantified for its DNA content.

		Column
		1 2 3 4 5 6 7 8 9
Row	1	2 3 4 5 6 7 8 9 1
	2	3 4 5 6 7 8 9 1 2
	3	4 5 6 7 8 9 1 2 3
	4	5 6 7 8 9 1 2 3 4
	5	6 7 8 9 1 2 3 4 5
	6	7 8 9 1 2 3 4 5 6
		Pool ID

Figure 9.4. Wrapping of wells in diagonal pools. The array shown here has dimensions of 6×9 and is sampled into nine pools, each containing six wells.

9.5 Forward and Reverse Integrated Physical-Genetic Mapping: Targeted DNA Marker Mapping

Several novel BAC fingerprinting techniques have been developed since the construction of BAC- based integrated maps became the standard in physical mapping (Coulson et al. 1986; Marra et al. 1997; Luo et al. 2003). Most of these methods focus on a higher information content of the individual BAC fingerprints, with the aim to have better and more reliable assembly of the BAC contigs with fingerprint-editing computer programs (Soderlund et al. 2000). After the generation of BAC fingerprints and the subsequent assembly of the BAC contigs, the daunting task of verification of each individual contig remains before the dataset can be released for further exploration. In most BAC fingerprinting datasets, the majority of the assembled BAC contigs are not linked to a genetic map position, and normally a vast number of single BACs do not assemble into any contig. The main reason for the large number of unlinked contigs is the relatively low marker density of most genetics maps and the relatively small physical span of most assembled BAC contigs. In the ideal dataset, the physical coverage of the BAC contigs enables the anchoring of most contigs to their corresponding genetic map position. However, the assembly of a standard fingerprinting project of a 12× BAC library of a 350-Mb genome (~36,000 clones), for example, generates at least 10% singletons and a significant group of smaller contigs containing 2–25 BACs. Therefore, high-density genetic maps with one marker per 50 kb on average are necessary to anchor all computer-assembled BAC contigs.

A straightforward approach for the anchoring of BAC contigs to the genetic map is the hybridization of mapped probes (RFLPs, STSs, ESTs, etc.) to high-density BAC filters. The pooling of overgo probes in the hybridization process facilitates the throughput and the identification of individual BACs and BAC contigs. Methods that link genetically mapped AFLP markers to BACs and BAC contigs are desirable (Klein et al. 2000). The six-dimensional BAC pooling strategy in combination with an AFLP screening enables the linkage of up to 40 genetic map positions to BAC contigs for each AFLP primer combination (Klein et al. 2000).

9.5.1
Integrated AFLP Mapping

Another method is generating high-information-content fingerprints of all individual BACs by applying the AFLP protocol on individual BAC DNA and excluding the selective base pairs on the primers in the amplification step. For example, in an amplification step with *Hin*dIII+0/*Mse*I+0, 50–60 fragments on average are generated on an individual BAC DNA, enabling the assembly of BAC contigs with a resolution similar to that achieved by other high-information-content methods (Jess et al., unpublished). The next step is the screening of the BAC pools with the same AFLP enzyme combination *Hin*dIII/*Mse*I, but then including the selective base pairs +2 on the *Hin*dIII-primer and +3 on the *Mse*I-primer. The selective base pairs should have the same extensions as the extensions that were used to construct the genetic map. Consequently, the *Hin*dIII+2/*Mse*I+3 AFLP fragments can be linked *in silico* to the corresponding *Hin*dIII+0/*Mse*I+0 fragment of the fingerprint of the individual BAC clone in the positive pool, and therefore the individual BAC(s) can be linked directly to the corresponding genetic map position. In summary, the crux of this method – called "KeyMaps" – is the *in silico* linkage of genetically mapped $+K/+L$ AFLP fragments to $+M/+N$ AFLP fragments of an individual BAC fingerprint, where K, L, M, and N represent the primer extensions and can be any integer between 0 and 10. In general, K=2 or 3, L=3, M=0, and N=0. The $+K/+L$ primer combinations applied for genetic map construction are also used for screening the BAC pools containing 0.05–0.25 genome equivalents. The advantage of this approach is that one fingerprinting technique (AFLP) is used for two applications. In the first step, BAC fingerprints are generated by AFLP, and in the second step AFLP markers are used to assign genetic markers to BAC contigs. In Figure 9.5, the feasibility of the KeyMaps concept is demonstrated.

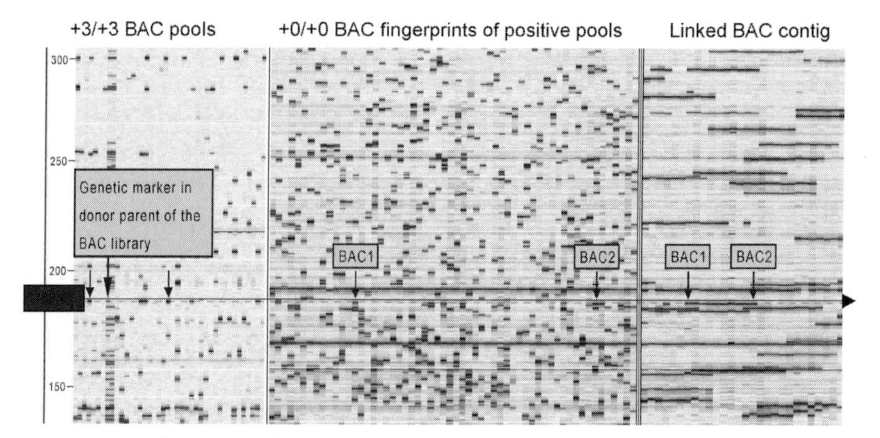

Figure 9.5. *In silico* linkage of +3/+3 and +0/+0 AFLP fragments. The genetic marker and a few positive BAC pools are indicated by the arrows. *Hin*dIII+0/*Mse*I+0 fingerprints of individual BACs of the corresponding positive pools.

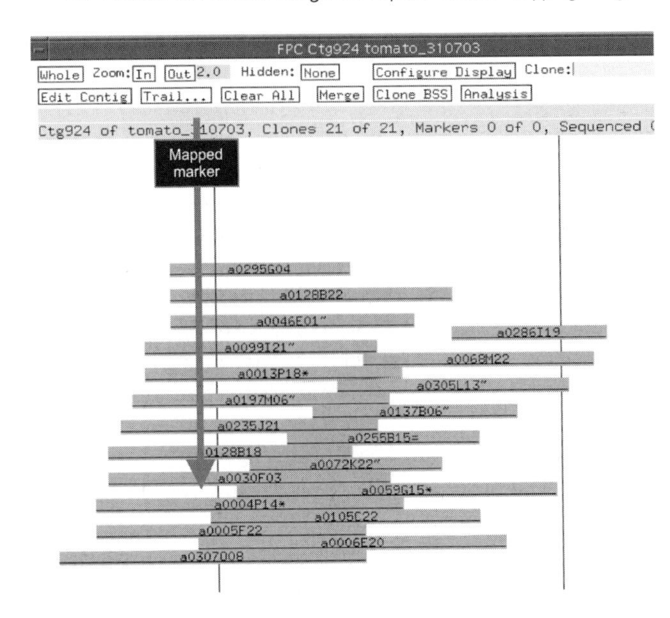

BAC Contig Matching:

1) Fingerprint a pooled genomic (BAC) library with genetic markers

2) Fingerprint the individual BACs in the pools

3) Contig assembly of the individual fingerprints (FPC)

4) Match the contig to the genetic markers in the fingerprint of the pool

5) Anchor the contig to the genetic map - Prediction of the presence/absence of individual bacs in the pools is based on the generated contig

Figure 9.6. Contig matching of the corresponding *Hind*III+0/*Mse*I+0 BAC clone anchored to the genetic map.

Besides a very efficient integrated map construction, the KeyMaps method facilitates the generation and addition of novel AFLP markers to the integrated genetic and physical map. Each *Hind*III+2/*Mse*I+3 fragment can be assigned to a BAC contig in the assembled contig dataset (Figure 9.6). If a particular BAC contig is already linked to the genetic map, a second and a third marker assigned to this contig are consequently mapped de novo to the position of the first marker. Finally, when sufficient AFLP primer combinations are screened on the BAC pools, the co-retention frequencies of these markers on the sub-genomic BAC pools enable the generation of de novo physical maps, analogous to HAPPY mapping and radiation hybrid mapping approaches (Thangavelu et al. 2003).

9.5.2 Targeted SSR Mapping

Although it is most common to use DNA markers to integrate a physical map, in genomes or targeted genome regions where additional DNA markers are needed, large-insert DNA clones can be used to generate DNA markers. The most common and easiest approach is the development of SSR markers (Cregan et al. 1999, Meksem et al. 2000).

Targeted SSR mapping relies on subcloning the large-insert DNA clones into small DNA insert clones, identifying tandem repeat sequences within the subclone using DNA probes designed to identify the repeats, sequence analysis of the positive subclones, and primer design to identify the polymorphic repeats.

9.5.2.1 **Subcloning of Large-insert DNA Clones**

Usually, 40 ng of a large-insert DNA clone from a library and a cloning plasmid vector (pBluescript II) are totally digested with 10 units of *Eco*RI restriction enzyme at 37 °C for 2 h. The digested DNA is precipitated with 0.6× volume of isopropanol and centrifuged at 10,621 *g* for 20 min. Precipitates are washed twice with 70 % (w/v) ethanol and dried using a CentriVAp Concentrator (LABCONCO Corporation, Kansas City, MO). Distilled water (13.5 μL) is added to each tube. Three units of T_4 DNA ligase and 1.5 μL of 10× T_4 DNA ligase buffer are then added to each tube. Next, the tubes are incubated at 16 °C overnight. On the following day, 1 μL of ligated DNA solution is electroporated into 20 μL of DH10B electro-competent cells, which are then cultured in 500 μL of SOC medium (Gibco BRL Life Technologies, Rockville, MD) at 37 °C in a shaker (New Brunswick Scientific, Edison, NJ) at 210 rpm for 50 min. Thirty microliters of the culture is then plated on an LB agar plate with the corresponding antibiotic (40 g of LB agar mix, 12.5 μg mL^{-1} chloramphenicol, 60 μg mL^{-1}, 5-bromo-4-chloro-3-indolyl-beta-D-galactoside [X-gal], 200 μg mL^{-1} isopropyl thiogalactoside [IPTG]). The plates are incubated at 37 °C for 16 h. Recombinant clones are picked and inoculated into a 384-well plate, which contains LB broth with 60 μg mL^{-1} ampicillin and freeze media (360 mM K_2HPO_4, 132 mM KH_2HPO_4, 17 mM sodium citrate, 4 mM Mg SO_4 × $7H_2O$, 68 mM $(NH_4)\ _2SO_4$, 44 % (w/v) glycerol in sterile H_2O), and incubated at 37 °C for 16 h. The plates are then stored at –80 °C.

9.5.2.2 **Colony Hybridization**

Subclones are inoculated with a 384-pin replicator onto a Hybond N^+ transfer membrane (Amersham Pharmacia Biotech Inc., Piscataway, MD). Sterilization of the replicator is done with 100 % (w/v) EtOH and flame. Once the replicator has cooled for a few seconds, it is dipped into the plate and placed down onto the membrane, which has been placed on an LB agar plate (60 μg mL^{-1} ampicillin, 60 μL mL^{-1} X-gal, 200 μL mL^{-1} IPTG). Cell colonies are incubated at 37 °C until they grow to 0.5–1.0 mm in diameter.

About 100 ng of each oligomeric repeat primer, $(AT)_{10}$, $(ATT)_{10}$, $(CG)_{10}$, $(CC)_{10}$, $(CT)_{10}$, and $(CTT)_{10}$ (Research Genetics, Huntsville, AL), is labeled by phosphorylating the 5′ end with 3 μL 3000 Ci/mM γ-^{33}P dATP (Fisher Scientific, Pittsburgh, PA) and 10 units of T4-DNA kinase (Amersham Pharmacia Biotech Inc., Piscataway, MD) at 37 °C for 1 h. The enzyme is inactivated by heating at 70 °C for 10 min. Filters are next incubated in pre-hybridization buffer (5× SSC, 0.5 % [w/v] SDS, and 5× Denhardt's reagent) at 39 °C for 2 h and then hybridized to the labeled probe at 39 °C in pre-hybridization buffer overnight. After hybridization, the filters are washed twice for 20 min at 37 °C in 2× SSC/0.2 % (w/v) SDS and twice for 1 h at 37 °C in 1× SSC/0.1 % (w/v) SDS. Plastic is wrapped around the filter to cover it, and the filter exposed to BioMax-MR X-ray film (Kodak, Rochester, NY) for 16 h. The film is then developed and analyzed. Once putative repeat subclones are identified, they are sequenced with the dideoxy chain-termination method using an ABI big dye cycle sequencing kit and an ABI377 automated DNA sequencer (Perkin Elmer, Foster City, CA). The sequence analysis allows the identification of tandem repeats (Figure 9.6). Primers for simple sequence repeat (SSR) markers are selected within 100 bp upstream and downstream from the sequences where repeats are identified, and the putative SSR marker is mapped in the appropriate population.

9.6 Bioinformatic Tools for an Integrated Physical-Genetic Map: Genome Browser

Information specifically related to a physical map is usually maintained using a community standard informatics tool called Generic Genome Browser (GBrowse) (Stein et al. 2002).

9.6.1 GBrowse

The GBrowse module is available for download at http://www.gmod.org, which provides tools that are maintained and updated during the process of integrated physical genetic mapping. One route of access is provided by "genetic markers" (e.g., SSR, RFLP, AFLP, hybridization probes), where the user enters GBrowse by typing a landmark name into the text field. Landmarks can be DNA markers, gene names, clone names, accession numbers, or any other identifier configured by the administrator. Once a region is selected, it is displayed in a detailed view that summarizes annotations and other genomic features. An overview panel and a navigation bar together allow the user to move from one place to another.

Once an interesting region of the genome is in view, the user can navigate through it in a variety of ways. A "web link" allows the genetic map of a particular linkage group to be displayed, selecting a marker within that group allows the contig(s) identified by that marker to be displayed with clone plate addresses, while

ACGTATCCACCAATACGGATATTTTATAATTGGAACGTACTCTTGGCATTATATTT

ATATATGTATAAAGTCATAGATTAGATAGGCACACTGATTTTGACCCCTTGTTTCC

TCCAACCACAACTCCAATCTATGTCC**CATTACAGCGGATGGAAC**AAAAAACATA

TATAGAACTGATTCTTGAAAAGTGAAAGCAAATTGGAATTTTGATTCTCAGCCAC

GTGTAATAATATATGATGTGTTAGCTGGTTGGGCCTTTCTT**TATATATATATATAT**

ATATATAGGATTGGTGGCTATATGCAAAAGAAGAAATGGGATAAAGGTTTTTTTT

TTAAAAAAAAAAGGAATGTTATGTTTATATTTTAAGTTCTATATTATATTTTGACT

TATT**CGGACAAATCTAGACTCTCAC**TATTCTC

Figure 9.7. Sequence of a plasmid clone from a soybean BAC containing putative SSR markers. The position of the tandem repeat is indicated in boldface and underlined. The position of the SIUC-SAT040 reverse primer used for the polymorphism test is also labeled in boldface.

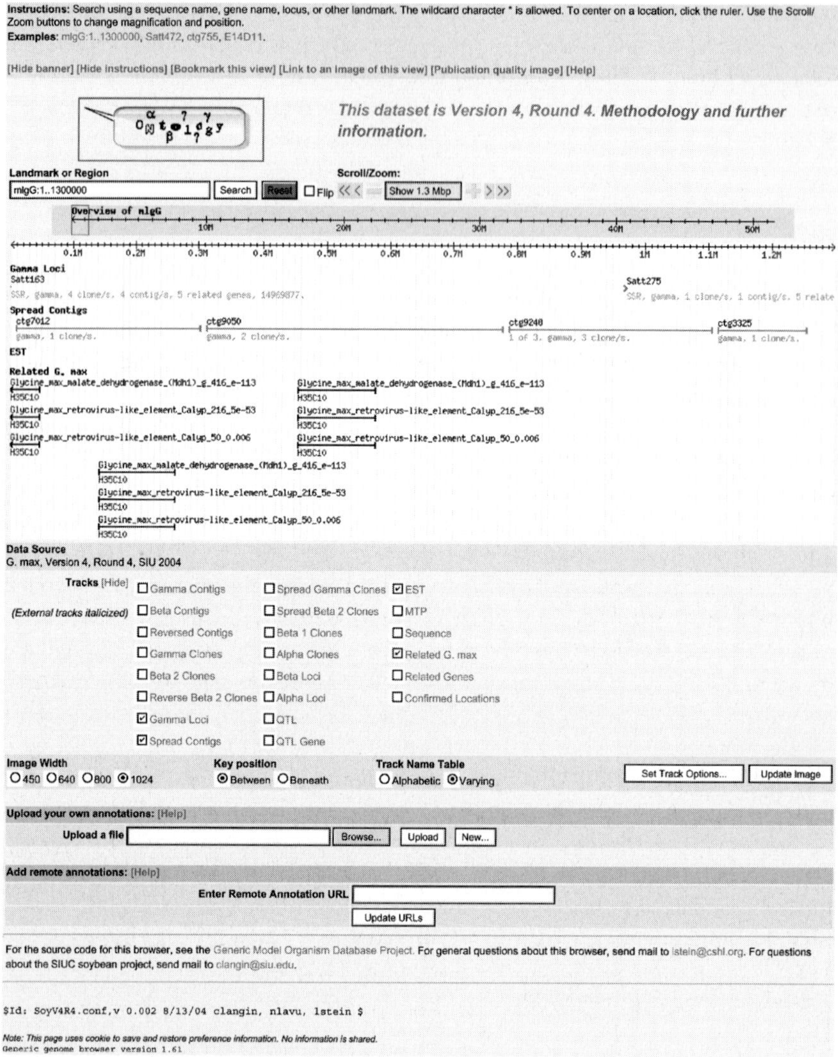

Instructions: Search using a sequence name, gene name, locus, or other landmark. The wildcard character * is allowed. To center on a location, click the ruler. Use the Scroll/Zoom buttons to change magnification and position.
Examples: mlgG:1..1300000, Satt472, ctg755, E14D11.

[Hide banner] [Hide instructions] [Bookmark this view] [Link to an image of this view] [Publication quality image] [Help]

This dataset is Version 4, Round 4. Methodology and further information.

Landmark or Region
mlgG:1..1300000 Search Reset Flip

Scroll/Zoom:
<< < Show 1.3 Mbp > >>

Overview of mlgG
10M 20M 30M 40M 50M

0.1M 0.2M 0.3M 0.4M 0.5M 0.6M 0.7M 0.8M 0.9M 1M 1.1M 1.2M

Gamma Loci
Satt163
SSR, gamma, 4 clone/s, 4 contig/s, 5 related genes, 14969877.
Satt275
SSR, gamma, 1 clone/s, 1 contig/s, 5 relate

Spread Contigs
ctg7012
gamma, 1 clone/s.
ctg9050
gamma, 2 clone/s.
ctg9248
1 of 3, gamma, 3 clone/s.
ctg3325
gamma, 1 clone/s.

EST

Related G. max
Glycine_max_malate_dehydrogenase_(Mdh1)_g_416_e-113
H35C10
Glycine_max_malate_dehydrogenase_(Mdh1)_g_416_e-113
H35C10
Glycine_max_retrovirus-like_element_Calyp_216_5e-53
H35C10
Glycine_max_retrovirus-like_element_Calyp_216_5e-53
H35C10
Glycine_max_retrovirus-like_element_Calyp_50_0.006
H35C10
Glycine_max_retrovirus-like_element_Calyp_50_0.006
H35C10
Glycine_max_malate_dehydrogenase_(Mdh1)_g_416_e-113
H35C10
Glycine_max_retrovirus-like_element_Calyp_216_5e-53
H35C10
Glycine_max_retrovirus-like_element_Calyp_50_0.006
H35C10

Data Source
G. max, Version 4, Round 4, SIU 2004

Tracks [Hide]
(External tracks italicized)

Gamma Contigs	Spread Gamma Clones	EST
Beta Contigs	Spread Beta 2 Clones	MTP
Reversed Contigs	Beta 1 Clones	Sequence
Gamma Clones	Alpha Clones	Related G. max
Beta 2 Clones	Beta Loci	Related Genes
Reverse Beta 2 Clones	Alpha Loci	Confirmed Locations
Gamma Loci	QTL	
Spread Contigs	QTL Gene	

Image Width 450 640 800 1024
Key position Between Beneath
Track Name Table Alphabetic Varying
Set Track Options... Update Image

Upload your own annotations: [Help]
Upload a file Browse... Upload New...

Add remote annotations: [Help]
Enter Remote Annotation URL
Update URLs

For the source code for this browser, see the Generic Model Organism Database Project. For general questions about this browser, send mail to lstein@cshl.org. For questions about the SIUC soybean project, send mail to clangin@siu.edu.

$Id: SoyV4R4.conf,v 0.002 8/13/04 clangin, nlavu, lstein $

Note: This page uses cookie to save and restore preference information. No information is shared.
Generic genome browser version 1.61

Figure 9.8. G. max, Version 4, Round 4, SIU 2004
Showing 1.3 Mbp from mlgG, positions 1 to 1,300,000

another link leads to the DNA sequences associated with that marker. The sequences associated with a contig represent a second route of entry to the physical map. A third route of access is provided by the plate addresses for clones that can be used to view individual BAC fingerprints. The user can jump rapidly from region to region by clicking in the overview panel, a schematic located at the top of the image that shows the entire chromosome (or, in the case of unfinished genomes, a contig) and a number of landmark features such as well-known genetic markers, cytogenetic bands, or sequence scaffolds. GBrowse provides multiple configurable levels of zoom and two scroll speeds. For fine adjustment, the user can click on the scale that appears at the top of the detailed view to center the displayed region at that point or can select the "+" and "" buttons to make fine adjustments in the zoom level.

9.6.2
iMap

iMap allows simultaneous viewing of the genetic map and associated BAC contigs that have been assigned to their respective genetic locations. The iMap software was developed based on the GIOT software written by the Rice Genome Project. iMap is a Web-based browser that allows genetic map data stored in databases with FPC-derived physical map data to be retrieved. The display features side-by-side views of the genetic map and associated physical map (Figure 9.9). The database search utility may extend to report information on map location, probe, sequence accession number, and contig number in the search results. For each locus, information about genetic marker type or contig can be displayed in popup windows. Details on genetic markers are retrieved via links to other databases. The iMap graphic contains the following components: header, selection, genetic map, marker type, physical object, physical map, and popup window(s).

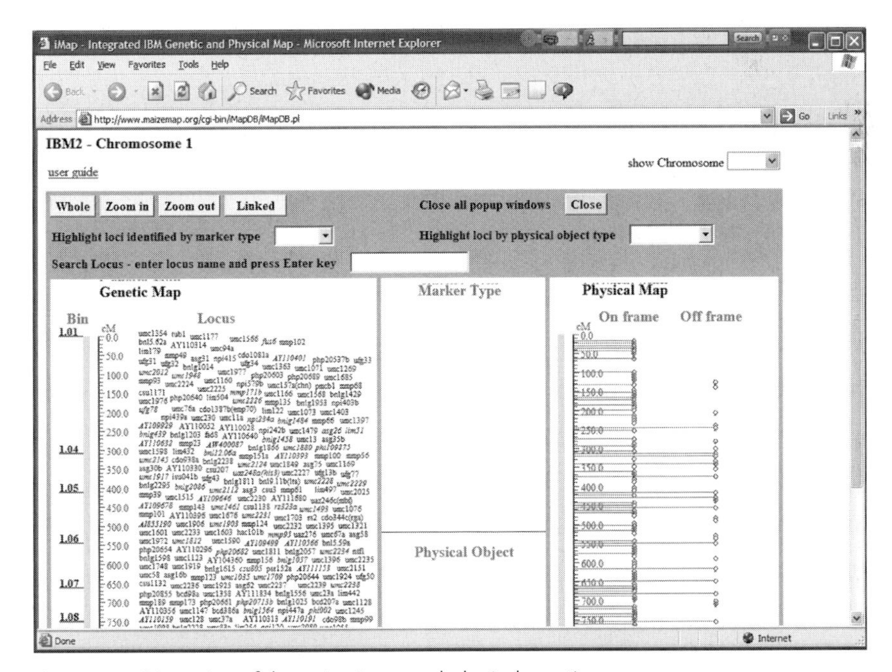

Figure 9.9. iMap: view of the maize integrated physical-genetic map.

9.7 Discussion

When a probe or a PCR DNA marker hits a large-insert DNA clone, if the clone is part of an assembled contig, the contig is then assigned a specific genetic location. However, this concept is not easy as it sounds in practice; factors such as genome duplication and complexity may make contig anchoring a challenging prospect. Using single-copy probes is an ideal strategy to anchor the physical map to the genetic map, where a contig can be assigned with high confidence to a single genetic map location. Unfortunately, the patterns of [probe–large-insert DNA clone–contig] association obtained from FPC contig assembly (See Chapter 12) may reveal possible conflicts. Sometimes, two probes corresponding to distinct genetic locations associate with a single contig; this may be due to BAC address identification, genetic mapping, contig assembly, or *in silico* determination of mapped marker identity. In this case, verifying the results of the hybridization screen and pool analysis, checking genetic mapping assignments, and manual editing of contigs to test the possibility that the contig is chimeric may solve the conflict. Testing of the DNA pools by PCR using primers derived from the same hybridization probes used for hybridization-based screening of the large-insert DNA filters is recommended.

Hybridization-based screens tend to identify large-insert DNA clones that could be a false positive, arising either by a scoring error or by spurious hybridization. In contrast, PCR validation experiments indicate that false positives occur rarely, if at all. Therefore, during hybridization-screen experiments, it is advisable to wash the hybridizing filters at high stringency and to adopt strict scoring criteria that require multiple positive hybridization signals (e. g., each double-spotted large-insert DNA clone must hybridize to multiplexed probes that are pooled in two or three dimensions).

Because of its short sequence (40 bp), the overgo hybridization method does have several advantages, including low background of hybridization, low rate of false positives, high throughput, and ease of handling compared with conventional cloned probes. It is crucial that selection criteria for uniqueness be quite stringent. Therefore, screening against repeat databases is not sufficient for overgo selection; BLAST searches against GenBank should be an integral part of selection criteria for overgos.

FPC V4.7 is the most commonly used software that allows markers to be used with the contigs (Soderlund et al. 2000; Chapter 12, this volume). There are several existing services available to view integrated physical-genetic maps via the Internet. Web FPC offers a limited view of physical maps similar to FPC for maps such as rice, maize, sorghum, and *Arabidopsis thaliana* (http://www.genome.arizona.edu/software/fpc/). Other sites such as Ensembl (http://www.ensembl.org/) and NCBI (http://www.ncbi.nlm.nih.gov/mapview/) also offer views of BAC maps integrated with sequence and other information.

The community foresaw a need for packaged software that provides a viewing system sufficient to satisfy most of the investigators who wished to browse the data and the maps built from them without the requirement and overhead of downloading and updating datasets. A Java-based application called iCE (Internet Contig Explorer) has been designed to provide views of physical maps and associated data, where users can search for and display individual clones, contigs, clone fingerprints, clone insert sizes, and markers (Fjell et al. 2003). To retrieve and display the integrated map data, iMap (Cone et al. 2002) provides a side-by-side view of the genetic map and associated contigs, where the database is searchable by DNA marker or genetic locus. GBrowse, a flexible Web-based application (Stein et al. 2002), is becoming a community standard for displaying genomic annotations and other features associated with physical maps.

References

Coe E, Cone K, McMullen M, Chen SS, Davis G, Gardiner J, Liscum E, Polacco M, Paterson A, Sanchez-Villeda H, Soderlund C, Wing R.(2002) Access to the maize genome: an integrated physical and genetic map. *Plant Physiol.* 128(1):9–12.

Cone KC, McMullen MD, Bi IV, Davis GL, Yim YS, Gardiner JM, Polacco ML, Sanchez-Villeda H, Fang Z, Schroeder SG, Havermann SA, Bowers JE, Paterson AH, Soderlund CA, Engler FW, Wing RA, Coe EH Jr.(2002) Genetic, physical, and informatics resources for maize. On the road to an integrated map *Plant Physiol.* (4):1598–605.

Coulson A, Sulston J, Brenner S, Karn J (1986) Toward a physical map of the genome of the nematode *Caenorhabditis elegans. Proc Natl Acad Sci USA* 83:7821–7825.

Cregan, P. B., J. Mudge, E. W. Fickus, L. F. Marek, D. Danesh, R. Denny, R. C. Shoemaker, B. F. Mattews, T. Jarvik, N. D. Young. (1999). Target isolation of simple sequence marker through the use of bacterial artificial chromosomes. *Theor Appl Genet* 98:919–928.

Fjell CD, Bosdet I, Schein JE, Jones SJ, Marra MA. (2003) Internet Contig Explorer (iCE)–a tool for visualizing clone fingerprint maps. *Genome Res.*(6A):1244–9.

Kanazin V, Marek LF, Shoemaker RC (1996) Resistance gene analogs are conserved and clustered in soybean. *Proc Natl Acad Sci U S A.* 93(21):11746–50.

Klein, P. E., R. R. Klein, S. W. Cartinhour, P. E. Ulanch, J. Dong, J. A. Obert, D. T. Morishige, S. D. Schlueter, K. L. Childs, M. Ale, and J. E. Mullet. 2000. A high-throughput AFLP-based method for constructing integrated genetic and physical maps: progress toward a sorghum genome Map. *Genome Res.* 10:789–807.

Luo MC, Thomas C, You FM, Hsiao J, Ouyang S, Buell CR, Malandro M, McGuire PE, Anderson OD, Dvorak J.(2003) High-throughput fingerprinting of bacterial artificial chromosomes using the snapshot labeling kit and sizing of restriction fragments by capillary electrophoresis. *Genomics.* 82(3):378–89.

Marra MA, Kucaba TA, Dietrich NL, Green ED, Brownstein B, Wilson RK, McDonald KM, Hillier LW, McPherson JD, Waterston RH.(1997) High throughput fingerprint analysis of large-insert clones. *Genome Res.* 7(11):1072–84.

Marra M, Kucaba T, Sekhon M, Hillier L, Martienssen R, Chinwalla A, Crockett J, Fedele J, Grover H, Gund C, McCombie WR, McDonald K, McPherson J, Mudd N, Parnell L, Schein J, Seim R, Shelby P, Waterston R, Wilson R. (1999) A map for sequence analysis of the *Arabidopsis thaliana* genome *Nature Genet* 22(3):265–70.

Meksem K, Zobrist K. Ruben E., Hyten D., Quanzhou T., Zhang H-B, D. A. Lightfoot (2000) Two large insert soybean genomic libraries constructed in a binary vector: Applications in chromosome walking and genome wide physical mapping. *Theor Appl Genet* 101 5/6: 747–755.

Soderlund C, Humphray S, Dunham A, French L.(2000) Contigs built with fingerprints, markers, and FPC V4.7. *Genome Res.* (11):1772–87.

Stein LD, Mungall C, Shu S, Caudy M, Mangone M, Day A, Nickerson E, Stajich JE, Harris TW, Arva A, Lewis S. (2002) The generic genome browser: a building block for a model organism system database. *Genome Res.* (10):1599–610.

Thangavelu M, James AB, Bankier A, Bryan GJ, Dear PH, Waugh R (2003) HAPPY mapping in a plant genome: reconstruction and analysis of a high-resolution physical map of a 1.9 Mbp region of *Arabidopsis thaliana* chromosome 4. *Plant Biotechnology Journal* (1):23–31.

Wu C, Sun S, Nimmakayala P, Santos FA, Meksem K, Springman R, Ding K, Lightfoot DA, Zhang HB (2004) A BAC- and BIBAC-based physical map of the soybean genome *Genome Research.* 14(2):319–26.

10
Positional Cloning of Plant Developmental Genes

Peter M. Gresshoff

Overview

Positional, or map-based, cloning is an experimental approach to isolate causative DNA sequences responsible for altered characteristics of a developmental or metabolic phenotype. More than any other approach to gene isolation, it generates new knowledge through a direct nexus of structure (the DNA sequence) and function (the mutant phenotype). Positional cloning relies on a symphony of techniques that allow the isolation of an unknown gene by its position in the genome. Molecular markers are mapped ("positioned") closely to the locus controlling the phenotype by the application of Mendel's laws and then are used to isolate large DNA molecules, usually in the form of bacterial artificial chromosomes (BACs), that are sequenced to yield candidate gene sequences. The causative gene is usually confirmed by transformation of the candidate DNA sequence into the mutant or wild-type background (depending on the dominant or recessive nature of the mutant allele) or by repeated sequencing of multiple alleles at the same locus and establishment of a strong correlation between mutant phenotype and molecular DNA change. The approach has been particularly successful in the model plant *Arabidopsis thaliana*, where numerous genes controlling plant development and hormone response were isolated. Recently, advances in the discovery of developmental genes have been achieved with members of the legume family (such as *Glycine max* [soybean], *Lotus japonicus* [related to the forage plant birdsfoot trefoil], and *Medicago truncatula* [a forage medic]), where several genes controlling nodulation and leading to symbiotic nitrogen fixation have been isolated and characterized. Legumes have broad agricultural uses as they produce seeds rich in vegetable oil and protein as well as pharmaceutically significant products such as antioxidants, phytoestrogens (as isoflavones), and saponins. Legumes also have a major ecological function through their ability to introduce nitrogen into the biosphere, and they often are pioneering species on nutrient-deficient soils. With the advance in genome-wide DNA sequencing and the development of physical DNA fragment maps, the approach of positional cloning, together with that of

The Handbook of Plant Genome Mapping.
Genetic and Physical Mapping. Edited by Khalid Meksem, Günter Kahl

ISBN 3-527-31116-5

gene insertions, will be instrumental in finding further genetic components of plant development

Abstract

Despite knowledge of the complete DNA sequence of two plant genomes (*Arabidopsis* and rice) and extensive databases of expressed sequences (ESTs) for many others, the technique of positional or map-based cloning of genes continues to contribute to the discovery of plant genes known only by their inheritance, mutant phenotype, and genetic map position. After the initial successes of positionally cloning genes in *Arabidopsis*, tomato, and rice, success has now been swift in legumes. Critical genes controlling the perception of the *Rhizobium*-synthesized nodulation factor and subsequent systemic regulation of cell proliferation were recently cloned in the model legumes *Lotus japonicus* and *Medicago truncatula* as well as in the more recalcitrant crop legume soybean (*Glycine max*). Key genes for the perception of the lipo-oligosaccharide nodulation (Nod) factor (namely *NFR1* and *NFR5* as well as *LYK3*) encode transmembrane receptor kinases characterised by LysM domains known to interact in other cases with oligosaccharide and peptidoglycan moieties. Positional cloning also isolated the *NORK/SYMRK* gene, that functions early in the Nod factor initiated signal transduction cascade but is different from *NFR1/NFR5/LYK3*, as loss of function mutations fail to establish symbioses with either *Rhizobium* nor mycorrhizal fungi. Likewise genes for a calmodulin-dependent protein kinase (*MtDMI3*), a nucleoporin (*LjSYM3*), a potassium ion channel (*MtDMI1/LjCASTOR*), a Lotus symbiosome transporter as well as a new shoot-controlled supernodulation gene (*LjDISTANS*) were cloned. The main gene controlling autoregulation of nodulation (AON), first recognized in supernodulation mutants in soybean, was positionally cloned and characterized to encode a leucine-rich repeat receptor kinase, termed NARK (nodule autoregulation receptor kinase), related to the *Arabidopsis* CLAVATA1 (AtCLV1). Loss of function mutations or fast neutron–induced deletions exhibit a supernodulation or hypernodulation phenotype characterized by abundant nodule numbers, increased proportion of the root that is nodulated, and nodulation tolerance to otherwise inhibitory levels of nitrate. Cloning the *NARK* gene in soybean was a challenge because of the ancestral tetraploidy of the species and its recalcitrance to efficient transformation. The discovery of a CLAVATA1-related kinase has opened the search for long-distance signals as AON is regulated through functions in the leaf and not the root. Conversely, AtCLV1 is functional within the shoot apical meristem, controlling cell proliferation and cell fate. Since AtCLV1 directly or indirectly interacts with a peptide ligand, the possibility exists that AON is regulated by long-distance peptide signals or molecules related in shape to these. By comparative genomics, gene discovery in the model systems allowed discovery of identical genes in other legumes, such as pea (*Pisum sativum*), hindered by larger genomes and lower transformation frequencies. The combination of reliable high-throughput marker technology, high-density molecular maps, well-characterized mutant phenotypes, sequenced artificial chromosomes, physical maps, and genetic complementation techniques continues to advance our mechanistic insights into complex biological networks underlying plant development and plant responses to the environment.

10.1
Introduction

The basic aim of biological research is to establish a linkage between structure and function. To do this, we ask questions such as "How do telomeres work? How is auxin recognized? How is development derailed in cancerous cells?" The fundamental knowledge required to answer these questions requires information about the involved molecules, their interaction, and the fluxes between different structures and locations. Intuitively one would assume that the directed analysis of biochemical and chemical processes would reveal such information. However, living systems are chemically complex, and too many compounds exist within the same mixture (more than 200,000 metabolites have been estimated to exist in legumes); more importantly, this is a "Gemisch" that requires extraction from different cell types, or intracellular locations, that may respond differentially and may be subject to extraction artifacts. Thus, additional tools of analysis are required that link structure to function. In the future these may come through direct intracellular monitoring of metabolite changes, optimally in a real-time platform. Early advances in micro-metabolomics are promising, and the future seems to be limited only by instrumentation and interactive creativity. At present, however, the major tools come from the field of genetics, specifically, molecular genetics and functional genomics, that permit the coupling of gene structures to biological functions, thus assigning causative roles to genes and their resultant polyribonucleotide (RNA) or protein sequences.

Plant development involves the coordinated expression of gene programs during the induction of *de novo* meristems or the decision-making process following their induction to develop. Plants are bipolar organisms involving two embryonic meristem growth centers which proliferate in a stem cell–like fashion to yield new cells used for both maintenance of the stem cell linage and differentiation of new, committed cells. Plant differentiation, in contrast to animals, is mainly post-embryonic, subject to external and internal factors during the adult growth phase of the organism. Thus, plants require sophisticated recognition systems of such signals, reflected in part by the abundance of genes encoding receptor-like kinases (610 are found in the *Arabidopsis* genome, 206 of these are LRR-RLKs) in plant genomes (as compared to animal genomes, about 50 found in *Drosophila*, human and nematode). Differentiation into *de novo* organ structures such as lateral roots, flowers, and nodules is commonly initiated either at the apical meristems (or its developmental duplication of axial primordia) or from an interesting tissue maintained throughout the plant as a cylinder of pluripotent stem cells, called the cambium (in the shoot) and the pericycle (in the root). These tissues harbor immense regenerative potential, as is seen in wound healing, grafting, callus induction, organogenesis, and embryogenesis of plants. Indeed it seems essential to propose a genetic suppression system that prevents uncontrolled proliferation of such cells during development. Thus, some processes such as the formation of nitrogen-fixing root nodules in legumes may be seen as a duplicate process of induction as well as de-repression of mitosis. One sees such duplicity in the control of flowering, with both a floral inducer and repressor being regulated (Beveridge et al. 2003), and the induction of legume nodules, where the oligosaccharide nodulation factor is needed for the induction of the nodule (Spaink 2000; Oldroyd 2001) while in parallel, an independent circuit-suppressing cell proliferation from pericycle cells, seemingly mutated in "spontaneous" nodulators, needs to be inactivated (Caetano-Anollés et al. 1991a).

The discovery of plant developmental genes promises the construction of a causative gene network that constitutes pathways of organ development or environmental, nutrient, or disease responses. Over the last two decades, the extensive application of molecular biological tools (as seen as an extension of biochemistry) has revealed thousands of gene sequences that are differentially expressed during such responses or differentiation (see Figure 10.1). Such response genes are frequently different from the genes that control the developmental cascade starting with induction, then differentiation, then growth, and finally response. Response genes are usually expressed abundantly and control genes scarcely, making the discovery of regulatory genes causative of developmental mutations difficult through biochemical technology.

A good example of this approach is the use of microarrayed EST libraries that are probed with mRNA (or cDNA derived from it). Maguire et al. (2002) used 4000 unique ESTs of soybean, arrayed on glass slides, and compared the abundance in root and shoot tissue of soybean. A large number of genes were differentially expressed, some being as much as 300-fold higher in one tissue compared to the other. Differences were verified by Northern hybridization as well as by quantitative RT-PCR. A strong correlation was established between RT-PCR and microarray expression levels. Despite these findings, little can be extrapolated from such results about the regulatory genes involved in root vs. shoot differentiation. It appears that the microarray approach is an excellent surveillance procedure but that it needs to be followed up by comparative analysis of mutant material or reverse genetics such as RNA interference (RNAi; Smith et al. 2000) to develop a nexus between function and gene.

The field of legumes nodulation gained significant insight into the types of genes that are expressed during early and late stages of the symbiosis through the screening of libraries for differential gene expression (cf. Appel et al. 1999; Maguire et al. 2002). Some of these "nodulins" were characterized in detail and represent genes involved in nitrogen metabolism (glutamine synthase, uricase), oxygen transport (leghemoglobins), transport (aquaporins such as nodulin26 in soybean), structure (ENOD2, a hydroxyproline-rich cell wall protein), or infection response (ENOD12) (Mylona et al. 1995). Their function has illuminated our understanding of symbiotic nitrogen fixation, but has not told us which gene is regulating the

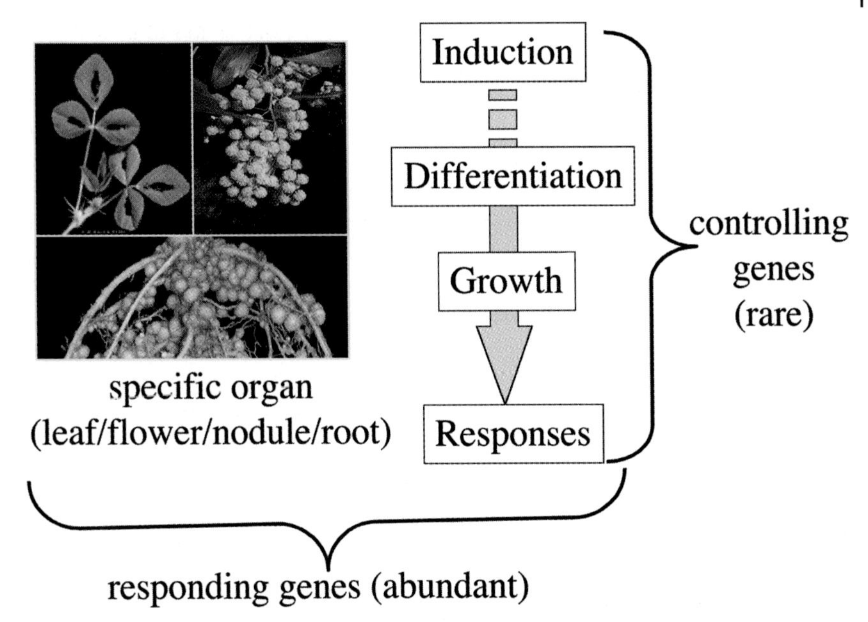

Figure 10.1. Genes control and respond during developmental events. Many molecular genetic approaches such as differential display and microarray analysis detect gene expression changes related to responses. These often are abundant and significant. Often the responses are not tissue-specific but tissue-accentuated. Control genes govern the development of the organ at different stages of commitment and often are rare and of low expression levels. Control genes frequently are transcription factors or receptor kinases (cf. *NARK, ETR1, ASTRAY, NIN1*). Shown are subterranean clover leaves and axial buds, Australia wattle flowers (a legume), and the root and nodule system of a NARK-deficient soybean.

symbiosis. Some uncharacterized nodulin genes, such as ENOD40 encoding two putative peptides, or regulatory RNA, or both (Röhrig et al. 2002), may represent regulatory functions. Many such "molecular markers for development" represent unknown functions, and despite extensive knowledge of their promoters, site of gene expression, and putative computer-based assignment to certain functions, their direct role, and indeed necessity, in the developmental step remains unproven. Recent technologies such as RNA interference (RNAi; see Smith et al. 2000) may help to reveal some function, although gene redundancy or biochemical compensation may prevent clear interpretation of results.

As yet it has not been possible to relate a known phenotypic alteration in symbiosis (as described by Kawaguchi et al. 2002; Szczyglowski et al. 1998; Carroll et al. 1985a, 1985b) with a differentially expressed gene discovered by the purely molecular biological (in contrast to molecular genetic) approach. The experimental approach is further complicated by the fact that most "tissue-specific" proteins/genes/RNAs are really not. Most nodulins are also expressed in root tissue and often in floral organs, perhaps indicating some requirement for functions (sinks, cell type differentiation). The *ENOD40* gene is one such example; most correctly, its characteristics should be defined as "nodule-enhanced" (Charon et al. 1997). Similarly, genes known to be specifically involved in photosynthetic function were shown to be expressed at very low levels (0.3 % compared to shoot controls) in complete dark-exposed roots of soybean (Maquire et al. 2002). Furthermore, a hydroxyproline rich cell wall protein was expressed 300 times higher in the root than the shoot. Classical Northern analysis in that case showed a complete "yes-no" result, illustrating that the term "specificity" is subject to the sensitivity of the analytical method (cf. Edwards and Coruzzi [1990]).

Furthermore, often RNA expression is observed in tissues that apparently do not 'use' that RNA; thus transcription control, in contrast to prokaryotic systems, may not be a major regulator of biological function in eukaryotes. An excellent example of this is seen for *Arabidopsis* CLAVATA1, where mutants display a clear developmental phenotype in the shoot apical meristem, leading to fasciation (fused or enlarged multiple meristems; Fletcher, 2002). Yet the *AtCLV1* gene is expressed in all tissues including leaves and roots (P. Gresshoff and P. Schenk, unpublished data) without an apparent root/leaf phenotype in otherwise strong mutants. Similar results were obtained for the soybean autoregulation receptor kinase gene *GmNARK* (D. Buzas, unpubl. data).

It is thus clear that genes involved in plant development (and indeed in all multicellular development) cannot be easily revealed and characterized through differential gene expression or through attempted genetic complementation of biochemical functions in lower eukaryotes such as yeast or *Chlamydomonas*, because (1) these organisms usually do not express the differentiation process in question or (2) the precise biochemical alteration is unknown.

10.2
Gene Discovery Through Insertional Mutagenesis

The approach of inserting a known piece of DNA into an unknown one, followed by the loss of function and subsequent analysis of the flanking regions (known as "insertional mutagenesis") works well and already has contributed substantially to plant, fly, worm, yeast, and bacterial gene discovery, but it will not be covered in this chapter. Numerous experimental avenues exist in plants, based mainly on the insertion of a T-DNA element from *Agrobacterium* or transposition of a transposable element (such as *Ac* or *Ds* from maize, or *TnT* from tobacco) from an appropriate launching site within the genome (Schauser et al. 1999). Some plants, especially *Arabidopsis thaliana*, possess high transformation efficiencies through floral dip transformation. Thus, tens of thousands of T-DNA insertion lines have been generated that now function as a resource of gene discovery. Naturally or induced phenotypes can be matched with insertion lines, flanking DNA can be characterized, and candidate genes are confirmed by transformation and complementation of the mutant condition. This works perfectly and is the "dream" of every developmental biologist: a fully insertionally mutated set of seeds and a completed genome. Presently, this is only possible for *Arabidopsis thaliana* and feasible, as their genome analysis nears completion, for the model legumes *Lotus japonicus* and *Medicago truncatula* in the near future.

In this context it may be noted that high-efficiency T-DNA insertion technology is paramount for this process. Several model legumes are transformable (Handberg and Stougaard 1992; Nolan and Rose, 1997; Jiang and Gresshoff 1997), but both require a labor-intensive cell culture step (Lohar et al. 2001; Stiller et al. 1997). The recent report of floral dip transformation of *M. truncatula* (Harrison et al., 2001), although extremely promising, apparently cannot be repeated. Our own attempts with seed or floral dip transformation of *L. japonicus* ecotype Gifu with *Agrobacterium tumefaciens* strain AGL1 carrying a GUS-intron reporter construct gave positive sectors in developing seedlings, but failed completely in the resultant progeny. Thus, we do not know enough about transformation in legumes or whether the soma-seed barrier, and issues as simple as the floral architecture may prevent floral dip transformation. *Arabidopsis* is somewhat special as the micropyle is open, allowing *Agrobacterium* penetration as the right time.

The introduction of a transposable element such as the maize *Ac* or *Ds* elements promises to reveal genes through insertional phenotypes. Schauser et al. (1999) were the first to utilize this approach in legumes and detected the nodule-initiation deficient (NIN) phenotype and gene. NIN appears to be a transcriptional factor, and inactivation of its gene causes termination of the infection process, leading to non-nodulation. Biochemical analysis of such mutant material may be complicated by insertionnal instability. NIN mutants frequently revert to wild type through excision of the element. Thus, developmental studies are feasible but biochemical analysis is complicated by possible chimerism.

As yet there is no large T-DNA insertion library of legumes. Few promising starts exist: Martirani et al. (1999) used *L. japonicus* and transferred *en large* T-DNA carrying a promoter-less β-glucoronidase (GUS) gene. This approach mirrored the one developed for *Arabidopsis* (Koncz et al. 1989; Kertbundit et al. 1991). Primary transformants were selected by GUS staining in nodules, nodule primordia, and roots. Examples are lines such as "Machinegun" (characterized by multiple staining foci representing all internal lateral root primordia), "Arrowhead" (expressed only in emerged root meristems including the root apical meristem [RAM]), "Hyaena" (with "dirty spots" on the root representing nodule primordia), "Cheetah" (expressing GUS in both persistent root meristems and emerging nodule meristems), "Fata morgana" (expressed in pericycle at the site of nodule initiation and the nodule interior), "Vasco" (expressing only in the nodule vasculature), and "Timpa" (extensive nodule interior GUS expression) (see Buzas et al. 2005). Disappointingly, many primary transgenic lines (50 %) were infertile, most likely through collateral damage during cell culture and the selection process. This may have been lessened through optimized transformation protocols involving BASTA (phosphinotricin) selection and shorter cell culture times (Lohar et al. 2001). Three lines were analyzed in detail: Vasco, Fata morgana, and Cheetah. In each case the T-DNA insertion seems to have occurred outside a "normal" gene or coding sequence. Detailed analysis of this material is in progress, but the essential lessons are that larger numbers of transgenics need to be screened and that not all tagged lines will yield genes. However, it is becoming clear that tagged lines are useful cytological markers for developmental processes and may be used as assay lines for critical factors. Using the same approach in *L. japonicus*, Webb et al. (2000) isolated a gene that appears to encode a calcium-binding protein.

One concludes that other methodologies for the isolation of plant developmental genes exist (Feldman 1991), having proven outcomes, but that many of these are limited to a specific organism or are limited by biological issues.

In parallel to the gene discovery technologies based on transformation- or molecular biology, another genetic approach, called positional or map-based cloning, proved fruitful. The remainder of this chapter will focus on the accomplishments and insights gained from this seemingly laborious, but progressively easier, approach.

10.3 Technology Requirements for Map-based Cloning

The overall strategy of map-based cloning relies on the successful combination and execution of several biological, genetic, and technical components (Table 10.1). Weakness at any one level will increase the workload and lessen the chances of success. It is essential that a clear and stable mutant phenotype exist for the targeted gene. Multiple alleles, preferably of variable phenotype, are advantageous for later investigations. It is easier to clone a gene that has monogenic recessive inheritance, as homozygous segregants are easily determined. Dominant mutant alleles require a further generation to verify dominant homozygotes. Attempts to clone genes that are part of a quantitative trait, even major genes, are more difficult to verify. This situation arises especially in disease resistance genes, where genes tend to work in compensating for gene networks, as well as in genes controlling environmental response traits (draft, salinity, heat).

Mutant individuals are crossed with a genetically divergent relative. For example, in the cloning of soybean (*Glycine max*) genes, we used crosses between the commercial soybean *Glycine max*, with its relevant recessive mutant locus, and *Glycine soja*, a congenic wild relative (Landau-Ellis et al. 1991). Issues of the genetic diver-

Table 10.1a. Steps of positional cloning in plants: the genetics phase.

1. Detect a clear mutant phenotype (preferably in a parent used to create a molecular map).
2. Determine inheritance (stable, distorted, monogenic, recessive?).
3. Cross mutant with genetically distinct wild type (preferably the second parent used to construct an existing molecular map).
4. Create F_2 individuals and score phenotype (about 300–500 individuals minimally needed).
5. Score same F_2 individuals for evenly spaced molecular markers.
6. Establish co-segregation of locus controlling the phenotype and molecular markers.
7. Establish map position and order relative to other nearby molecular markers.
8. Screen for further markers using arbitrary markers in bulked segregant analysis

Table 10.1b. Steps of positional cloning in plants: the genomics phase.

9. Isolate BAC clone using closely linked molecular probes by either filter hybridization or PCR screening.
10. Discover candidate BAC and isolate end clone.
11. Determine polymorphism for end clone and map in F_2 mapping population.
12. Use end clone to isolate neighboring BACs again from filters or pools.
13. Construct contig in the region.
14. Sequence candidate BAC to determine coding potential.
15. Select putative candidate genes.
16. Sequence candidate genes for mutant alleles and wild type.
17. Test allele-specific primer for co-segregation with mutant locus.
18. Carry out genetic complementation of mutant locus with wild-type allele.
19. Characterize the candidate gene for regulation and function.

(a) The genetic phase: A stable phenotype is detected and tested for inheritance. In parallel, a molecular map is used to provide molecular markers spaced perhaps 30 cM from each other. The mutant phenotype-expressing plant is crossed with the distant partner of the molecular map population, and an F_2 mapping population is created containing perhaps 200–300 individual families. Homozygous recessives are selected by phenotype. If a dominant locus is pursued, an F_3 population is needed to detect homozygous dominants based on their lack of segregation of recessive phenotypes. Flanking molecular markers, detected by linkage disequilibrium, are used to guide further tests with closer-linked markers. If marker density is low in the desired region, extra markers, perhaps based on AFLP or DAF technology, may be needed.
(b) The molecular phase: Closely linked markers are used as probes to detect a high-molecular-weight DNA clone (usually a BAC or TAC). End clones are isolated and tested for polymorphism between the two mapping parents. End clones are mapped and must co-segregate with the mutant locus. BAC clones are characterized after shotgun sequencing. Candidate genes are selected and tested by high-quality sequencing in wild-type and mutant genomes. Detected polymorphisms are further verified by sequencing of different alleles at the same locus. Complementation of the mutant phenotype is achieved by transformation either into stable transgenics or "hairy root" tissues.

sity of the mapping parents need to be considered in light of either scarcity of markers or interference in recombination. It is of advantage to isolate the original mutation in a genotype that has been used to construct a molecular map and to utilize the original mapping partner in the current positional cloning population. From this cross, an F_2 (or F_3) generation is created by selfing and is screened for expected segregation of the mutant and wild-type phenotypes. Homozygotes are assigned according to phenotype and are used to determine co-segregation with known molecular markers from an existing map. Variations of approach are discussed below in detail. Association of a molecular marker with the locus of interest is further refined either by using the map more extensively or by searching for new molecular beacons using, e. g., arbitrary marker technology in procedures such as bulk segregant analysis (BSA) (Kolchinsky et al. 1997).

The aim is to achieve close linkage (or association) between a molecular marker and the locus controlling the 'gene-associated' phenotype. But what do we call "close linkage"? There is no arbitrary guide to this, as genetic linkage is a derived value based on the frequency of recombination. Theoretically, the genetic distance between two markers (measured in the unit of centimorgans (cM; after the great *Drosophila* geneticist Morgan) varies inversely to the physical distance. But in reality, there are hot and cold spots of recombination. Centromeres and associated heterochromatin are known to restrict recombination, as evidenced by clustering of molecular markers on genetic maps. Likewise, there are recombinational hotspots. Mapping of polymorphic ends from sized characterized BAC chromosomes showed that 1 cM in tomato may span between 40 and 4000 kb of actual DNA. If the mapping population was recreated with distant relatives, as, for example, in an interspecific cross, then suppression of recombination will occur because of inversions and translocations (Sandal et al. 2002). Thus, choosing the right parents is an essential aspect of a successful mapping approach.

Usually the average physical distance per centimorgan ranges around the 200-kb range for genes in transcriptionally active regions. This was first shown for soybean in a physical mapping of a cluster of RFLP-defined loci on molecular linkage group H (MLG H; Funke et al. 1993). High-molecular-weight DNA was digested with rare cutters such as *Not*I and separated by pulse-field gel electrophoresis (CHEF-PFGE); the gel lane was then digested in the restriction nuclease that was known to release the polymorphic fragment of known size. The gel lane was then placed perpendicular onto a normal agarose gel, electrophoresed, and blotted/probed with the appropriate probe. The physical sizes of the original *Not*I fragments together with joint hybridization signals revealed that in this region of the soybean genome, one centimorgan was equivalent to about 330 kb. Interestingly, the same conversion factor was calculated for the entire genetic map if one divided the total genome size of soybean (1100 Mb per haploid genome) by the size of the entire map (averaging contractions and expansions).

Thus, the answer to the question "What is close and what is close enough?" depends on the chromosomal region and quality of segregation of flanking markers. Derivation from expected ratios would suggest chromosomal causes of altered recombination. Usually, the best attempt is to be within about 0.1–1.0 cM, which makes the chromosome walk feasible. Of course, the direct "hit", where the marker turns out to be the gene of interest, is the most optimal scenario!

Increased mapping of functional genes as defined by ESTs also permits gene isolation. Single-nucleotide polymorphisms (SNPs) stemming from such EST tags are valuable and are starting to enrich genetic maps based predominantly on non-defined markers such as random RFLP clones (see Landau-Ellis et al. 1991), microsatellites (SSRs; Cregan and Quigley 1997), and arbitrary markers such as AFLPs, DAFs, or RAPDs (Vos et al. 1995; Caetano-Anollés et al. 1991; Bassam et al. 1991; Hayashi et al. 2001; Sandal et al. 2002). Progressively, as sequencing costs decrease, RFLP probes are being sequenced. For example, many databases (such as TIGR; www.tigr.org) provide convenient tools for exploring the depth of information coupling DNA sequence, map positions, and recombination distances.

The two closest flanking molecular markers are used to isolated high-molecular-weight (HMW) DNA clones usually in the form of bacterial artificial chromosomes (BACs; original BAC reference 1992; see Men et al. 2001). A variant of BACs are transformation-competent chromosomes (TACs), which combine elements of the *Agrobacterium* T-DNA and normal replicative function in their cloning vector (Meksem et al. 2002). While these appear to facilitate potential gene cloning and library screening through complementation of recessive mutants, their application in this fashion is limited. Yeast artificial chromosomes (YACs) as a cloning vector because of their potential to carry larger DNA inserts (Pillai et al. 1996; Zhu et al. 1996). YACs, however, present additional problems in both their construction and maintenance. The homologous recombination system of yeast also tends to create undesirable deletions yeast transformation is also variable, as it prefers to transform with smaller inserts.

BAC libraries now exist for a larger range of plants and no longer are a limiting step to research (Tomkins et al. 1999; Meksem et al. 2000; Men et al. 2001). Scanning the abstracts of recent conferences such as the Plant and Animal Genome (PAG) meetings in San Diego reveals common approaches and common BAC library properties. Average BAC size is around 100 kb, with an even distribution of repeated and coding DNA. Chloroplast DNA inserts are usually present at low abundance. HMW DNA is usually extracted from isolated nuclei from plant leaves treated for several days in darkness to remove starch. Partially digested DNA (as verified by CHEF pulse-field gel electrophoresis) is ligated into a binary vector such as pBACBeLo11 or V41 (Figure 10.2), which is then transformed into *E. coli*. For TAC libraries, the clones were mated en masse into *Agrobacterium tumefaciens* (Men et al. 2001). BAC/TAC clones remain stable, as verified by size determinations and presence of markers. BAC libraries are maintained in multititer plates as well as in pools and multidimensional superpools. Detection and isolation of a candidate BAC clone occur either by hybridization of labeled probes to nylon membrane–spotted BACs (Figure 10.3a) or by PCR screening of BAC pools and superpools.

For some organisms such as soybean and *Lotus japonicus*, physical libraries are becoming available. These are based on BAC fingerprinting (Figure 10.3b), end

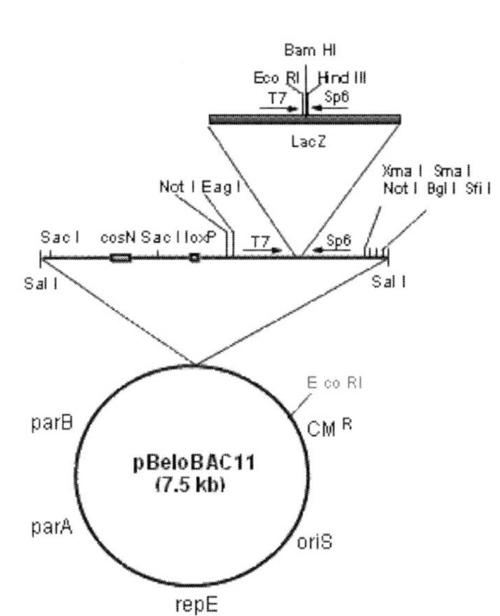

Figure 10.2. Example of a BAC clone vector. The vector usually contains a *Hin*dIII cloning site for insertion of partially digested plant DNA. Usually, the cloning site is inside the *E. coli LacZ* gene, allowing easy "blue-white" selection of inserted clones. Selectable markers and replication origin as well as components of the *Agrobacterium tumefaciens* mini-Ti plasmid are present.

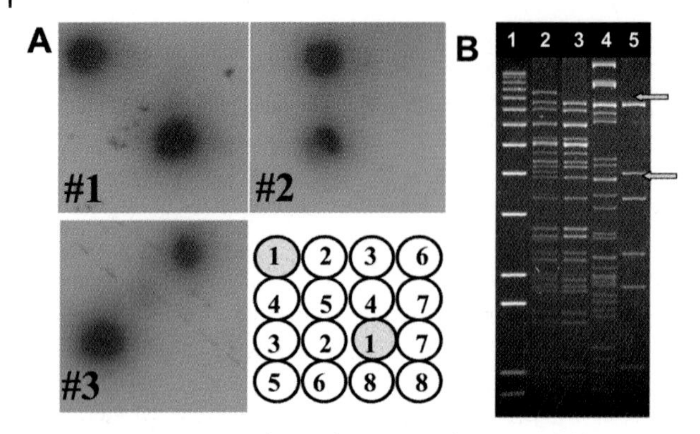

Figure 10.3. Detection of BAC clones. (a) Filter-spotted libraries. The BAC library was spotted robotically onto nylon membranes in a spatially distinct twin-spot pattern. Following hybridization with the labeled probe, X-ray film or phosphoimaging instrumentation was used to detect twin spots. The arrangement of spots helps with the determination of the candidate BAC clone. High-intensity signals usually indicate a direct "hit." However, attention needs to be paid to multiple gene copies or pseudo-genes. (b) PCR-based pool screening. BAC clones from multititer plates are pooled according to an X-Y-Z dimensional scheme. Superpools are screened by PCR. Positive pool overlaps are further tested with more detailed pools and eventually single clones.

clone sequencing, and mapping of end clones. The availability of a physical map will increase the speed of map-based cloning.

Once a candidate BAC is detected and confirmed in its position by sequencing and then mapping polymorphic end clones, the BAC is subjected to shotgun sequencing to determine its coding content. In many plant species, regions are duplicated and one may find that the candidate BAC represents another genomic region. Great care is needed at this stage. For example, we, in collaboration with the Australian Genome Research Facility in Brisbane, sequenced 135 kb of soybean BAC17107 (c.f., Searle et al., 2003); a total of 2,300 reads were needed for over 900 nebulization-derived fragments of about 1 kb in size (A. Men; pers. comm.); the BAC contained numerous repeat elements, or parts thereof, from retro-transposons and transposons. Annotation revealed many putative genes found to be collinear to *Arabidopsis thaliana* as well as *Lotus japonicus* and *Medicago truncatula*.

The DNA sequence of the candidate BAC is then annotated for candidate genes. For example, soybean BAC17107 contains 11 genes with high confidence values, supported by EST, protein and genomic data. When looking within that set of candidates for the "correct" gene, bias always plays into this stage, as researchers tend to "favor" genes involved in control functions (e. g., transcriptional factors, receptors, kinases). Household genes such as rRNA, ribosomal proteins, or metabolic enzymes often are overlooked at this early stage. Likely candidates are sequenced carefully using multiple sequencing runs; mutant alleles are also sequenced to determine whether a SNP can be correlated with the phenotype of interest. Candidate alleles are verified by mapping the SNP through allele-specific PCR amplification. Complete co-segregation in all mutant progeny is expected. If the SNPs co-map and provide an explanation for the phenotype through an altered protein structure, then the absolute verification of the candidate gene status is complementation of the mutant allele by the wild type. This is achieved by a variety of methods with different levels of rigor and effort. The positionally cloned, dominant *Pst1* disease resistance gene was complemented by transient bombardment. The early infection gene *LYK3*, known to encode a LysM domain receptor kinase (Limpens et al. 2003), was complemented by *Agrobacterium rhizogenes* hairy root cultures, representing the first positional cloning in the model legume *Medicago truncatula*; this was acceptable, as the phenotype was root-controlled. Many researchers are worried about the use of hairy roots in nodulation research because the transgenic tissue is altered in its phytohormonal balance. This appears to be of minimal consequence, as studies in several legumes (Beach and Gresshoff 1988; Martirani et al. 1999; Stiller et al. 1997; Bond et al. 1996) show that nodulation pattern, nodule number, and ability to respond to autoregulation are maintained in hairy roots. Moreover, nodule-related gene expression (*ENOD40* and leghemoglobin *LbhA* were tested) is maintained in the same cell-specific manner as in non-transgenic roots (Martirani et al. 1999). The only unusual phenotype associated with hairy roots was observed in strain K599 transformed roots of soybean Bragg, which nodulated normally but developed carrot-shaped nodules on about 5–15 % of transgenic roots (J. Bond and P. Gresshoff, unpublished data). It is possible that in roots with a high Ri-T-DNA copy number, the developmental program leading to cessation of cell proliferation in determinate nodules is altered to create meristematic nodules. The hypernodulation gene *Har1* (see Wopereis et al. 2000), encoding an LRR receptor kinase (Nishimura et al. 2002a), was complemented by stable transformation using *A. tumefaciens* in *L. japonicus*. This was needed, as the *Har1*-controlled phenotype (hypernodulation) is shoot-controlled (see Jiang and Gresshoff 2002; Krusell et al. 2002).

10.4 Positional Cloning Successes in Legume Nodulation Genes

10.4.1 Nodulation Biology

The process of nodulation encapsulates multiple plant developmental features so that its analysis provides factual and conceptual background for many developmental processes in plants (Figure 10.4). For example, lateral root formation shares many common features with nodulation, such as the involvement of the pericycle. Conversely, many features are distinct, such as the inverse regulation by ethylene (Lohar et al. 2005; Penmetsa et al. 2003). Legume nodules are controlled outgrowths from the cortex and pericycle opposite from the xylem poles. Nodules are induced by soil bacteria broadly known as *Rhizobium*. The *Rhizobium* genome harbors about 50 nodulation-related genes that are transcriptionally activated by root exudates usually belonging to the flavone/isoflavone family. The nodulation

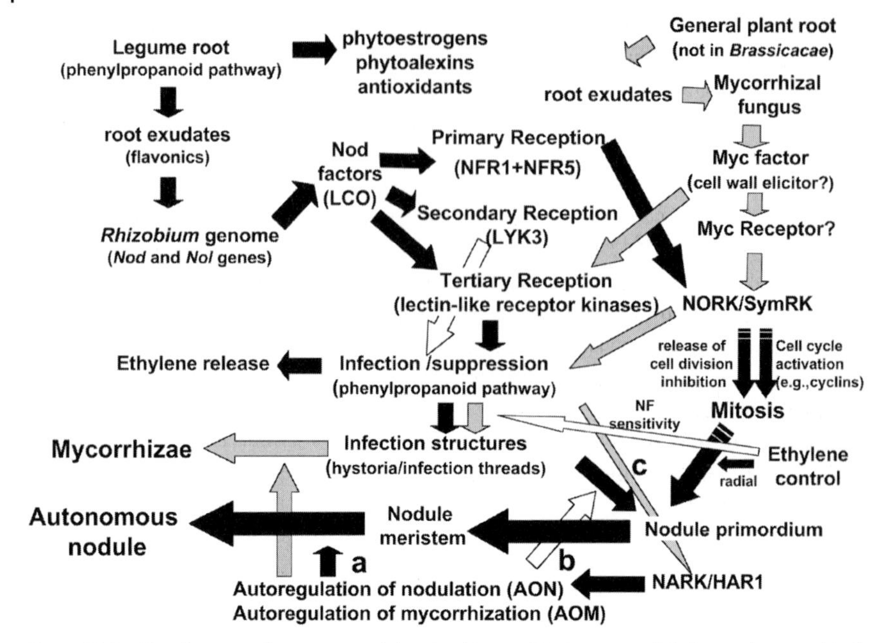

Figure 10.4. Developmental gene network involved in symbiotic root nodulation and mycorrhizal development. Gray arrows refer to the mycorrhizal pathway, and black arrows imply the nodule pathway. The target of autoregulation is still unclear and may differ between legumes and in the degree of regulation. For example, for alfalfa, *b* is greater than *a* activity, while in soybean, *a* is greater than *b*. It is still unknown whether successful "colonization" of the primordium is required for the transition of the primordium to the autonomous nodules. Autoregulation of mycorrhization and its dependence on GmNARK was recently demonstrated (Vierheilig and Gresshoff, unpublished data). It is not clear what signal is perceived by NARK/HAR1; thus arrow *c* is a hypothesis based on the common infection-related features. Although the mycorrhizal and rhizobial signals are perceived in part by a common cascade, and cross-regulation occurs, distinction is maintained, as neither mycorrhizal nor other fungi (perhaps acting through the common "tertiary receptor" recognizing chitin-like elicitors) induce cell division leading to nodulation. Ethylene controls nodulation by altering the sensitivity of the root to Nod factor and also by inhibiting cortical and pericycle cell divisions close to phloem bundles (Lohar et al. 2004).

genes cooperate to synthesize a novel class of plant growth regulators called "Nodulation (Nod) factors". These are oligosaccharides consisting of three to five β-(1-4)-linked *N*-acetyl-D-glucosamine units modified by a long fatty acid at the C2 position of the non-reducing end sugar and several minor decorations involving sulfate, acetate, or fucose (Oldroyd 2001). For example, the major Nod factor for *Lotus japonicus* is a pentameric *N*-acetylglucosamine carrying a *cis*-vaccenic acid and a carbonyl group at the non-reducing terminal residue together with a 4-*O*-acetylfucose on the reducing terminal residue (Niwa et al. 2001). As such, the Nod factor is related to chitin, commonly seen in insect exoskeletons and fungal cell walls. The role of the decorations appears to be transport and survival from degradation (presumably from chitinases). Different *Rhizobium* species produce different decorations and thus provide one level of host specificity. Some rhizobia, (such as strain NGR234) produce a large cocktail of factors and are able to nodulate multiple hosts.

Nod factors are perceived by a specific recognition machinery recently elucidated through positional cloning of genes controlling non-nodulation phenotypes (Figure 10.5). It was known that some legume mutants (such as *Ljsym1* and *Ljsym5*) uniquely lack the ability to respond to the *Rhizobium*, while others (such as *Ljsym2*, *Mtdmi2*, and *MsMN1007*) showing the same non-nodulation phenotype also lacked the ability to form symbioses with mycorrhizal fungi. Thus, the *Rhizobium* and mycorrhizal symbioses share common components, reflecting evolutionary modification of the ancient fungal symbiosis for the more modern nodule symbiosis. It is still unknown how the original signaling occurs for the fungal symbiosis, but it is likely that the fungus releases a Nod factor equivalent, perhaps similar to other fungal elicitors known from the field of plant pathogenicity (Kosata et al. 2003). The lack of axenic fungal culture and the absence of mycorrhizal genetics hinders progress in this area.

10.4.2
Non-nodulation Genes

Positional cloning has revealed a large set of critical genes involved in nodulation (Table 10.2). Often the isolated gene was used to find orthologs in related species, illustrating that legume research can help plant biology in general. The first legume gene isolated by map-based cloning was the non-nodulation gene from the MN1008 mutant originally isolated by Barnes in alfalfa (*Medicago sativa*). This mutant is special, as alfalfa is tetraploid. In a monumental effort, Kiss and colleagues in Hungary made crosses and transferred the mutation to a diploid alfalfa species. Then, using RAPD markers (not really the most robust or high-throughput), they managed to find the causative gene and labeled it *NORK* (nodulation receptor kinase; Endré et al. 2002). This kinase contains LRR domains and a nucleotide-binding site (NBS). Interestingly, the MN1008 mutant is known to form spontaneous nodules, (i.e., nodules in the absence of *Rhizobium* inoculation), thus, the mutatio must act downstream from the *NORK* gene. In parallel M. Parniske's group in Britain used *Lotus japonicus* and isolated a *NORK* ortholog from the *Ljsym2* mutant (Stracke et al. 2002). The *Lotus* gene was labeled *SYMRK* (symbiosis receptor kinase), reflecting its role in both *Rhizobium* and mycorrhizal symbioses (Figure 10.4). Genetics predicted that *LjSym1* and *LjSym5* acted upstream of *LjSym2*.

The group of J. Stougaard in Denmark achieved the cloning of a putative Nod factor receptor late in 2003, opening a major avenue of investigation (Radutoiu et al. 2003; Madsen et al. 2003). *LjSym1* and *LjSym5* code for two related LysM domain receptor kinases named NFR1 and NFR5. LysM domains have long been known in several gene products with putative oligosaccharide- or peptidoglycan-binding activities. It is likely that the two proteins interact in the plasma membrane to form a heterodimer, which in turn may directly or indirectly interact

Table 10.2. Examples of positionally cloned plant genes.

Gene name	*Organism*	*Biochemical function*	*Approach*	*Reference*
NORK	*Medicago sativa*	NBS-LRR-receptor kinase	MBC	Endré et al. 2002
SymRK	*Lotus japonicus*	NBS-LRR-receptor kinase	MBC	Stracke et al. 2002
NARK	*Glycine max*	LRR receptor kinase	MBC	Searle et al. 2002,2003
NARK	*Glycine max*	LRR receptor kinase	homology	Nishimura et al. 2002a
HAR1	*Lotus japonicus*	LRR receptor kinase	MBC	Krusell et al. 2002
HAR1	*Lotus japonicus*	LRR receptor kinase	MBC	Nishimura et al. 2002a
NFR1	*Lotus japonicus*	LysM receptor kinase	MBC	Radutoiu et al. 2003
NFR5	*Lotus japonicus*	LysM receptor kinase	MBC	Madsen et al. 2003
LYK3	*Medicago truncatula*	LysM receptor kinase; NFR1 like	MBC	Limpens et al. 2003
ASTRAY	*Lotus japonicus*	Transcription factor; Hy5 like	homology	Nishimura et al. 2002b
Nin	*Lotus japonicus*	Transcription factor	insertion	Schauser et al. 1999

MBC = map-based cloning; homology = usually based on an *Arabidopsis gene*; insertion = using the Ac/Ds system

with the LjSYMRK/NORK protein. *NFR5* is the ortholog of the pea gene *Pssym10* and lacks an activation loop, suggesting that it lacks kinase activity. Presumably, the heterodimer functions through shared receptor and kinase activity. Direct Nod factor binding data for LjNFR1/5 do not exist as yet, although it was shown that the mutations affect intracellular calcium spiking, a direct response to Nod factor application. The two papers from 2003 couple the positional cloning of the Nod factor receptor genes to a cellular phenotype, namely, the inability of the mutant to respond to the Nod factor through membrane depolarization and extracellular alkalinization.

A similar LysM domain gene was isolated from *M. truncatula* by the group of T. Bisseling in the Netherlands (Limpens et al. 2003), who originally discovered a host strain–specific, non-nodulation phenotype in a natural accession from Afghanistan. This mutant, although well characterized, lacks the wild-type control. Therefore, Limpens et al. (2003) found tightly linked molecular markers to the *Pssym2* genes then isolated a BAC from *M. truncatula* based on extensive microsynteny. The *M. truncatula* BAC was sequenced and, contained seven different LysM-type receptor kinase genes in a small multigene cluster. (*NB*: Since Nod factor specificities in legumes seem to evolve rapidly, as evidenced by significant structural requirements in closely related legumes, it is likely that a *LysM* gene cluster could function in the generation of diversity through intragenic recombination, perhaps resembling immunoglobulin or plant disease gene evolution.) Allelic sequencing was not a feasible approach to discover a candidate gene, as in pea there was no wild type and in *M. truncatula* there was no mutant. Thus, brilliantly, Limpens et al. (2003) used transgenic hairy roots transformed with RNAi constructs for the multiple candidates; two of these significantly lowered or eliminated nodulation, suggesting that they are the candidate genes. The *LYK3* gene found in *M. truncatula*, although related, is not equivalent to the *NFR1* or *NFR5* genes discovered in *Lotus*. Indeed the mutant phenotype is different, as *Pssym2* allows infection thread formation that is then aborted, while NFR1/5 mutants are completely blocked in infection thread formation. This suggests that there are multiple "gatekeepers" as the infection progresses and signals back and forth. Using the above information as well as extensive mapping data, we determined that the soybean *rj1/nod49* locus is mutated at the *NFR1* gene (Indrasumunar et al., in preparation). Two alleles have been reliably sequenced and lead to frame shift mutations with severe consequences on protein structure (premature termination). Interestingly nod49 mutants do respond to *Bradyrhizobium* inoculation at high cell density with pseudoinfections (cell divisions in the sub-epidermal cell layer) and occasional nodules (Mathews et al., 1989). Why would an apparent strong knock-ouit mutant maintain a responsiveness to high elicitor levels? The answer may lie with parallel receptors of weaker affinity and receptor cross-talk. It is clear that the new field of post-genomic research will focus on the substrates of these kinases and the nature of their complexes. Technologies involving protein biochemistry, based on enriched protein expression, protein-protein interaction analysis by Surface Plasmon Resonance (SPR; see Biacore 3000 instrumentation), mass spectroscopy, phospho-proteomics, site-directed mutagenesis, as well as reverse genetics aimed at functional analysis (such as RNA interference (Jorgensen, 2002), and TILLING (Perry et al., 2003) will add new dimensions to our genetic and transcriptional analysis.

10.4.3 Autoregulation of Nodulation Genes

Multicellular organisms need to control the proliferation of cells. This is achieved at the cellular level, through cell cycle control, as well as at the tissue (localized) and organismal (systemic) level. This extent of control is paramount for plant development where pluripotent stem cells are maintained in the cambium, the pericycle, and the apical domes. How does a plant control the cell division potential and fate of resultant cells from such regenerative tissues? Clearly, multiple mechanisms overlap, one of which is the CLAVATA complex of the shoot apical meristem illustrated beautifully in *Arabidopsis* (see Fletcher 2002; Clark et al. 1997 and references therein). A similar system may function in other meristems, as is suggested by broad expression profiles of *AtCLV1* (Gresshoff and Schenk, unpublished data), the abundance of CLV1-related receptor kinases in the *Arabidopsis* genome, and the recent discovery of autoregulation mutants in legumes.

Autoregulation of nodulation (AON) is a feedback system in which young infection stages restrict development of new infection stages (Caetano-Anollés and Gresshoff 1991; Gresshoff 1993). Loss-of-function mutants in AON lead to abundant nodulation termed either hyper- or supernodulation (Carroll et al. 1985a, 1985b). The regulation functions by either blocking initiation of new cell division clusters (early-acting as in alfalfa) or blocking the advance of induced clusters (late-acting as in soybean) (Mathews et al. 1989; Caetano-Anollés and Gresshoff 1991). The regulation is systemic, as split root systems (Olssen et al. 1989) and reciprocal grafts (Delves et al. 1986, 1992; Jiang and Gresshoff 2002; Krusell et al. 2002) illustrated that the shoot controls autoregulation. By using rooted trifoliate leaves of soybean, it was possible to maintain the nodulation phenotype governed by the source plant, suggesting that the leaf itself is capable of controlling autoregulation (Bano and Harper 2002; Li and Gresshoff, unpublished data). Intergrafts of wild-type stems onto supernodulation mutant stems do not suppress nodulation (Li and Gresshoff, unpublished data). AON mutants have been isolated in soybean, pea, *Lotus japonicus*, and *Medicago truncatula* (see Searle et al. 2003 and references therein) and generally possess the same characteristics of increased nodule number, nitrate tolerance in nodulation, reduced root systems when nodulated, and shoot control of AON. The large range of mutants coupled with similar mutations in other LRR receptor kinases have allowed the definition of conserved structural components for the kinase activity, such as conserved APE and DFG sites, as well as the active site for phosphorylation (Diévart et al. 2003). Some hypernodulation mutants such as *Ljhar1-1* also have an altered lateral root phenotype under specific growth conditions, suggesting that the same gene controls nodule and lateral root initiation and outgrowth (Wopereis et al. 2000). Two exceptions exist to the generality of leaf-controlled autoregulation; in pea there is a mutant *nod3*, which is not an allele of *Pssym29* (the *GmNARK* ortholog of pea) and which is root-controlled. Similarly, Lotus mutant *astray* (Nishimura et al. 2002b) is characterized by increased nodulation and was found by an "educated guess" to be the ortholog of the *Arabidopsis Hy5* gene, known to increase the number of lateral roots. Interestingly the *Ljastray* mutant does not have an altered lateral root phenotype. *LjASTRAY* encodes a bZIP-type transcription factor and has an ortholog in *GmSTF1*. *Ljastray* also functions in the root. Real-time PCR expression analysis of *GmSTF1* shows that the gene is broadly expressed in soybean tissue (Davis 2003).

Map-based cloning was successful in finding the major leaf-active AON gene of legumes. Landau-Ellis et al. (1991) were the first to map an AON mutant locus (namely, *nts382* of soybean) on the newly emerging soybean map. Closely linked markers were used to isolate a candidate BAC, which revealed the CLAVATA1-related receptor kinase gene GmNARK (nodule autoregulation receptor kinase) (Searle et al. 2003; Men et al. 2002). Using multiple allelic mutants at the locus, a direct correlation was established between the mutations and the mutant phenotype. Most mutations turned out to be nonsense mutations, suggesting that minor structural changes caused by missense mutations, unless close to critical residues, result in a minor phenotype not easily detected in mutant screens. An excellent example in soybean is the mutant *nts1116*, which exhibits hypernodulation (three times that of wild type) and is mutated in V837A (Searle et al. 2003). The minor missense mutation changing a valine to an alanine is detected only phenotypically because the alteration is right next to the active site of the kinase domain. In parallel, Krusell et al. (2002) and Nishimura et al. (2002a) cloned a related gene, *LjHAR1* in *Lotus japonicus*. *GmNARK* is highly expressed as RNA in leaves and stems but only weakly in shoot apical meristem or nodules (Searle et al. 2003; Gresshoff et al. 2004). The high level of RNA expression but lack of biological activity of *GmNARK* in root tissue presents a paradox that still needs to be resolved by detailed biochemical studies involving tissue-specific partner molecules such as AtCLV2 in the *Arabidopsis* system, RNA turnover, or availability of specific ligands or downstream response elements (such as ROP, KAPP, WUSCHEL, or POLTERGEIST in *Arabidopsis*; cf. Fletcher 2002; Gresshoff et al. 2003). *GmNARK* is duplicated in the soybean genome as *GmCLV1A* (Men et al. 2002), which is 95 % identical at the DNA level. This duplicate locus does not possess autoregulation activity, as is evidenced by the fast neutron deletion mutant FN37 of *G. soja* (Men et al. 2002) that lacks *GmNARK* and possesses *GmCLV1A* yet is autoregulation-deficient. *GmCLV1A* shares an RNA expression profile with *GmNARK* and *AtCLV1*, with moderate expression in the SAM (10–30 times higher than *GmNARK*) and ubiquitous expression in leaves, roots, and flowers (D. Buzas, A. Hayward, S. Nontachaiyapoom, unpublished data; Gresshoff et al. 2003). Is it possible that *GmCLV1A* functions in the SAM similarly to *ATCLV1*? Does GmCLV1A possess some residual AON-related activity permitting older tissues in a *GmNARK* mutant (such as *nts1007*) to express some form of autoregulation? Does a *GmCLV1A* mutant possess a phenotype related to fasciation and pod development? Such questions can now be answered because positional cloning has revealed some key members in the AON cascade. However, analyses at the DNA and RNA levels are limited, as they do not extend into the molecular physiology, which, after all, is "where the action is." Coupling genomic data with those of the phenome will clearly be a major challenge of developmental biology in the next decade.

Discovery of new long-distance signals in plant development, especially those controlling genes involved in cell proliferation and cell fate, may induce a paradigm shift in plant development from small signal substances (auxins, cytokinins, etc.; see Beveridge et al. 2003) to larger molecules such as RNA (Jorgensen 2002; Tang et al. 2003) and peptides (Takayama and Sakagami 2002; Matsubayashi et al. 2002; Ryan et al. 2002). The purified kinase or LRR domains of *GmNARK* may allow the isolation of a specific antibodies that could lead to immune precipitation of the NARK complex and resultant detection of GmNARK interactors.

The fascinating aspects of the NARK system are its involvement in long-distance regulation of cell proliferation and its similarity to the CLAVATA complex. There are further receptor kinases related to the regulation of cell proliferation. The phytosulfokine receptor perceives different small sulfated peptides and regulates cell division in plant cells maintained in cell culture (Lorbiecke and Sauter 2002). The major soybean cyst nematode resistance genes (*Rhg1* and *Rhg4*) are LRR receptor kinases (K. Meksem, unpublished data). Nematode and *Rhizobium* interactions with a legume, despite broad differences in the biology, share common steps such as the activation of the Cheetah promoter fusion in *L. japonicus* (D. Lohar, unpublished data). Several fungal elicitor receptors also are LRR receptor kinases, as are components of the animal cell division machinery involving EGF (epider mal growth factor). The interesting cross-activity of NARK to control AON as well as autoregulation of mycorrhization, mirroring the overlap during the early induction of infection stages, suggest that AON perhaps also is a derivation from an ancestral fungal mechanism controlling establishment/infection of the microsymbiont in the plant tissue. One can assume that mycorrhizal fungi and *Rhizobium* induce a common plant response (possibly through the NORK/SYMRK step; see Figure 10.5) that results in a translocated signal that is recognized by the leaf and interpreted in a common way to produce the shoot-derived inhibitor that now blocks further invasion success of the microbe. If such persistence in the tissue is essential for further nodule proliferation, then regulation of invasion could regulate cell proliferation. One thus wonders which components of the systemic signaling involved in SAR and general fungal responses were "pirated" for the autoregulation of nodulation.

10.5 Conclusions

Positional cloning of plant developmental genes is now a reality actively used in several model plant systems. We can take advantage of a decade-long technology development in molecular marker technology, molecular maps, BAC library construction, data sorting, and high-speed and low-cost DNA sequencing. EST and SNP information grows daily and enriches the information pool needed to detect candidates. Plant transformation, although still lengthy in many species, is a common procedure in hundreds of laboratories.

It is likely that the next five years will see the complete or near-complete genome sequence of many more plants. Tomato, soybean, tobacco, *Lotus japonicus*, and *Medicago truncatula* are just some examples that spring to mind. Physical maps, even with numerous gaps, will sort BAC clones into larger contigs. T-DNA insertion collections will exist for several plant species, allowing rapid isolation of flanking sequences related to mutant phenotypes. Reverse genetics, involving either the search of a mutant based on a gene sequence (by T-DNA collection screening or TILLING; Perry et al. 2003) or RNA interference will enrich the search for genetic components of plant developmental phenomena. In all this explosion of knowledge, the need for physiological integration, the recognition of redundancy and lack of function, and the complexities of the "protein world" and the resultant metabolic pools that make the marvels of plant life should be foreseen.

Acknowledgments

The Australian Research Council, the University of Queensland through its Strategic Initiative Funds and other central funds, and the Queensland State Government are thanked for providing support for the ARC Centre of Excellence. I thank past and present colleagues, staff, and students for their interest in and support of nodulation biology.

References

Appel, M., Bellstedt, D. and Gresshoff, P. M. (1999) Differential display of eukaryotic mRNA: meeting the demands of the new millennium? *J. Plant Physiol.* 154:561–570.

Bano, A., and Harper, J.E. (2002) Plant growth regulators and phloem exudates modulate root nodulation of soybean. Funct. Plant Biol. 29:1299–1307.

Bano, A., Harper, J.E., Augé, R.M., and Neuman, D.S. and (2002) changes in the phytohormones levels following inoculation of two soybean lines differeing in nodulation. Funct. Plant Biol. 29:965–974.

Bassam, B. J., Caetano-Anollés, G. and Gresshoff, P. M. (1991). A fast and sensitive silver-staining for DNA in polyacrylamide gels. *Analytical Biochemistry* 196:80–83.

Beach, K. and Gresshoff, P. M. (1988). Characterization and culture of *Agrobacterium rhizogenes* transformed roots of forage legumes. *Plant Science* 57:73–81.

Beveridge, C. A., Gresshoff, P. M., Rameau, C., and Turnbull, C. G.N. (2003) Many more signals needed to orchestrate development. *J. Plant Growth Regulators* 22:15–24.

Bond, J. E., McDonnell, R. E. and Gresshoff, P. M. (1996) The susceptibility of nodulation mutants of *Glycine max* to *Agrobacterium tumefaciens*. *J. Plant Physiol.* 148:684–692.

Buzas, D. M., Lohar, D., Stiller, J. and P. M. Gresshoff, P.M. (2005) Lotus japonicus promoter trapped lines provide markers for developmental domains. (in preparation).

Carroll, B. J., McNeil, D. L. and Gresshoff, P. M. (1985a) Isolation and properties of soybean mutants which nodulate in the presence of high nitrate concentrations. *Proc. Natl. Acad. Sci. USA* 82:4162–4166.

Carroll, B. J., McNeil, D. L. and Gresshoff, P. M. (1985b). A supernodulation and nitrate tolerant symbiotic (nts) soybean mutant. *Plant Physiol.* 78:34–40.

Caetano-Anollés, G., Joshi, P. A., and Gresshoff, P. M. (1991a). Spontaneous nodules induce feedback suppression of nodulation in alfalfa. *Planta* 183:77–82.

Caetano-Anollés, G., Bassam, B. J. and Gresshoff, P. M. (1991). DNA amplification fingerprinting using very short arbitrary oligonucleotide primers. *Bio/Technology* 9:553–557.

Chang, C., Kwok, S. F., Bleecker, A. B. and Meyerowitz, E. M. (1993) *Arabidopsis* ethylene-response gene *ETR1*: Similarity of product to two-component regulators. *Science* 262:539–545.

Charon, C., Johansson, C., Kondorosi, E., Kondorosi, A. and Crespi, M. (1997) *enod40* induces dedifferentiation and division of root cortical cells in legumes. *Proc. Natl. Acad. Sci. USA,* 94:8901–8906.

Clark, S. E., Williams, R. W., and Meyerowitz, E. M. (1997) The *CLAVATA1* gene encodes a putative receptor kinase that controls shoot and floral meristem size in Arabidopsis. *Cell* 89:575–585.

Cregan, P. B. and Quigley, C. V. (1997) Simple sequence repeat DNA marker analysis. In: DNA Markers: protocols, applications and overviews. eds. G. Caetano-Anollés and P. M. Gresshoff. Wiley and Sons, New York. pp. 173–185.

Davis, P. (2003) Regualtion of nodule and lateral root related genes in soybean. Honours thesis, Botany Department, Univ. of Queensland, Brisbane, Australia.

Delves, A. C., Mathews, A., Day, D. A., Carter, A. S., Carroll, B. J. and Gresshoff, P. M. (1986). Regulation of the soybean-*Rhizobium* symbiosis by shoot and root factors. *Plant Physiol.* 82:588–590.

Delves, A. C., Higgins, A. and Gresshoff, P. M. (1992). The shoot apex is not the source of the systemic signal controlling nodule numbers in soybean. *Plant, Cell and Environment* 15:249–254.

Diévart, A. and Clark, S. E. (2003) Using mutant alleles to determine the structure and function of leucine-rich repeat receptor-like kinases. *Curr. Opin. Plant Biol.* 6:507–516.

Edwards, J. W. and Coruzzi, G. M. (1990) Cell-specific gene expression in plants. *Ann. Rev. Genet.*, 24:275–303.

Endré, G., Kereszt, A., Kevei, Z., Mihacea, S., Kalo, P. and Kiss, G. P. (2002) A receptor kinase regulating symbiotic nodule development. *Nature* 417:962–966.

Feldman, K. A. (1991) T-DNA insertion mutagenesis in *Arabidopsis*: mutational spectrum. *Plant J.*, 1:71–82.

Fletcher, J. C. (2002) Shoot and floral meristem maintenance in *Arabidopsis. Annu. Rev. Plant Biol.* 53:45–66.

Funke, R., Kolchinsky, A. and Gresshoff, P. M. (1993). Physical mapping of a region in the soybean (*Glycine max*) genome containing duplicated sequences. *Plant Mol. Biol.* 22:437–446.

Gresshoff PM (1993) Molecular genetic analysis of nodulation genes in soybean. *Plant Breeding Reviews* 11:375–401.

Gresshoff, P. M. (2001) *Glycine max* (soybean). *Encyclopedia of Genetics*. Eds. S. Brenner and J. Miller, Academic Press, London, New York, pp. 884–885.

Gresshoff, P. M. (2003a) Post-genomic insights into nodulation. *Genome Biology* 4:201.

Gresshoff, P. M. (2003b) Genetics and genomics of nodulation and symbiotic nitrogen fixation nodulation. *Handbook of Plant Biotechnology*, J. Wiley & Sons (London), eds. P. Christou and H. Klee, pp. 246–265.

Gresshoff, P. M., Rose, R. J., Singh, M., and Rolfe, B. G. (2003) Symbiosis signals. *Today's Life Science* May/June 2003, 30–33.

Gresshoff, P. M., Buzas, D. M., Laniya, T. S., Men, A., Jiang, Q., Schenk, P., Hayward, A., Kam, J., Li, D. X., Miyahara, A., Nontachaiyapoom, S., Indrasumunar, A., Brcich, T., Gualtieri, G., Davis, P. and Carroll, B. J. (2004) Systemic regulation of nodulation by a leaf-controlled LRR-receptor kinase. Mol. Plant Microbe Interactions St. Petersburg; eds. I. Tikhanovich pp. 233–236.

Handberg, K. and Stougaard, J. (1992) *Lotus japonicus*, an autogamous, diploid legume species for classical and molecular genetics. *Plant J.* 2:487–496.

Hayashi, M., Miyahara, A., Sato, S., Kato, T., Yoshikawa, M., Taketa, M., Hayashi, M., Pedrosa, A., Onda, R., Imaaizumi-Anraku, H., Kawasaki, S., Kawaguchi, M., and Harada, K. (2001) Construction of a genetic linkage map of the model legume *Lotus japonicus* using an intraspecific F_2 population. *DNA Research* 8:301–310.

Jiang Q., and Gresshoff, P. M. (1997) Classical and molecular genetics of the model legume *Lotus japonicus. Molecular Plant-Microbe Interactions* 10:59–68.

Jiang, Q. and Gresshoff, P. M. (2002) Shoot-control and genetic mapping of the har1-1 (hypernodulation and aberrant root formation) mutant of *Lotus japonicus. Funct. Plant Biol.* 29:1371–1376.

Jorgensen, R. A. (2002) RNA traffics information systemically in plants. *Proc. Natl. Acad. Sci.* USA 99:11561–11563

Kawaguchi, M., Imaizumi-Anraku, H., Koiwa, H., Niwa, S., Ikuta, A., Syono, K., and Akao, S. (2002) Root, root hair and symbiotic mutants of the model legume *Lotus japonicus. Mol. Plant. Microbe Interact.* 15:17–26.

Kertbundit, S., De Greve, H., DeBoeck, F., van Montagu, M. and Hernalsteens, J.-P. (1991) *In vivo* random β-glucuronidase gene fusions in *Arabidopsis thaliana. Proc. Natl Acad. Sci. USA,* 88:5212–5216.

Kolchinsky, A., Landau-Ellis, D. and Gresshoff, P. M. (1997). Map order and linkage distances of molecular markers close to the supernodulation (*nts-1*) locus of soybean. *Molec. Gen. Genetics* 254:29–36.

Koncz, C., Martini, N., Mayerhofer, R., Koncz-Kalman, Z., Korber, H., Redei, G. P. and Schell, J. (1989). High frequency T-DNA-mediated gene tagging in plants. *Proc. Natl Acad. Sci. USA,* 86:8567–8571.

Krusell, L., Madsen, L. H., Sato, S., Aubert, G., Genua, A., Szczyglowski, K., Duc, G., Kaneko, T., Tabata, S., de Bruijn, F., Pajuelo, E., Sandal, N., and Stougaard, J. (2002) Shoot control of root development is mediated by a receptor-like kinase. *Nature* 420:422–426.

Kosata, S., Chabaud, M., Lougnon, G., Gough, C., Dénarié, J. Barker, D. G., Bécard, G.(2003) A diffusible factor from arbuscular mycorrhizal fungi induces symbiosis-specific *MtENOD11* expression in roots of *Medicago truncatula*. *Plant Physiology* 131:952–962.

Landau-Ellis, D., Angermüller, S. A., Shoemaker, R., and Gresshoff, P. M. (1991). The genetic locus controlling supernodulation cosegregates tightly with a cloned molecular marker. *Mol. Gen. Genetics* 228:221–226.

Limpens. E., Franken, C., Smit, P., Willemse, J., Bisseling, T., and Geurts, R. (2003) LysM domain receptor kinases regulating rhizobial nod factor-induced infections. *Science* 302:630–633. (published online: 28 August 2003;10.1126/science.1090074.

Lohar, D., Schuller, K. A., Buzas, D., Gresshoff, P. M., and Stiller, J. (2001) Efficient transformation of *Lotus japonicus* using selection with the herbicide BASTA. *J. Exp. Botany* 43:857–863.

Lohar, D. P., Stiller, J., Hababunga, S., Kam, J. H.W., Tabata, S., Sato, S., Dunlap, J., Stacey, G., and Gresshoff, P. M. (2004). Ethylene insensitivity conferred by the *Arabidopsis ETR1-1* receptor gene alters nodulation control and lateral root formation in transgenic *Lotus japonicus*. Submitted for publication.

Lorbiecke, R. and Sauter, M. (2002) Comparative analysis of PSK peptide growth factor precursor homologs. *Plant Science* 163:321–332.

Madsen, E. B., Madsen, L. H., Radutoiu, S, Olbryt, M., Rakwalska, M., Szczyglowski, K., Sato, S., Kaneko, T., Tabata, S., Sandal, N., and Stougaard, J. (2003) A receptor kinase gene is involved in legume perception of rhizobial signals. *Nature* 425:637–640.

Maguire, T. L., Grimmond, S., Forrest, A., Iturbe-Ormaetxe, I., Meksem, K., Gresshoff, P. M. (2002) Tissue-specific gene expression monitored by cDNA microarray analysis of soybean (*Glycine max*). *Journal of Plant Physiology* 159:1361–1374.

Martirani, L., Stiller, J., Mirabella, R., Alfano, F., Lamberti, A., Radutoiu, S. E., Iaccarino, M., Gresshoff, P. M., and Chiurazzi, M. (1999) T-DNA tagging of nodulation and root related genes in *Lotus japonicus*: expression patterns and potential for promoter trapping and insertional mutagenesis. *Mol. Plant-Microbe Interact.* 12:275–284.

Mathews, A., Carroll, B. J., and Gresshoff, P. M. (1989) Development of *Bradyrhizobium* infections in supernodulating and non-nodulating mutants of soybean (*Glycine max* (L.) Merrill). *Protoplasma* 150:40–47.

Matsubayashi, Y., Ogawa, M., Morita, A., and Sakagami, Y. (2002) An LRR receptor kinase involved in perception of a peptide plant hormone, phytosulphokine. *Science* 296:1470–1472.

Meksem, K., Zobrist, K., Ruben, E., Hyten, D., Quanzhou, T., Zhang, H. B., and Lightfoot, D. A. (2000). Two large-insert soybean genomic libraries constructed in a binary vector: applications in chromosome walking and genome wide physical mapping, *Theor Appl Genet* 101:747–753.

Men, A. E., and Gresshoff, P. M. (2001) DAF yields a cloned marker linked to the soybean (*Glycine max*) supernodulation *nts-1* locus. *J Plant Physiol* 158:999–1006.

Men, A. E., Laniya, T. S., Searle, I. R., Iturbe-Ormaetxe, Hussain, A. K.M., Gresshoff, I., Jiang, Q., Carroll, B. J., and Gresshoff, P. M. (2002) Fast neutron mutagenesis produces a supernodulating mutant containing a large deletion in linkage group H of soybean (*Glycine soja* L.). *Genome Letters* 1:147–155.

Mylona, P., Pawlowski, K. and Bisseling, T. (1995) Symbiotic nitrogen fixation. *Plant Cell*, 7:869–885.

Nishimura, R., Hayashi, M., Wu, G.-J., Kouchi, H., Imaizumi-Anraku, H., Murakami, Y., Kawasaki, S., Akao, S., Ohmori, M., Nagasawa, M., Harada, K., and Kawaguchi, M. (2002a) HAR1 mediates systemic regulation of symbiotic organ development. *Nature* 420:426–429.

Nishimura, R., Ohmori, M., Fujita, H., and Kawaguchi, M. (2002b). A *Lotus* basic leucine zipper protein with a RING-finger motif negatively regulates the developmental program of nodulation. *Proc. Natl. Acad. Sci. (USA)* 99:15206–15210.

Niwa, S., Kawaguchi, M., Imaizumi-Anraku, H., Chechetka, S. A., Ishizaka, M., Ikuta, A and Kouchi, H. (2001) Responses of a model legume *Lotus japonicus* to lipochitin oligosaccharide nodulation factor purified from *Mesorhizobium loti* JRL501. *Mol. Plant Microbe Interact.* 14:848–856.

Oldroyd, G. E.D. (2001) Dissecting symbiosis: developments in nod factor perception. *Annals of Botany* 87:709–718.

Olsson, J. E., Nakao, P., Bohlool, B. B. and Gresshoff, P. M. (1989). Lack of systemic suppression of nodulation in split root systems of supernodulating soybean (*Glycine max* (L.) Merr.) mutants. *Plant Physiol.* 90: 1347–1352.

Perry, J. A., Wang, T. L., Welham, P. J., Gardner, S., Pike, J. M., Yoshida, S., and Parniske, M. (2003) A TILLING reverse genetics tool and web-accessible collection of mutants of the legume *Lotus japonicus. Plant Physiol.* 131:866–871.

Pillai, S., Funke, R. P., and Gresshoff, P. M. (1996) Yeast and bacterial artificial chromosomes (YAC and BAC) clones of the model legume *Lotus japonicus. Symbiosis* 21:149–164.

Penmetsa, R. V., Frugoli, J. A., Smith, L. S., Long, S. R., and Cook, D. R. (2003) Dual genetic pathways controlling nodule number in *Medicago truncatula. Plant Physiology* 131:1–11.

Radutoiu, S, Madsen, L. H., Madsen, E. B., Felle, H. H., Umehara, Y., Gronlund, M., Sato, S., Nakamura, Y., Tabata, S., Sandal, N., and Stougaard, J. (2003) Plant recognition of symbiotic bacteria requires two LysM receptor-like kinases. *Nature* 425:585–592.

Röhrig, H., Schmidt, J., Miklashevichs, E., Schell, J., and John, M. (2002) Soybean ENOD40 encodes two peptides that bind to sucrose synthase. *Proc. Natl. Acad. Sci.* (USA) 99:1915–1920.

Ryan, C. A., Pearce, G., Scheer, J., and Moura, D. S. (2002) Polypeptide hormones. *Plant Cell* 14:S251–264.

Sandal, N. et al. (2002) A genetic linkage map of the model legume *Lotus japonicus* and strategies for fast mapping of new loci. *Genetics* 161:1673–1683.

Schauser, L., Roussis, A., Stiller, J. and Stougaard, J. (1999) A plant regulator controlling development of symbiotic root nodules *Nature* 402:191–195.

Searle, I. R., Men, A. M., Laniya, T. S., Buzas, D. M., Iturbe-Ormaetxe, I., Carroll, B. J., and Gresshoff, P. M. (2003) Long distance signalling for nodulation requires a CLAVATA1-like receptor kinase. *Science* 299:108–112. (first published on line 31. Oct. 2002. 10.1126/science.1077937).

Smith, N. A., Singh, S. P., Wang, M. B., Stoutjesdijk, P., Green, A. and Warerhouse, P. M. (2000) Total silencing by intron-spliced hairpin RNAs. *Nature* 407:319–320.

Spaink, H. P. (2000) Root nodulation and infection factors produced by rhizobial bacteria. *Annual Review of Microbiology* 54:257–288.

Stiller, J., Martirani, L., Tuppale, S., Chian, R.-J., Chiurazzi, M. and Gresshoff, P. M. (1997) High frequency transformation and regeneration of transgenic plants in the model legume *Lotus japonicus.* J. exp. Botany 48:1357–1365.

Stracke, S., Kistner, C., Yoshida, S., Mulder, I., Sato, S., Kaneko, T., Tabata, S., Sandal, N., Stougaard, J., Szczyglowski, K. and Parniske, M. (2002) A plant receptor-like kinase required for both bacterial and fungal symbiosis. *Nature* 417:959–962.

Szczyglowski, K., Shaw, R. S., Wopereis, J., Copeland, S., Hamburger, D., Kasiborski, B., Dazzo, F. B. and de Bruijn, F. J. (1998). Nodule organogenesis and symbiotic mutants of the model legume *Lotus japonicus. Mol. Plant-Microb. Inter.*, 11:684–697.

Takayama, S. and Sakagami, Y. (2002) Peptide signalling in plants. *Current Opin. in Plant Biol.* 5:382–387.

Tang, G., Reinhart, B. J., Bartel, D. P., and Zamore, P. D. (2003). A biochemical framework for RNA silencing in plants. *Genes Dev.* 17: 49–63.

Tomkins, J. P. Mahalingam, R., Smith H, Goicoechea, J. L., Knap, H. T. and Wing, R. A. (1999). A bacterial artificial chromosome library for soybean PI 437654 and identification of clones associated with cyst nematode resistance. *Plant Mol Biol* 41:25–33.

Vos, P., Hogers, R., Bleeker, M., Reijans, M., van de Lee, T., Hornes, M., Frijters, A., Pot, A. J., Peleman, J., Kuiper, M., and Zabeau, M. (1995). AFLP: a new technique for DNA fingerprinting, *Nucleic Acids Res* 23:4407–4412.

Webb, J. et al. (2000) *Mesorhizobium loti* increases root-specific expression of a calcium-binding protein homologue identified by promoter tagging in *Lotus japonicus. Mol. Plant Microbe Interact.* 13:606–616.

Wopereis J., Pajuelo E., Dazzo F. B., Jiang Q., Gresshoff P. M., de Bruijn F. J., Stougaard J., and Szczyglowski K. (2000) Short root mutant of *Lotus japonicus* with a dramatically altered symbiotic phenotype. *Plant Journal* 23:97–114.

Zhu, T., Shi, L., Funke, R. P., Gresshoff, P. M. and Keim, P. (1996) Characterization and application of soybean YACs to molecular cytogenetics. *Mol. Gen. Genetics* 252:483–488.

11
Whole-genome Physical Mapping: An Overview on Methods for DNA Fingerprinting

Chengcang Wu, Shuku Sun, Mi-Kyung Lee, Zhanyou Xu, Chengwei Ren, Teofila S. Santos, and Hong-Bin Zhang

Overview

Whole-genome physical mapping with bacteria-based, large-insert clones is becoming an active research area of genome research. Previous studies have demonstrated that whole-genome clone-based physical maps are crucial to many aspects of advanced genomics and molecular biology research, including large-scale genome sequencing, large-scale gene mapping, gene cloning and characterization, and long-range genome analysis. DNA fingerprinting has become the method of choice for large-scale, bacteria-based large-insert clone analysis and generation of whole-genome, clone-based physical maps. To facilitate whole-genome physical mapping with bacteria-based large-insert clones, several fingerprinting methods have been developed. Using these methods supplemented with other tools, whole-genome physical maps have been developed from bacteria-based large-insert clones for several species of economic and/or biological importance. These physical maps have greatly promoted the comprehensive genome research of these and related species. The effective use of the techniques of clone DNA fingerprinting will greatly expedite whole-genome physical mapping and thus rapidly advance the genome research of plants, animals, humans, insects, and microbes. This chapter provides an overview of the fingerprinting methods and their applications in whole-genome physical mapping and explains the detailed procedures for generating fingerprints from bacteria-based large-insert clones.

Abstract

DNA fingerprinting combined with the technologies of large DNA fragment cloning in bacteria has become the most widely used approach to whole-genome physical mapping with large-insert clones. To facilitate effective use of this approach for whole-genome physical mapping, this chapter provides an overview of the major fingerprinting methods developed to date, including the agarose gel-based

The Handbook of Plant Genome Mapping.
Genetic and Physical Mapping. Edited by Khalid Meksem, Günter Kahl

ISBN 3-527-31116-5

restriction fingerprinting method, the manual sequencing or polyacrylamide gel-based restriction fingerprinting method, and the automatic sequencing gel- and capillary electrophoresis-based restriction fingerprinting method. The procedures of the fingerprinting methods that have been used or that have potential for wider utilization are detailed. Analysis of the whole-genome physical maps developed to date by using the fingerprinting methods and the most recent comparative studies among different fingerprinting methods have shown that quality contig map construction is significantly affected by fingerprinting methods. A capillary electrophoresis-based, two- or three-enzyme fingerprinting method seems to be the most efficient for whole-genome physical mapping, although any of the fingerprinting methods could be used to generate fingerprints and to construct contig maps from large-insert clones. It is also observed from these studies that other tools such as DNA markers of genetic maps are needed to assist in the construction of robust physical maps consisting of one or a few contigs for each chromosome from the fingerprints of the source clones generated by the fingerprinting methods. New techniques or strategies are also needed to construct contig maps for the genomic regions that are abundant in tandem repeats such as centromeric and rDNA repeats.

11.1
Introduction

11.1.1
Inception and Development of DNA Fingerprinting

DNA fingerprinting or profiling refers to a method of detecting unique restriction patterns of the DNA of an individual organism or a DNA fragment cloned in a cloning host. The uniqueness of the organism's DNA or the cloned DNA fragment allows distinction of one organism or clone from another. The term DNA fingerprinting was first described by Jeffreys et al. (1985) using hybridization of DNA with mini-satellite DNA probes. Since then a number of DNA fingerprinting systems have been developed based on DNA hybridization, PCR amplification, and/or restriction enzyme digestion. Hybridization- and PCR-based DNA fingerprinting systems have been used extensively in forensic and legal medicine and in plant research for molecular marker development, genetic mapping, breeding, and diagnostics (Weising et al. 1995).

The focus of this chapter is on various methods of fingerprinting bacteria-based large-insert clones cloned in bacterial artificial chromosomes (BACs), bacteriophage P1-derived artificial chromosomes (PACs), plant transformation-competent binary BACs (BIBACs), plasmid-based large-insert bacterial clones (PBCs), and transformation-competent artificial chromosomes (TACs) (see Chapter 9) using restriction enzyme(s), and their applications in the generation of whole-genome integrated physical and genetic maps of large, complex genomes.

11.1.2
Clone-based Whole-genome Physical Mapping

The goal of clone-based whole-genome physical mapping is to reconstruct the entire genome of an organism from large-insert, arrayed DNA libraries. In general, it involves an orderly arrangement of the clones of large-insert DNA libraries that represent the entire genome and are overlapped using different methods (see below). The overlapping clones that are expected to derive from the same genomic region are grouped by computers based on their similarities in nucleotide sequences and/or restriction profiles, forming an overlapping clone group known as a "contig." The continuously overlapping clone contigs spanning a genomic region or the entire genome are referred to as clone-based physical maps for the genomic region or whole genome. The physical map contigs can be further landed to a genetic linkage map using DNA markers mapped to the genetic map. As more mapped markers are used, all contigs of the physical map can be landed to the genetic linkage map, forming a robust, integrated physical/genetic map (Figure 11.1). Similarly, the large-insert clones containing DNA markers can be used as probes to anchor the integrated physical/genetic map to individual chromosomes by fluorescent *in situ* hybridization (FISH), forming an integrated physical, genetic, and cytogenetic map.

Integrated physical and genetic maps are the centerpieces of genomics research. They provide not only revolutionary platforms and organizational frameworks for all genomics research activities and consequences – including genome sequence, genetic linkage maps, DNA markers, genes, quantitative trait loci (QTLs), expressed sequence tags (ESTs), regulatory sequences, and repeat elements – but also a "freeway" for structural, functional, and evolutionary genomics research (Figure 11.1).

Bacteria-based large-insert clone fingerprinting and contig assembly have become the methods of choice for whole-genome physical mapping. To facilitate construction of a whole-genome physical map from bacteria-based large-insert clones, several restriction fingerprinting methods have been developed, including the agarose gel-based method (Olson et al. 1986; Marra et al. 1997), the manual polyacrylamide gel- or sequencing gel-based method (Coulson et al. 1986; Zhang and Wing 1997), and the automatic sequencing gel- (Gregory et al. 1997; Ding et al. 1999, 2001) or capillary electrophoresis-based methods (Xu et al. 2002, 2004). The agarose gel-based method and the manual sequencing gel-based method have been successfully used in the development of whole-genome, large-insert, clone-based physical maps of several species, including *Arabidopsis* (Marra et al. 1999; Chang et al. 2001), *Drosophila* (Hoskins et al. 2000), human (IHGMC 2001), indica rice (Tao et al. 2001), mouse (Gregory et al. 2002), japonica rice (Chen et al. 2002; Li et al. 2004), soybean (Wu et al. 2004), and chicken (Ren et al. 2003). The automatic sequencing gel- or capillary electrophoresis-based methods represent the latest techniques for whole-genome physical mapping with bacteria-based large-insert clones. Because of their high resolution, high throughput, automation, and amenability to multiplex fingerprint genera-

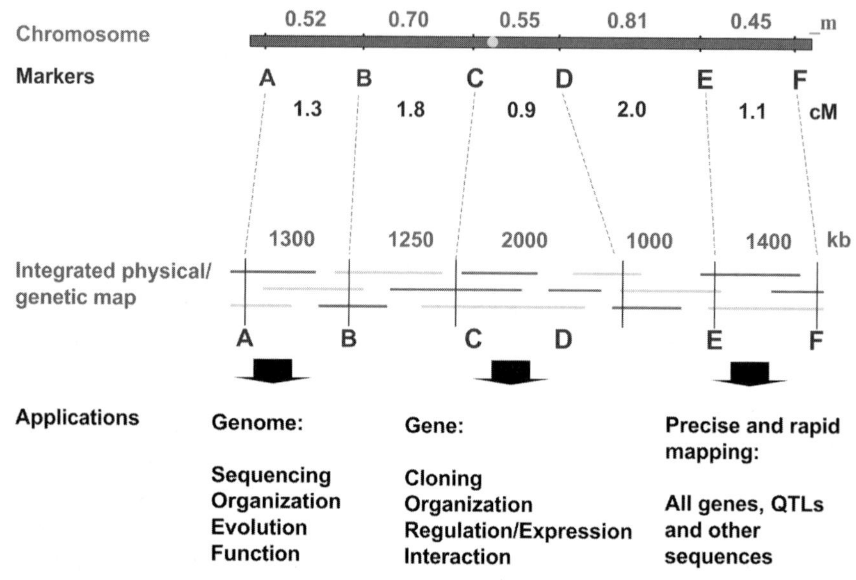

Figure 11.1. A conceptual integrated physical, genetic, and molecular cytogenetic map of a plant chromosome and its applications. The overlapping dark grey and light grey bars in the integrated physical/genetic map are large-insert genomic DNA clones such as BACs/BIBACs. The minimally overlapping tiling path of clones is highlighted in light grey. Letters A to F indicate DNA markers selected from a genetic linkage map. The markers are integrated into the physical map by screening its source BAC libraries. The integrated physical/genetic map is mapped to the chromosome by FISH of the marker-associated BACs/BIBACs. The distances between the markers are measured in both kilobases and centimorgans in the integrated physical/genetic map and in micrometers on the molecular cytogenetic map. The integrated physical, genetic, and molecular cytogenetic map provides a powerful platform for all genomics research activities.

tion, these methods are being used to construct whole-genome physical maps of several species.

11.1.3
Source DNA Libraries for Whole-genome Physical Mapping

Bacteria-based large-insert arrayed DNA libraries are desirable substrates for whole-genome physical mapping by restriction fingerprinting and contig assembly. They are essential not only for whole-genome physical mapping but also for large-scale sequencing and analysis of large, complex genomes (Zhang and Wu 2001; see also Chapter 9). Bacteria-based large-insert DNA libraries cloned in BAC, PAC, BIBAC, PBC, or TAC represent the state-of-the-art technology for large DNA fragment cloning in bacteria. Unlike the earlier DNA libraries cloned in cosmid, fosmid, P1, and YAC, the bacteria-based large-insert DNA libraries are capable of cloning DNA fragments as large as 300 kb in bacteria, are stable in the host and low in chimeric clones, and are amenable to high-throughput DNA purification.

Bacteria-based large-insert DNA libraries that are desired for whole-genome physical mapping should have the following features (see Chapter 9). First, they should have large insert sizes with fewer insert-empty clones. Based on the studies of Soderlund et al. (2000), although libraries with larger average inserts better facilitate construction of quality whole-genome physical maps, libraries with average insert sizes of 160 kb or larger are desired for whole-genome physical mapping. Second, the source libraries should ideally represent the entire target genome. Because almost all of the bacteria-based large-insert DNA libraries were constructed from genomic DNA partially digested with restriction enzymes and because the distribution of the restriction sites of a restriction enzyme is uneven along a genome, several libraries constructed with different restriction enzymes are needed. Moreover, multiple libraries cloned in different vectors and/or host systems will further increase their representation for the whole genome. The use of different vectors and/or host systems in the construction of the libraries enhances the capacity of cloning the sequences that are difficult to be cloned in one or another vector/host. In fact, almost all of the whole-genome physical maps reported to date were developed from two or more DNA libraries constructed with different restriction enzymes (see above). Third, because one of the major goals of genome research in crop plants is to improve the crop plants using the tools provided by genome research, the libraries constructed in transformation-competent vectors such as BIBAC and TAC streamline the utility of the resultant physical maps for functional analysis and genetic engineering of the genome through genetic transformation. Finally, the genome coverage is another important aspect to be considered. Studies have shown that the number of clones equivalent to approximately 10× haploid genomes seems to be the most efficient for global genome physical mapping (Xu et al. 2004). This conclusion is strongly supported by the whole-genome shotgun sequencing of both microorganisms (Fleischmann et al. 1995; Fraser et al. 1995) and higher organisms (Hoskins et al. 2000), in which 6–10× genome coverage of clones were sequenced.

11.2 Techniques and Methodologies

Restriction fingerprinting of bacteria-based large-insert clones relies upon DNA digestion with one or more restriction enzymes, followed by the measurement of the sizes of the resulting fragments. Although the clone DNA could be digested with a single restriction enzyme and fractionated by agarose gel electrophoresis (Olson et al. 1986; Marra et al. 1997) – or with two or more restriction enzymes, end-labeled with radioactive or fluorescent nucleotides, and fractionated by polyacrylamide gel (Coulson et al. 1986; Carrano et al. 1989; Gregory et al. 1997; Hong 1997; Zhang and Wing 1997; Ding et al. 1999, 2001) or capillary electrophoresis (Xu et al. 2002, 2004), leading to different restriction fingerprinting methods – the procedure of whole-genome physical mapping by fingerprint analysis of bacteria-based large-insert clones includes the following steps.

1. Clones are selected from the source DNA libraries and grown in the medium containing appropriate antibiotics, and clone DNA is isolated and purified.
2. DNA is subjected to restriction digestion with one or more restriction enzymes, and the restricted fragments are either directly fractionated on agarose gel matrices or end-labeled and then fractionated on sequencing gel matrices or in capillaries.
3. The restriction profiles or fingerprints of the clones are visualized and digitized in computers.
4. Overlapping clones are assembled into contigs using computer software called Fingerprinted Contig (FPC) (http://www.sanger.ca.uk/Software/fpc/) based on the similarity of the restriction patterns or fingerprints of the clones.

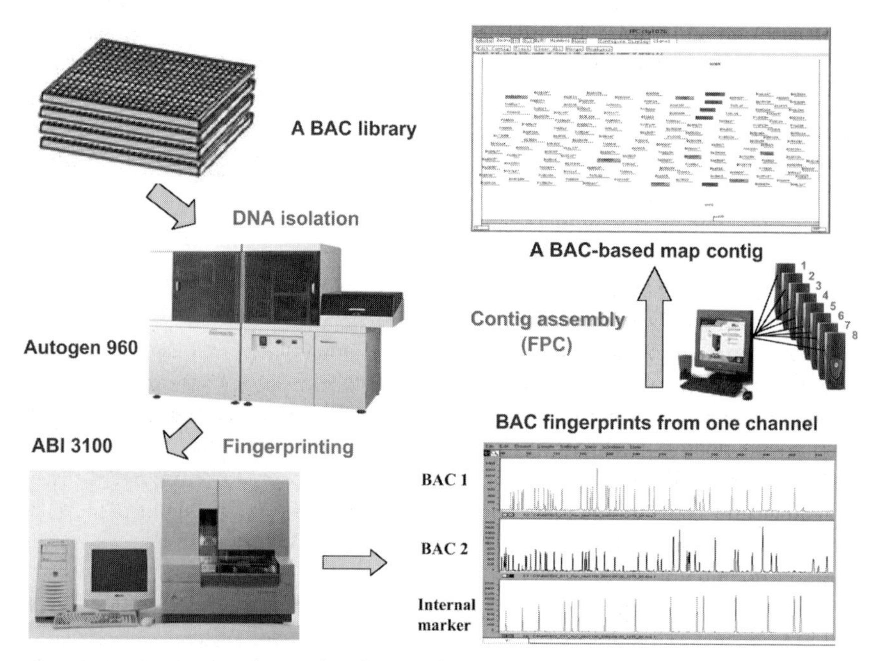

Figure 11.2. An automated procedure for large-insert clone fingerprinting and contig map assembly (modified from Xu et al. 2002 and http://hbz.tamu.edu - Analysis Tools – Automated). The source BAC library is stored in 384-well microtiter plates. BAC DNA is isolated by the Auto-Genprep 960 Robotic Workstation, digested with *Hin*dIII, *Bam*HI, and *Hae*III, and end-labeled at the *Hin*dIII and *Bam*HI sites with HEX-ddATP (BAC 1) or NED-ddATP (BAC2) in a single step. The restricted and end-labeled DNA fragments are fractionated with an internal molecular weight marker on the ABI 3100 Genetic Analyzer. The fingerprints of two BACs, along with one internal standard marker (orange), are generated from a single capillary channel of the ABI 3100. The fingerprints are collected with the GeneScan installed in the ABI 3100, transferred into a high-performance calculating cluster of computer workstations, and assembled into a contig, a segment of a whole-genome physical map, using a computer program called FingerPrinted Contig (FPC). The contig is finally anchored to a genetic linkage map with one or more mapped DNA markers shown below the contig.

Figure 11.2 shows an automated procedure of genome physical mapping with bacteria-based large-insert clones using the DNA isolation robotic workstation AutoGenprep 960 (AutoGen, Inc. USA), the automatic DNA capillary sequencer ABI 3100 (Applied Biosystems, USA), and a high-performance calculating cluster of computer workstations (Dell, USA). This chapter will depict in detail fingerprint generation and fingerprinting method comparison and its applications, whereas fingerprint editing and contig map construction are described in Chapter 12.

11.2.1 Preparation of DNA from Bacteria-based Large-insert Clones

Several protocols have been used to prepare DNA from bacteria-based large-insert clones for fingerprinting. All of the protocols are based on the alkaline lysis method developed by Birnboim and Doly (1979). Introduced here are three of the methods: the clone-by-clone method and the manual and fully automated 96-well format mini-preparations. In comparison, the manual and fully automated 96-well format mini-preparations have a throughput several-fold higher than the clone-by-clone approach. Using the latter methods, DNA can be routinely isolated from 1200–2400 clones in one scientist day.

11.2.1.1 Isolation of DNA from Bacteria-based Large-insert Clones Using the Clone-by-Clone Approach

1. Using sterile toothpicks, inoculate individual clones from the source-arrayed DNA library into 2 mL of Luria broth (LB) or Terrific broth (TB) supplemented with appropriate antibiotic in 10-mL polypropylene or glass tubes. Incubate the cells at 37 °C with shaking at 250 rpm for at least 16 h.
2. Harvest the cells by centrifugation at 2500 rpm (1200 *g*) for 10 min and discard the supernatant. Thoroughly remove the remaining culture medium because this may interfere with subsequent endonuclease digestion.
3. Add 0.3 mL of solution I (15 mM Tris-HCl, pH 8.0, 10 mM EDTA, 100 μg mL^{-1} RNase A) and resuspend the cell pellet by vortex and incubate on ice for 5 min.
4. Add 0.3 mL of freshly prepared solution II (0.2 N NaOH, 1% SDS) and gently rotate the tube to thoroughly mix the contents. Keep on ice for 5 min. The cell lysis is considered to be successful when the turbid suspension becomes translucent or clear.
5. Slowly add 0.3 mL of solution III (3 M KOAc, pH 5.5) to the tube and gently mix the contents. Thick, white *E. coli* cell debris and genomic DNA will be visible. Keep the tube on ice for at least 15 min.
6. Spin at 3250 rpm (1900 *g*) for 30 min at 4 °C to collect the cell debris and chromosomal DNA. Transfer all supernatant to a 1.5-mL Eppendorf tube. While transferring, try to avoid any white precipitated materials. Add 0.8 mL isopropanol, mix well, and keep the tube at room temperature for at least 10 min. Otherwise, the sample may be left at –20 °C overnight.

7. Precipitate DNA in a tabletop microcentrifuge for 10 min. Pour off the supernatant and wash the DNA pellet twice with 0.5 mL of 70% ethanol. Spin at room temperature for 5 min.
8. Discard the supernatant and air-dry the pellet at room temperature and resuspend in 30 μL of TE (10 mM Tris.HCl, 1 mM EDTA, pH 8.0). The DNA can be used for both agarose gel-based and polyacrylamide gel-based restriction fingerprinting.

11.2.1.2 Isolation of DNA from Bacteria-based Large-insert Clones in 96-well Format Using a Manual Approach

1. Using a 96-pin clone replicator (Fisher Scientific, USA), inoculate the clones stored in a 384-well microtiter dish of the source library into four 96-deep well plates (Beckman, USA), with each well containing 1 mL of TB or LB medium supplemented with appropriate antibiotics. Seal the plates with aluminum foil and incubate at 37 °C with shaking at 325 rpm for 20 h.
2. Pellet the cells by centrifugation at 2500 rpm (1200 g) for 10 min in a GS-6KR centrifuge with microplate carriers (Beckman, USA) and discard the supernatant by inverting the plates.
3. Add 40 μL of solution I (15 mM Tris-HCl, pH 8.0, 10 mM EDTA, 100 μg mL^{-1} RNase A) to each well using an 8- or 12-channel pipette, vortex to resuspend the cell pellets, and incubate on ice for 5 min.
4. Add 80 μL of freshly prepared solution II (0.2 N NaOH, 1% SDS) to each well and mix gently by tapping the microplate on top of lab bench at least 20 times. Incubate at room temperature for 5 min and then on ice for 5 min. The suspension should change from turbid to translucent, suggesting that the cells are successfully lysed.
5. Slowly add 60 μL of solution III (3 M KOAc, pH 5.5) to each well and gently mix using a vortex. A white precipitate of *E. coli* cell debris and genomic DNA will be seen. Keep the plates on ice for at least 15 min.
6. Spin at 3250 rpm (1900 g) for 30 min at 4 °C to pellet the cell debris and chromosomal DNA, and transfer 100 μL of the supernatant with an 8- or 12-channel pipette into eight 0.2-mL tube strips or 96-well 0.2-mL microtube plates (Applied Biosystems, USA). Try to avoid any white precipitated materials during transferring.
7. Add 100 μL of isopropanol to each sample, mix well, and incubate at room temperature for at least 10 min. At this stage, the samples can be kept at –20 °C overnight.
8. Precipitate DNA by centrifugation at 3250 rpm (1900 g) for 15 min. Pour off the supernatant and wash DNA pellets twice with 0.1 mL of 70% ethanol and invert the plates several times each wash. Spin at 3250 rpm for 5 min.
9. Discard the supernatant, air dry the pellets at room temperature, and resuspend each sample in 20 μL of TE. The DNA can be used for both agarose gel- and polyacrylamide gel-based restriction fingerprinting.

11.2.1.3 Isolation of DNA from Bacteria-based Large-insert Clones in 96-well Format Using the AutoGenprep 960 Robotic Workstation

1. Using a 96-pin clone replicator, inoculate the clones stored in a 384-well microtiter dish of the source library into four 96-deep well plates (Autogen, Inc., USA) containing 1.4 mL of TB (or LB) medium supplemented with appropriate antibiotic in each well, seal the plates with aluminum foil, and grow at 37 °C with shaking at 350 rpm for 24 h.
2. Isolate the clone DNA using the AutoGenprep 960 DNA Isolation Robotic Workstation according to the standard procedure recommended by its manufacturer.
3. Resuspend the DNA in 40 μL of TE with 100 μg mL^{-1} of DNA-free RNase, vortex for 20 s, incubate at 37 °C for 10 min, and spin down. The DNA then can be used for restriction fingerprinting.

11.2.2 Fingerprinting the Clone DNA Using Restriction Enzymes

Several methods have been developed to generate fingerprints from bacteria-based large-insert clones using restriction enzymes for whole-genome physical mapping. Below is a summary of each of the methods.

11.2.2.1 Agarose Gel-based, Restriction Fingerprinting Method

The agarose gel-based method (Marra et al. 1997) generates fingerprints from bacteria-based large-insert clones by digesting the clone DNA with a single 6-bp restriction enzyme such as *Hin*dIII or *Eco*RI, followed by fractionating the restricted fragments on agarose gel. The fingerprints of the clones are visualized by staining the gel with Vistra or SYBR Green and are captured into a computer for analysis using a FluorImager. This method uses one restriction enzyme to generate the fingerprint fragments because the agarose matrix can allow only the restricted fragments ranging from 200 bp to 20 kbp that are typically generated by a 6-bp cutting restriction enzyme to be fractionated. Figure 11.3 shows examples of human PAC fingerprints generated with *Hin*dIII and fractionated on an agarose gel. The following procedure is outlined according to Marra et al. (1997).

Equipment and Materials

1. 20×25-cm UV transparent trays and Model H4 electrophoresis units (Life Technologies, USA) with a refrigerated re-circulator (Model 1170, VWR Scientific, USA).
2. Molecular Dynamics FluorImager SI (Amersham Biosciences, USA).
3. SYBR Green (FMC BioProducts, USA) or Vistra Green (Molecular Probes, USA).
4. 10× TAE: 400 mM Trizma base, 200 mM glacial acetic acid, 10 mM EDTA pH 8.0.

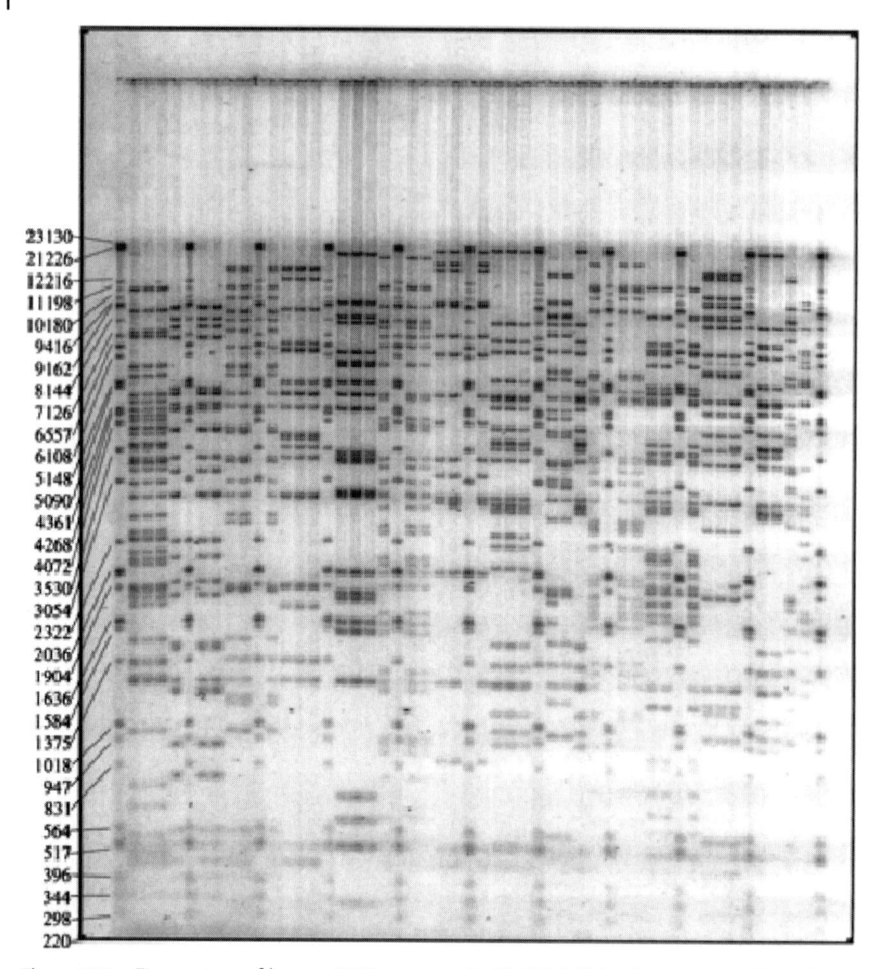

Figure 11.3. Fingerprints of human PACs generated with *Hin*d III by the agarose gel-based restriction fingerprinting method (Marra et al. 1997). DNA size standards are present in the first lane from the left and every fifth lane, and the sizes of each marker fragment are indicated in base pairs in the first lane. The fingerprints in the remaining lanes are derived from human PACs. Note that the identical fingerprints in different lanes are the duplication of the same PAC (from Marra et al. 1997 with permission).

5. 6× loading dye: 0.25 % bromophenol blue, 0.25 % xylene cyanol, 15 % Ficoll.
6. Standard marker mixture: 0.83 μL of 1 μg μL^{-1} 1 kb ladder (Life Technologies, USA), 3.33 μL of 250 ng μL^{-1} marker II, and 3.33 μL of 250 ng μL^{-1} marker III (Boehringer-Mannheim, Germany), 92.51 μL of TE, and 25 μL of 6× loading dye.

Clone DNA Restriction Digestion and Standard Marker Preparation

1. Set up the digestions of BAC DNA in a 10-μL reaction per sample in 96-micro-tube plates:

ddH_2O	3.75 μL
10× reaction buffer	1.00 μL
40 U μL^{-1} *Hin*dIII	0.25 μL
DNA	5.00 μL (1.2–1.5 μg)
Total volume	10.00 μL

2. Incubate at 37 °C for 3 h in a water bath or a thermocycler.
3. After digestion, collect DNA at the bottom of the wells by brief centrifugation, add 4 μL of the 6× loading dye, mix, and collect the DNA sample by brief centrifugation. Seal the plates with foil tape and keep on ice or store at 4 °C before loading gels.
4. Immediately before electrophoresis, remove 20 μL of the standard marker mixture to a separate tube, dilute by adding 17 μL of TE and 3 μL of 6× loading dye, and incubate at 60 °C for 5 min.

Restricted Fragment Fractionation by Agarose Gel Electrophoresis and Data Acquisition

1. Prepare 1% agarose gels in 1× TAE. Cool down the molten agarose to 46 °C in a water bath with occasional swirling. Pour approximately 200 mL of the molten agarose into a 20×25-cm UV transparent tray and insert a comb of appropriate size (e. g., 2 mm wide by 1 mm thick).
2. After the gel is solidified, remove the comb and place in electrophoresis unit with enough pre-cooled (14 °C) 1× TAE buffer to submerge the gel.
3. Load 1.75 μL of the DNA-restriction enzyme digestion/loading dye mixture into each well, with 1 μL of the standard marker in the first and every fifth well.
4. Electrophorese the samples at 56 V at a constant temperature of 14 °C for 14 h.
5. Stain the gel with 500 mL of a 1:10,000 dilution of Vistra Green in 1× TAE and agitate in the dark for 10–15 min. Two gels can be stained at one time and up to four gels can be stained using the same stock of dye. Store the diluted stains in a foil-covered container at 4 °C.
6. After staining, view the gel using a Fluorimager with the following scan settings: pixel size at 200 μm, digital resolution at 16 bits, detection sensitivity at high, PMT voltage at 950 V, and filter at 530 nm. Crop the gel images and convert them from the proprietary 16-bit Molecular Dynamics format to 8-bit TIFF images.
7. Transfer by ftp to Unix workstations (SUN Microsystems) for band calling and contig building (see Chapter 13).

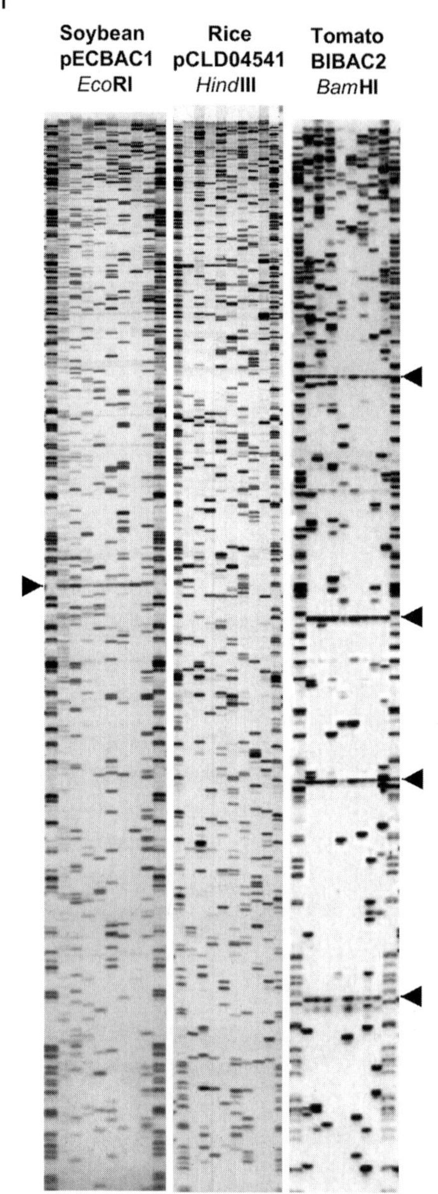

Figure 11.4. Fingerprints of soybean BACs cloned in pECBAC1, rice PBCs cloned in pCLD04541, and tomato BIBACs cloned in BIBAC2 generated with *Hind*III/*Hae*III by the manual sequencing gel-based restriction fingerprinting method. The first and last lanes of each type of sample are λ/*Sau*3AI DNA markers end-labeled with ^{33}P-dATP, and the remaining lanes are plant DNA BAC, PBC, or BIBAC clones. The DNA of BAC, PBC, and BIBAC was purified manually in 96-well format according to Chang et al. (2001) and Tao et al. (2001). The digestion of the BAC, PBC, and BIBAC DNA with *Hind*III and *Hae*III and end labeling of the *Hind*III sites with ^{33}P-dATP were conducted in a single step in an eight-microtube strip. No vector band was observed in the pCLD04541 clone fingerprints, one vector band was observed in the pECBAC1 clone fingerprints, and four vector bands were observed in the BIBAC2 clone fingerprints, each being indicated by arrowheads.

11.2.2.2 Polyacrylamide (Sequencing) Gel-based, Restriction Fingerprinting Method

The polyacrylamide gel-based restriction fingerprinting method (Coulson et al. 1986; Hong 1997; Zhang and Wing 1997); generates fingerprints from bacteria-based large-insert clones by digesting the clone DNA with a sticky-end, 6-bp restriction enzyme such as *Hind*III; end-labeling with a radionucleotide such as ^{32}P-dATP, ^{33}P-dATP or ^{35}S-dATP; and digesting with a 4-bp restriction enzyme (a blunt-ended enzyme is preferable) simultaneously or separately, followed by fractionating the restricted fragments on DNA sequencing gels. The fingerprints of clones are visualized by autoradiography and converted into image files by scanning the fingerprints into a computer. The sticky-end 6-bp restriction enzyme is used to digest the DNA, control the number of restricted fragments, and provide sticky ends for labeling by a filling-in reaction, whereas the 4-bp restriction enzyme is used to further digest the DNA fragments into smaller ones so that they can be fractionated in a range from 50 to 1000 nucleotides on a denaturing DNA sequencing gel. Figure 11.4 shows examples of BAC, PBC, and BIBAC fingerprints generated using this method. The below procedure is prepared according to Zhang and Wing (1997), Chang et al. (2001) and Tao et al. (2001), in which DNA digestion and end-labeling are completed in a single step.

Equipment and Materials

1. Sequi-gen GT sequencing system (Bio-Rad, USA).
2. Double-up gel drying system (Bio-Rad, USA).
3. 40 % polyacrylamide solution (Bio-Rad, USA).
4. 10× TBE: 0.89 M Tris base, 0.89 M boric acid, 0.02 M EDTA, pH 8.0.
5. 3× loading dye: 98 % deionized formamide, 0.3 % bromophenol blue, 0.3 % xylene cyanol, and 10 mM EDTA, pH 8.0.
6. 8 M urea.
7. Kodak Biomax MR film (35×43 cm).

Standard DNA Marker Preparation

The standard molecular weight DNA markers are prepared by digesting λ DNA and end-labeling the restricted fragments in a reaction buffer containing 0.1 M potassium acetate, 25 mM Tris-acetate (pH 7.5), 10 mM magnesium acetate, and 1 mM DTT.

1. Set up the following reaction of DNA marker preparation:

dd H_2O	36 μL
0.5 μg μL^{-1} Lambda DNA (Biolabs, USA)	2 μL
10× reaction buffer (see above)	5 μL
10 mM dGTP (Sigma, USA)	2 μL
10 mM ddTTP (Sigma, USA)	2 μL
10 U μL^{-1} *Sau*3AI (Promega, USA)	1 μL
10 U μL^{-1} AMV-Reverse transcriptase (Promega, USA)	1 μL
^{33}P-dATP (>2500 Ci mM^{-1}, Amersham, USA)	1 μL
Total volume	50 μL

2. Mix well and incubate at 37 °C for 2 h. Stop the reaction by adding 25 μL of 3× loading dye and store at –20 °C.
3. Heat at 95 °C for 5 min and transfer on ice before use. Load 1.5~3.0 μL per lane.

DNA Restriction Digestion and End Labeling With a Radioactive Nucleotide

Each DNA sample is digested and labeled at a single step in a reaction buffer containing 27.5 mM Tris-acetate, pH 7.8, 110 mM potassium acetate, 11 mM magnesium acetate, 1 mM DTT, 110 μg mL^{-1} BSA, and 7 μM ddGTP (as necessary, 110 μg mL^{-1} DNase-free RNase A is added to the buffer) in 8- or 12-microtube strips.

1. Set up the reaction of sample DNA digestion and labeling:

BAC DNA	3 μL (100–200 ng)
2× fingerprinting reaction buffer (see above)	2.79 μL
50 U μL^{-1} *Hind*III (Promega, USA)	0.05 μL
50 U μL^{-1} *Hae*III (Promega, USA)	0.05 μL
10 U/50 U μL^{-1} AMV-reverse transcriptase (Promega, USA)	0.06 μL
^{33}P-dATP (>2500 Ci mM^{-1}, Amersham)	0.05 μL
Total volume	6.00 μL

2. Mix well and incubate at 37 °C for 2 h. Stop the reaction by adding 3 μL of 3× loading dye.
3. Heat at 95 °C for 5 min and transfer on ice before use. Load 1.5~3.0 μL per lane.

Restricted Fragment Fractionation by Denaturing Sequencing Gel Electrophoresis and Data Acquisition

1. Carefully clean up a pair of sequencing plates, set up the sequencing gel cast unit, and prepare 100 mL of 4% polyacrylamide gel solution. To prepare the solution, mix well 10 mL of 40% polyacrylamide, 90 mL of 8 M urea in 1× TBE, 40 μL TEMED, and 500 μL of 10% APS.
2. Pour the gel solution carefully into the gel cast and use a 0.4-cm, 73-well shark-tooth comb (Bio-Rad, USA) and allow the gel to completely solidify.
3. Pre-run the gel for 1 h at 90 W in 1× TBE buffer.
4. Rinse the top of the gel thoroughly with 1× TBE, insert a shark-tooth comb (0.4-cm, 73-well) and rinse all the wells thoroughly using a plastic syringe.
5. Load 1.5~3.0 μL of each DNA fingerprinting sample per lane prepared above and 1.5~3.0 μL of the lambda/*Sau*3AI standard marker per lane, beginning with the first lane, every ninth lane, and the last two lanes.
6. Run the gel at 90 W for about 100 min until the bromophenol blue dye is about 4 cm from the bottom of the gel.
7. Lift the gel on 3 MM chromatograph paper (Whatman, USA), wrap with Saran Wrap, and vacuum dry at 80 °C using a vacuum gel dryer.
8. Expose the gel for 1–2 days to the Kodak Biomax MR film and develop the film.
9. Scan the fingerprints on the autoradiograph into image files using a UMAX Mirage D-16L scanner and transfer to a Unix workstation (SUN Microsystems) for further analysis (see Chapter 13).

11.2.2.3 Automatic Sequencing Gel- and Capillary Electrophoresis-based Restriction Fingerprinting Methods

The automatic restriction fingerprinting methods using the polyacrylamide gel-based (Gregory et al. 1997; Ding et al. 1999, 2001) or capillary electrophoresis-based (Xu et al. 2002, 2004) automatic sequencers generate fingerprints from bacteria-based large-insert clones by digesting the DNA with one or more sticky-end 6-bp restriction enzymes, end-labeling with fluorescent dideoxynucleotide dyes such as ddATP-HEX or ddATP-NED, and then digesting with one 4-bp, blunt-end restriction enzyme. In the manual sequencing gel-based restriction fingerprinting method, the sticky-end 6-bp restriction enzyme is used to digest the DNA, control the number of restricted fragments, and provide sticky ends for labeling by a filling-in reaction, whereas the 4-bp restriction enzyme is used to further digest the DNA fragments into smaller ones so that they can be fractionated in a range from 35 to 600 nucleotides on a sequencing gel- or capillary electrophoresis-based automatic sequencer. The fragments are separated on DNA sequencing gels or sequencing capillaries. The fingerprints of clones are directly visualized on automatic DNA sequencers such as ABI PRISM 377 (Applied Biosystems, USA) and ABI 3100 and captured into a computer with the GeneScan software installed in the sequencer. Figure 11.5 shows examples of the fingerprints of BAC clones digested with three restriction enzymes and labeled with one fluorescent color as well as the fingerprints of BAC clones digested with five different restriction enzymes, labeled with four fluorescent dyes (SNaPshot kit, Applied Biosystems, USA), and fractionated in one capillary of the ABI 3100 sequencer.

Equipment and Materials

ABI PRISM 377, ABI 3100, or any advanced model of the automatic DNA sequencers with supplemental materials for DNA genotyping analysis can be used.

Restriction Digestion and End-labeling With Fluorescent Nucleotides

In this fingerprinting method, each BAC DNA can be digested with (1) two enzymes such as *Hin*dIII/*Sau*3AI and end-labeled at the *Hin*dIII sites with a single fluorescent dye such as ddATP-HEX, ddATP-NED, or ddATP-TET (Gregory et al. 1997); (2) one type IIS and one type III restriction enzyme such as *Hga*I/*Rsa*I and end-labeled at the unknown *Hga*I sites with a fluorescent dideoxy terminator kit (Applied Biosystems, USA) containing ddATP-R6G, ddT(U)TP-ROX, ddGTP-R110, and ddCTP-TEMRA (Ding et al. 2001); (3) three enzymes such as *Hin*dIII/*Bam*HI/*Hae*III and end-labeled at the *Hin*dIII, *Bam*HI, or both enzyme sites with one fluorescent dye such as ddATP-HEX, ddATP-NED, or ddATP-TET (Applied Biosystems, USA) (Xu et al. 2002, 2004); (4) three pairs of restriction enzymes such as *Hin*dIII/*Rsa*I, *Hin*dIII/*Dpn*I, and *Hin*dIII/*Hae*III separately and end-labeled at the *Hin*dIII sites with three different fluorescent dyes such as ddATP-HEX, ddATP-NED, or ddATP-TET, respectively (Ding et al. 1999); or (5) five restriction enzymes such as *Hin*dIII/*Bam*HI/*Xba*I/*Xho*I/*Hae*III and end-labeled at the *Hin*dIII, *Bam*HI, *Xba*I, and *Xho*I sites using the SNaPshot kit con-

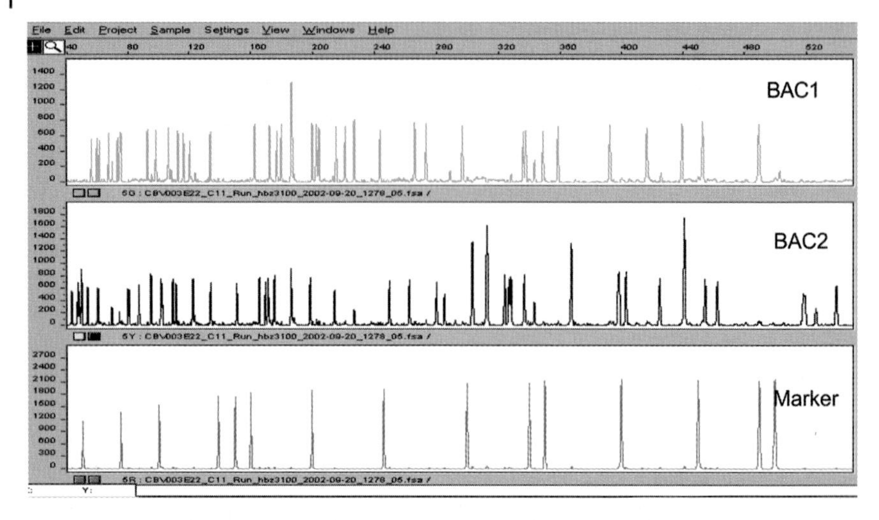

A: Two BAC fingerprints from one channel of the ABI 3100 sequencer

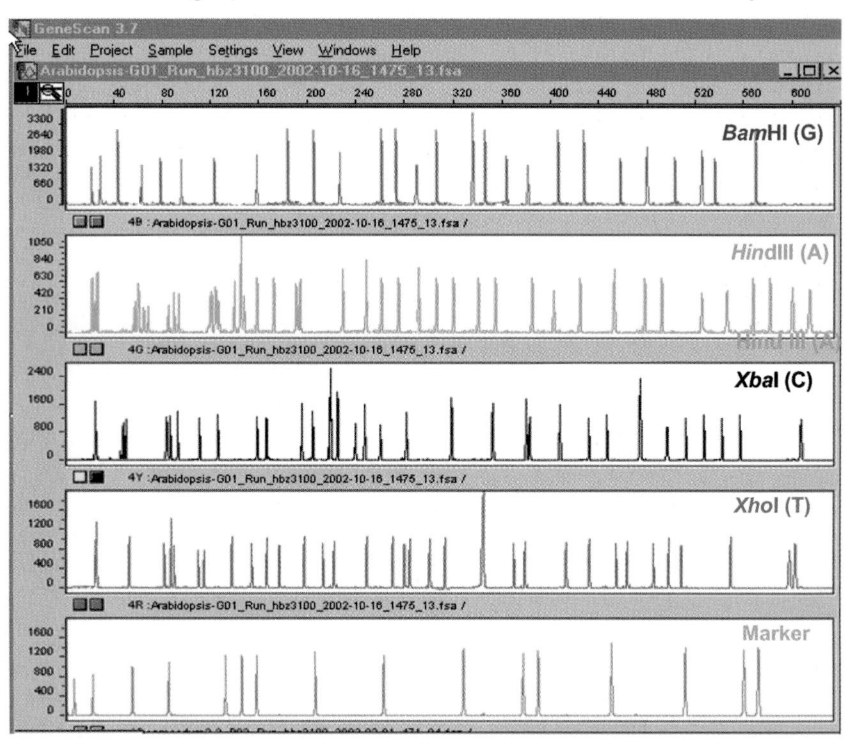

B: One BAC fingerprint from one channel of the ABI 3100 sequencer

◀ **Figure 11.5.** Fingerprints of *Arabidopsis* BACs generated by the capillary electrophoresis-based restriction fingerprinting methods (Xu et al. 2002, 2004; Figure 11.2). BAC DNA was isolated with the AutoGenprep 960 Robotic Workstation in 96-well format. (A) BAC DNA was digested with *Hind*III/*Bam*HI/*Hae*III and end-labeled with HEX-ddATP (BAC1) or NED-ddATP (BAC2), and the BACs labeled with different colors are combined. (B) BAC DNA was digested with *Hind*III/*Bam*HI/*Xba*I/*Xho*I/*Hae*III and end-labeled at the *Hind*III, *Bam*HI, *Xba*I, and *Xho*I sites with the SNaPshot kit containing R6G-ddATP (green), R110-ddGTP (blue), TAMRA-ddCTP (black), and ROX-ddT(U)TP (red). The restricted and end-labeled fragments are mixed with the internal molecular weight DNA markers (orange) and fractionated on the ABI 3100 Genetic Analyzer. The fingerprint of each BAC was captured by the GeneScan software (from Xu et al. 2004).

taining ddATP- R6G, ddGTP- R110, ddCTP- TAMRA, and ddT(U)TP- ROX (Xu et al. 2002, 2004). Note that the number of restriction enzymes used to generate fingerprints depends on the number of bands that are preferred to be generated per clone. According to Xu et al. (2004), for the fixed range of band size (35~500 bases) of an automatic sequencer-based method, the number of bands per clone ranging from 50 to 70 is the most desired for accurate, large contig assembly using the FPC program (Soderlund et al. 2000). Both Ding et al. (1999) and Xu et al. (2004) showed that although a certain number of bands per clone are required for its identity and assembly into a contig, too many bands in each clone make it troublesome for the FPC program to assemble contigs accurately. Below, several procedures for large-insert clone restriction digestion and end labeling with fluorescent nucleotides according to Xu et al. (2000, 2004) are detailed.

1. For DNA digestion with *Hind*III/*Hae*III and end labeling at the *Hind*III sites with a single fluorescent dye, set up the reaction in a reaction buffer containing 25 mM Tris-acetate (pH 7.8), 100 mM potassium acetate, 10 mM magnesium acetate, and 1 mM DTT:

dd H_2O	6.425 μL
DNA	20.00 μL (200–500 ng)
10× reaction buffer	3.00 μL
10 mg mL^{-1} BSA	0.30 μL
80 U μL^{-1} *Hind*III	0.037 μL
80 U μL^{-1} *Hae*III	0.038 μL
NED-ddATP or HEX-ddATP	0.10 μL
8 U μL^{-1} *Taq* DNA polymerase FS	0.10 μL
Total volume	30.00 μL

For DNA digestion with *Hind*III/*Bam*HI/*Hae*III and end labeling at the *Hind*III and *Bam*HI sites with a single fluorescent dye, set up the reaction in a reaction buffer containing 25 mM Tris-acetate (pH 7.8), 100 mM potassium acetate, 10 mM magnesium acetate, 1 mM DTT, and 6 μM dGTP:

dd H_2O	6.388 μL
DNA	20.00 μL (200–500 ng)
10× reaction buffer	3.00 μL
10 mg mL^{-1} BSA	0.30 μL
80 U μL^{-1} *Hind*III	0.037 μL
80 U μL^{-1} *Bam*HI	0.037 μL
80 U μL^{-1} *Hae*III	0.038 μL
NED-ddATP or Hex-ddATP	0.10 μL
8 U μL^{-1} *Taq* DNA polymerase FS	0.10 μL
Total volume	30.00 μL

For DNA digestion with *Hind*III/*Bam*HI/*Xba*I/*Xho*I*Hae*III and end labeling at the *Hind*III, *Bam*HI, *Xba*I, and *Xho*I sites using the SNaPshot kit containing ddATP-R6G, ddGTP-R110, ddCTP-TAMRA, ddT(U)TP-ROX, and *Taq* DNA polymerase FS, first set up the reaction in reaction buffer I containing 50 mM NaCl, 10 mM Tris-HCl (pH 7.9), 10 mM magnesium acetate, and 1 mM DTT:

dd H_2O	4.45 μL
DNA	35.00 μL (200–500 ng)
10× reaction buffer I	4.50 μL
10 mg mL^{-1} BSA	0.45 μL
20 U μL^{-1} *Hind*III	0.15 μL
20 U μL^{-1} *Xba*I	0.15 μL
20 U μL^{-1} *Xho*I	0.30 μL
Total volume	45.00 μL

After incubating at 37 °C for 2 h, heat at 65 °C for 15 min, precipitate the DNA with isopropanol, wash in 70 % ethanol, air dry, dissolve in 45 μL dd H_2O, and then set up the reaction in reaction buffer II containing 13.8 mM Tris-acetate (pH 7.8), 37.5 mM NaCl, 25 mM potassium acetate, 10 mM magnesium acetate, and 1 mM DTT.

dd H_2O	5.80 μL
DNA	45.00 μL
10× reaction buffer II	6.00 μL
10 mg mL^{-1} BSA	0.60 μL
80 U μL^{-1} *Bam*HI	0.05 μL
80 U μL^{-1} *Hae*III	0.05 μL
The SnaPshot kit	2.50 μL
Total volume	60.00 μL

2. Mix well and incubate at 37 °C for 2 h and then at 65 °C for an additional 45 min.
3. Combine multiple clones labeled with different colors as needed, precipitate with 100 % ethanol (previously stored at –20 °C), wash with ice-cold 70 % ethanol, and air-dry the pellets.

4. Dissolve the pellets in 9.75 μL Hi-formamide, add 0.25 μL Liz 500 size standard, vortex, spin down, and keep at 4 °C overnight.
5. Heat at 95 °C for 5 min and transfer on ice before use. Load a 1.0-μL sample per lane on an ABI PRISM 377 or fractionate the DNA on the ABI 3100 Genetic Analyzer (sequencer) using the Run Module.
6. Collect the fingerprint data using the GeneScan software for physical map contig assembly (see Chapter 13).

11.2.3 Applications of DNA Fingerprinting for Genome Physical Mapping

11.2.3.1 Whole-genome Physical Maps Generated from Large-insert Bacterial Clones by Fingerprint Analysis

Whole-genome physical maps have been successfully generated from bacteria-based large-insert clones for several plant, animal, and microbe species by using the restriction fingerprinting methods followed by contig assembly using the computer program FPC (Sulston et al. 1988; Soderlund et al. 1997, 2000). They are the physical maps of *Arabidopsis thaliana* (Marra et al. 1999; Chang et al. 2001), indica rice (Tao et al. 2001), japonica rice (Chen et al. 2002; Li et al. 2004), soybean (Wu et al. 2004), *Drosophila melanogaster* (Hoskins et al. 2000), human (IHGMC 2001), mouse (Gregory et al. 2002), chicken (Ren et al. 2003), and *Bradyrhizobium japonicum* (Tomkins et al. 2001). Table 11.1 summarizes the basic features of the physical maps. These studies show at least the following points. First, the restriction fingerprinting methods can be used to construct reliable whole-genome, contig-based physical maps from random bacteria-based large-insert clones. Physical maps could be constructed by using either the agarose gel-based restriction fingerprinting method (Marra et al.1997) or the manual sequencing gel-based restriction fingerprinting method (Zhang and Wing 1997; Tao and Zhang 1998). It should be noted that the construction of all of the physical maps by these two methods does not imply that they are the most powerful for whole-genome physical mapping; it indicates only that they are the first to be used for physical map construction. The automatic sequencing gel- and capillary electrophoresis-based fingerprinting methods are now being used to generate physical maps in several species. Second, whole-genome physical maps by the fingerprinting methods are not affected significantly by the sizes and complexities of the genomes, as indicated by the fact that quality physical maps were constructed from species with genome sizes ranging from several million base pairs to 3200 Mb. Third, DNA libraries constructed from two or more restriction enzymes facilitate quality physical map construction, as indicated by the fact that most of the physical maps were constructed from such libraries. Fourth, source clones equivalent to 6.9×–77× were used to construct the physical maps, suggesting that studies are needed to determine how many equivalents of clones are needed to construct high-quality physical maps efficiently. Finally, the fingerprinting methods are only able to construct overlapping clone contigs each spanning less than one million base pairs, suggesting that other tools, such as mapped DNA markers, and clone end sequences and syn-

Table 11.1. Physical maps constructed from bacteria-based large-insert clones by fingerprint analysis.

Species	*Genome size (Mb)*	*Fingerprinting methods*	*Source libraries*	*Supplemental tools**	*No. of automatic contigs*	*No. of edited contigs*	*Genome coverage*	*Reference*
Plants:								
Arabidopsis	125	Agarose gel	BAC (two, 17×)	NA	372	169	NA	Marra et al. 1999
		Polyacrylamide gel	BAC/BIBAC (three, 8.1×)	77 markers	268	194	96.9 %	Chang et al. 2001
Indica rice	430	Polyacrylamide gel	BAC (three, 6.9×)	83 markers	585	298	97.4 %	Tao et al. 2001
Japonica rice	430	Agarose gel	BAC (two, 25.6×)	3199 ESTs, 1704 markers, 110,438 STCs, sequences	1069	284	90.6 %	Chen et al. 2002
Soybean	1115	Polyacrylamide gel	BAC/BIBAC (three, 9.6×)	388 markers	5488	2905	NA	Wu et al. 2004
Animals:								
Drosophila melanogaster	97 (Chr. 2, 3)	Agarose gel	BAC (one, 14×)	1226 markers, 697 STSs, sequences	378	9	97.9 %	Hoskins et al. 2000
Human	3200	Agarose gel	BAC (five, 15×)	69,507 markers/ STSs, sequences	7700	1246	96.0 %	IHGMC 2001
Mouse	2800	Agarose gel	BAC (two, 15×)	16,992 markers, sequences, human syntenic sequences	7587	296	90.0 %	Gregory et al. 2002
Chicken	1100	Polyacrylamide gel	BAC (three, 8.2×)	367 markers	5543	2331	NA	Ren et al. 2003
Microbes:								
Bradyrhizobium japonicum	8.7	Agarose gel	BAC (one, 77×)	194 markers, 1152 BAC ends	6	2	99.7 %	Tomkins et al. 2001

* NA indicates that data are not available, and "sequences" indicates that the majority of the genome that was used to assist in physical map assembly was sequenced.

tenic sequences, are needed to further extend the contigs into larger ones and finally construct robust physical maps consisting of fewer contigs per chromosome.

11.2.3.2 Comparison of Different Fingerprinting Methods

Table 11.2 summarizes the major features of the restriction fingerprinting methods that have been developed so far to generate fingerprints from bacteria-based large-insert clones using restriction enzymes. Each of these methods has its advantages and disadvantages over the others. The agarose gel-based method (Marra et al. 1997) is relatively simple and economical and is able to fractionate DNA fragments up to 20 kb, thus being amenable to fragment size estimation. These features facilitate large-scale clone fingerprinting. However, it has a very low resolution of fragments, ranging from 4 bp at the 600-bp zone to 112 bp at the16-kb zone. Moreover, because only one restriction enzyme is used to generate the fingerprint from each clone, the agarose gel-based method generates fewer fingerprint bands per clone than do the polyacrylamide gel- and automatic sequencer-based methods (Figure 11.3). Furthermore, because the samples and standard molecular weight markers are loaded in separate lanes, the accuracy of band calling is much lower than in the automated sequencer-based method. These shortcomings all could reduce the accuracy of large contig assembly. In addition, because the samples need to be loaded manually and the method is still not suited for multiplex analysis, it is not amenable to automation, and the throughput is lower than in the automatic DNA sequencer-based method.

The manual polyacrylamide gel-based method (Zhang and Wing 1997; Tao and Zhang 1998) is also relatively simple and economical. The similarity of the method ensures that the restriction digestion is completed and that the digestion and labeling can be conducted in a single step. This method has a high resolution of one nucleotide, which allows the fragments differing by one nucleotide to be separated into different bands. Furthermore, this method generates fragments ranging from 58 to 773 bases that are well suited for contig map construction, and the number of bands per clone can be readily adjusted along the insert sizes of the source clones by adjusting the number of restriction enzymes used in the digestion and labeling. These features are crucial to large-scale, quality fingerprint generation and contig map assembly. Nevertheless, because the fragments are visualized by fragment end labeling with radioactive nucleotides and the fractionation gels are loaded manually, this method is also not suited for automation and high-throughput fingerprint generation. The separate loading of the samples from the standard molecular weight markers likewise reduces the accuracy of band calling.

The automatic DNA sequencer-based methods (Gregory et al. 1997; Ding et al. 1999, 2001; Xu et al. 2002, 2004) are promising for the rapid development of whole-genome physical maps for large, complex genomes despite the fact that no whole-genome physical maps developed by these methods have been reported so far. These methods allow generation of fingerprints at a resolution of 0.5 or 0.2 base, depending on the DNA sequencers used. They were developed to facilitate automation of the fingerprinting process and high-throughput generation of fin-

Table 11.2. Comparison among different restriction fingerprinting methods.

Methods	*Equipment*	*No. of enzymes used (e. g.)*	*Usable fragment size range*	*Resolution*[1]	*Size standard*	*Visualization methods*	*Automation*	*Throughput*[2]	*Cost*[3] *($/BAC)*	*Reference*
Agarose gel-based	Conventional	One (*Hind*III)	600 bp~16 kb	4~112 bp	Separate lanes	Vistra or SYBR	No	384	1.20	Marra et al. 1997
Manual polyacrylamide gel-based	Conventional	Two (*Hind*III/*Hae*III)	58~773 base	1 base	Separate lanes	^{32}P, ^{33}P or ^{32}S	No	384	0.50	Tao et al. 2001
Automatic DNA sequencer-based:										
Gel matrix	ABI 377	Two (*Hind*III/*Hae*III)	75~500 base	0.5 base	Inside sample	Fluorescence	Yes	1536	1.30	Gregory et al. 1997
Gel matrix	ABI 377	Three pairs (*Hind*III/*Rsa*I, *Hind*III/*Dpn*I and *Hind*III/*Hae*III)	75~500 base	0.5 base	Inside sample	Fluorescence	Yes	384	3.00	Ding et al. 1999
Gel matrix	ABI 377	A pair of type IIS and type II (*Hga*I/*Rsa*I)	75~500 base	0.5 base	Inside sample	Fluorescence	Yes	384	3.00~4.00	Ding et al. 2001
Capillaries	ABI 3100	Three (*Hind*III/*Bam*HI/*Hae*III)	35~500 base	0.2 base	Inside sample	Fluorescence	Yes	1536	1.30	Xu et al. 2004
Capillaries	ABI 3100	Five (*Hind*III/*Bam*HI/*Xba*I/*Xho*I/*Hae*III)	35~500 base	0.2 base	Inside sample	Fluorescence	Yes	384	3.00~4.00	Xu et al. 2004

[1] The resolution of agarose gels is variable, ranging from 4.2 bp at the 600-bp zone to 112 bp at the 16-kb zone of the gel, whereas the resolution of the DNA sequencing gels or capillary electrophoresis is fixed according to its manufacturers.

[2] The throughput was based on the output of two persons in a working day, from clone cells to fingerprints. Note that only the two-enzyme method of Gregory et al. (1997) and the three-enzyme method of Xu et al. (2004) can be multiplexed; up to four clones can be analyzed per gel lane or capillary channel.

[3] Estimates are based on the methods used in the publications, including the cost for DNA isolation and fingerprinting but not including labor.

gerprints. Generated fingerprints can be directly visualized, automatically captured into a computer during electrophoresis, edited, and converted into an FPC file by a customer-devised software. Labeling different clones with different migration rate–comparable fluorescent dyes following the method of Gregory et al. (1997) and the three-enzyme method of Xu et al. (2004) allows up to four clones to be fractionated on a single gel lane or capillary channel, which could further increase the throughput of fingerprint production by fourfold relative to the other fingerprinting methods. The loading of standard molecular weight markers with samples in the same lane or capillary channel has significantly enhanced the accuracy of band calling. As in the manual sequencing gel-based method, the number of bands per clone can be adjusted readily, corresponding to the insert sizes of the source clones. According to the method of Ding et al. (1999, 2001) and the five-enzyme method of Xu et al. (2004), up to five restriction enzymes are used and a single higher number of bands are generated per clone. Although the increased number of bands per clone may not enhance the quality of the physical map constructed (see below), the increased number of restriction enzymes and labeling processes would complicate the fingerprint generation, including more steps and/or higher potential in nonspecific digestions. This could not only influence the quality of contigs but also increase the cost of the fingerprinting reactions. Nevertheless, the automatic DNA sequencer-based method with type IIS and type III restriction enzymes is not cost-efficient but is more accurate in detecting small overlaps, and thus is useful for gap closing for large-scale fingerprinting and physical mapping of an entire genome (Ding et al. 2001).

The development of various fingerprinting methods has provided different tools for and facilitated whole-genome physical mapping rapidly and efficiently. However, no comparative study has been conducted among the methods until recently (Xu et al. 2004), although the agarose gel-based method (Marra et al. 1997) and the manual sequencing gel-based method (Zhang and Wing 1997) have been used in the development of whole-genome physical maps of several species. Xu et al. (2004) first systematically studied influences of different fingerprinting methods on quality physical map construction *in silico* and experimentally. These methods included the agarose gel-based, one-enzyme method (Marra et al. 1997), the manual sequencing gel-based, two-enzyme method (Zhang and Wing 1997), the capillary electrophoresis-based, three-enzyme method (Xu et al. 2002, 2004), the automated sequencing gel-based four-enzyme method (Ding et al. 1999), and the capillary electrophoresis-based five-enzyme method (Xu et al. 2002, 2004). They found that different quality physical maps were constructed using different fingerprinting methods. The results showed that the manual sequencing gel-based, two-enzyme method consistently generated larger and more accurate contigs, followed by the capillary electrophoresis-based, three-enzyme method, the agarose gel-based, one-enzyme method, the capillary electrophoresis-based five-enzyme method, and the automated sequencing gel-based four-enzyme method in descending order. This is because although a certain number of bands per clone are required for its identity and assembly into a contig, too many bands in each clone make it troublesome for the FPC program to assemble contigs accurately due to the increase in the chances of band mismatching. The results also showed that fingerprint band resolution played an important role in the quality map of construction: the resolutions that are one nucleotide or higher facilitate quality map construction. Xu et al. (2004) also studied the influence of genome coverage of source clones, ranging from 5× to 15× genomes, on quality map construction. The results revealed that as the number of clones increased from 5× to 10×, the contig length rapidly increased for all methods studied. However, when the number of clones was increased from 10× to 15× coverage, the contig length at best increased at a lower rate or even decreased. Therefore, they concluded, along with considerations of fingerprinting throughput and automation, that the capillary sequencing-based, three-enzyme method (Xu et al. 2002, 2004) or a capillary sequencing-based, two-enzyme method with approximately 10× coverage clones is the most efficient for quality map construction. It is worth noting that both of the fingerprinting methods are well amenable to multiplex fingerprint production, which adds further advantages to these methods over others for whole-genome physical mapping.

11.3 Discussion

DNA fingerprinting is a powerful tool for whole-genome physical mapping from bacteria-based large-insert clones. This has been demonstrated in the past few years in the physical mapping of several species, including human, plants, animals, insects, and microbes (Table 11.1). If a capillary electrophoresis-based, two- or three-enzyme fingerprinting method (Xu et al. 2004) is used, the fingerprints of approximately 70,000 clones equivalent to 10× genome coverage of an organism could be generated within four scientist months, which are sufficient for assembly of a whole-genome physical map of a species with a genome size of 1000 Mb. Among the DNA fingerprinting methods developed in the past few years (although each has its advantages and disadvantages over the others and any of them might be able to generate contigs), selection of the method that is most suited to a physical mapping project is of significance for constructing quality physical maps rapidly and efficiently. The studies of Xu et al. (2004) suggested that a capillary electrophoresis-based, two- or three-enzyme fingerprinting method is the most efficient for quality physical map construction because it is simple, fully automated, and amenable to multiplex and high-throughput fingerprint production. In addition, the number of bands generated per clone is the most desired for physical map contig assembly using the automatic DNA sequencer-based method.

Whole-genome physical mapping by fingerprint analysis is, however, a complicated process. In addition to the use of the most suited fingerprinting method, quality physical map construction from the fingerprints of clones is influenced by many other factors. These factors include the average insert size, genome representation, and genome coverage of the source clones (see Chapter 9; Soderlund et al. 1997). It might also be affected by the computer program used for map contig assembly from the fingerprints. The FPC program is the most widely used program for whole-genome physical map contig assembly from large-insert clone fingerprints. It was initially developed by Sulston et al. (1988) to construct the contig map of *Caenorhabditis elegans* from fingerprints generated from cosmids using the manual sequencing gel-based method (Coulson et al. 1986). Later, the program was further modified two or more times by Soderlund et al. (1997, 2000) to facilitate contig map construction from fingerprints generated from large-insert clones using the agarose gel-based method (Marra et al. 1997) and the manual sequencing gel-based method (Hong 1997; Zhang and Wing 1997) (see Chapter 13). Although the improved version of the FPC program provides several advanced features over the original version of the program, it still needs further improvements, especially for construction of contig maps from fingerprints generated with different automatic sequencer-based fingerprinting methods. Currently, large and accurate contig assembly using the FPC program is still largely influenced by human experiences and intervention.

Multiple tools are needed to develop a robust physical map consisting of one or a few contigs for each chromosome. Previous physical mapping (Marra et al. 1999; Hoskins et al. 2000; Chang et al. 2001; IHGMC 2001; Tao et al. 2001; Chen et al. 2002; Gregory et al. 2002; Ren et al. 2003; Wu et al. 2004) showed that the DNA fingerprinting methods allowed only the contigs shorter than 1.0 Mb to be assembled. Additional tools are needed to extend or merge the neighboring contigs into larger ones or a "contig scaffold." DNA markers mapped to genetic linkage maps provide powerful means in this regard. Additionally, anchoring of contigs to genetic linkage maps with DNA markers could also further verify the contigs and increases the utility of the whole-genome physical map. Other tools that have been used to assist in high-quality physical map construction include assignment of STSs derived from the fingerprinted BAC end sequences along syntenic genomic sequences and FISH mapping. It is also apparent from previous studies (Wu et al. 2004) that novel technologies are needed to challenge the construction of physical maps for the heterochromatic portions of the genomes, especially those containing tandem repeat elements such as centromeric repeats and rDNA. This effort is essential to obtaining a physical map that covers the entire genome and to studying genomic organization, function, and evolution.

Whole-genome physical mapping by fingerprint analysis will continue to play an important role in genome research. First, although whole-genome physical maps have been constructed for several important species (Table 11.1), physical maps will be needed for many other species or genotypes for advanced genomics research. Second, in the past decades DNA sequencing technology has been revolutionized, and thus the sequencing cost of large, complex genomes has become affordable. The genomes of more and more human health–related, scientifically, economically, agriculturally, and/or environmentally important plant, animal, insect, and microbe species are undergoing sequencing or will be sequenced in the near future. No matter which of the sequencing strategies – clone-by-clone or whole-genome shotgun – is used, bacteria-based large-insert clone maps are indispensable for the accurate assembly and finishing of the entire genome sequence. The clone-based physical maps also provide an integrated resource for analysis of genome structure, organization, function, and evolution, and use of the genome sequencing results in plant genetic improvement.

Acknowledgements

This article was supported in part by the National Science Foundation Plant Genome Program (award numbers 9872635 and 0077766), the USDA/CSREES National Research Initiative (award numbers 99-35205-8566 and 2001-52100-11225), and the Texas Agricultural Experiment Station (Acc. #8536-203104).

References

Birnboim, H. C. and J. Doly (1979) "A rapid alkaline extraction procedure for screening recombinant plasmid DNA" Nucleic Acids Res. 7:1513–1523.

Carrano, A. V., J. Lamerdin, L. K. Ashworth, B. Watkins, E. Branscomb, T. Slezak, M. Raff, P. J. de Jong, D. Keith, L. McBride, S. Meister, and M. Kronick (1989) "A high-resolution, fluorescence-based, semiautomated method for DNA fingerprinting." Genomics 4:129–136.

Chang, Y. L., Q. Tao, C. Scheuring, K. Meksem, and H.-B. Zhang (2001) "An integrated map of *Arabidopsis thaliana* for functional analysis of its genome sequence." Genetics 159: 1231–1242.

Chen, M., G. Presting, W. B. Barbazuk, J. L. Goicoechea, B. Blackmon, G. Fang, H. Kim, D. Frisch, Y. Yu, S. Sun, S. Higingbottom, J. Phimphilai, D. Phimphilai, S. Thurmond, B. Gaudette, P. Li, J. Liu, J. Hatfield, D. Main, K. Farrar, C. Henderson, L. Barnett, R. Costa, B. Williams, S. Walser, M. Atkins, C. Hall, M. A. Budiman, J. P. Tomkins, M. Luo, I. Bancroft, J. Salse, F. Regad, T. Mohapatra, N. K. Singh, A. K. Tyagi, C. Soderlund, R. A. Dean, and R. A. Wing (2002) "An integrated physical and genetic map of the rice genome." Plant Cell 14:537–545.

Coulson, A., J. Sulston, S. Brenner, and J. Karn (1986) "Toward a physical map of the genome of the nematode *Caenorhabditis elegans*." Proc. Natl. Acad. Sci. USA 83: 7821–7825.

Ding, Y., M. D. Johnson, R. Colayco, Y. J. Chen, J. Melnyk, H. Schmitt, and H. Shizuya (1999) "Contig assembly of bacterial artificial chromosome clones through multiplexed fluorescence-labeled fingerprinting." Genomics 56:237–246.

Ding, Y., M. D. Johnson, R. Colayco, Y. J. Chen, J. Melnyk, H. Schmitt, and H. Shizuya (2001) "Five-color-based high-information-content fingerprinting of bacterial artificial chromosome clones using type IIS restriction endonucleases." Genomics 74: 142–154.

Fleischmann, R. D., M. D. Adams, O. White, R. A. Clayton, et al., (1995) "Whole-genome random sequencing and assembly of *Haemophilus influenzae Rd.*" Science 269(5223): 496–512.

Fraser, C. M., J. D. Gocayne, O. White, M. D. Adams, R. A. Clayton, R. D. Fleischmann, C. J. Bult, A. R. Kerlavage, G. Sutton, J. M. Kelley, J. L. Fritchman, J. F. Weidman, K. V. Small, M. Sandusky, J. Fuhrmann, D. Nguyen, T. R. Utterback, D. M. Saudek, C. A. Phillips, J. M. Merrick, J.-F. Tomb, B. A. Dougherty, K. F. Bott, P.-C. Hu, T. S. Lucier, S. N. Peterson, H. O. Smith, C. A. Hutchison III, and J. C. Venter (1995) "The minimal gene complement of *Mycoplasma genitalium*." Science 270(5235):397–403.

Gregory, S. G., G. R. Howell, and D. R. Bentley (1997) "Genome mapping by fluorescent fingerprinting." Genome Res. 7: 1162–1168.

Gregory, S. G., M. Sekhon, J. Schein, S. Zhao, K. Osoegawa, C. E. Scott, R. S. Evans, P. W.

Burridge, T. V. Cox, C. A. Fox, et al. (2002) "A physical map of the mouse genome." Nature 418(6899):743–750.

Hong, G. (1997) "A rapid and accurate strategy for rice contig map construction by combination of fingerprinting and hybridization." Plant Mol. Biol. 35:129–133.

Hoskins, R. A., C. R. Nelson, B. P. Berman, T. R. Laverty, R. A. George, L. Ciesiolka, M. Naeemuddin, A. D. Arenson, J. Durbin, R. G. David, P. E. Tabor, M. R. Bailey, D. R. DeShazo, J. Catanese, A. Mammoser, K. Osoegawa, P. J. de Jong, S. E. Celniker, R. A. Gibbs, G. M. Rubin, and S. E. Scherer (2000) "A BAC-based physical map of the major autosomes of *Drosophila melanogaster.*" Science 287(5461):2271–2274.

(IHGMC) International Human Genome Mapping Consortium (2001) "A physical map of the human genome." Nature 409(6822): 934–941.

Jeffreys, A. J., V. Wilson, and S. L. Thein (1985) "Hypervariable 'minisatellite' regions in human DNA." Nature 314(6006):67–73.

Li, Y., T. Uhm, T. S. Santos, C. Ren, C. Wu, M.-K. Lee, B. Yan, F. Santos, A. Zhang, Z. Xu Z., D. Luo, and H.-B. Zhang (2004) "An integrated map of rice *japonica* for functional analysis of its genome sequence." Genetics (submitted).

Marra, M. A., T. A. Kucaba, N. L. Dietrich, E. D. Green, B. Brownstein, R. K. Wilson, K. M. McDonald, L. W. Miller, J. D. McPherson, and R. H. Waterston (1997) "High throughput fingerprint analysis of large-insert clones." Genome Res. 7:1072–1084.

Marra, M. A., T. A. Kucaba, M. Sekhon, L. Hiller, R. Martienssen, A. Chinwalla, J. Crockett, J. Fedele, H. Grover, C. Gund, W. R. McCombie, K. McDonald, J. McPherson, N. Mudd, L. Parnell, J. Schein, R. Seim, P. Shelby, R. H. Waterston, and R. Wilson (1999) "A map for sequence analysis of the *Arabidopsis thaliana* genome." Nat. Genet. 22: 265–270.

Olson, M. V., J. E. Dutchik, M. Y. Graham, G. M. Brodeur, C. Helms, M. Frank, M. MacCollin, R. Scheinman, and T. Frank (1986) "Random-clone strategy for genomic restriction mapping in yeast." Proc. Natl. Acad. Sci. USA 83:7826–7830.

Ren, C., M.-K. Lee, B. Yan, K. Ding, B. Cox, M. N. Romanov, J. A. Price, J. B. Dodgson, and H.-B. Zhang (2003) "A BAC-based physical map of the chicken genome." Genome Res. 13:2754–2758.

Soderlund, C., I. Longden, and R. Mott (1997) "FPC: a system for building contigs from restriction fingerprinted clones." Comput. Appl. Biosci. 13:523–535.

Soderlund, C., S. Humphray, A. Dunham, and L. French (2000) "Contigs built with fingerprints, markers, and FPC V4.7." Genome Res. 10:1772–1787.

Sulston, J., F. Mallett, R. Staden, R. Durbin, T. Horsnell, and A. Coulson (1988) "Software for genome mapping by fingerprinting techniques." Comput. Appl. Biosci. 4:125–132.

Tao, Q., Y. L. Chang, J. Wang, H. Chen, M. N. Islam-Faridi, C. Scheuring, B. Wang, D. M. Stelly, and H.-B. Zhang (2001) "Bacterial artificial chromosome-based physical map of the rice genome constructed by restriction fingerprint analysis." Genetics 158:1711–1724.

Tomkins, J. P., T. C. Wood, M. Stacey, J. T. Loh, A. Judd, J. L. Goicoechea, G. Stacey, M. J. Sadowsky, and R. A. Wing (2001) "A marker-dense physical map of the *Bradyrhizobium japonicum* genome." Genome Res. 11:1434–1440.

Weising, K., H. Nybom, K. Wolff, and W. Meyer (1995) DNA Fingerprinting in Plants and Fungi, CRC Press, Inc., pp. 157–226,

Wu, C., S. Sun, P. Nimmakayala, F. Santos, R. Springman, K. Ding, K. Meksem, D. Lightfoot, and H.-B. Zhang (2004) "A BAC and BIBAC-based physical map of the soybean genome." Genome Res. 14:319–326.

Xu, Z., Y. L. Chang, K. Ding, L. Covaleda, S. Sun, C. Wu, and H.-B. Zhang (2002) "Automation of the procedure for whole-genome physical mapping from large-insert random BACs and BIBACs." In: Proceedings of the Plant & Animal Genomes XI Conference, San Diego, California, pp. 123 (abstract).

Xu, Z., S. Sun, L. Covaleda, K. Ding, A. Zhang, C. Wu, C. Scheuring, and H.-B. Zhang (2004) "Genome physical mapping with large-insert bacterial clones by fingerprint analysis: methodologies, source clone genome coverage and contig map quality." Genomics (submitted).

Zhang, H.-B. and C. Wu (2001) "BAC as tools for genome sequencing." Plant Physiol. Biochem. 39:1–15.

12
Software for Restriction Fragment Physical Maps

William Nelson and Carol Soderlund

Overview

Over a decade ago, the first contigs built by restriction fragment fingerprints were published. Coulson et al. [1] mapped *C. elegans* using a two-enzyme method, which produced small fragments that could be run on an acrylamide gel. The result was many high-precision fragments covering a subset of the clones. Olson et al. [2] mapped yeast using a complete-digest method, which produced large fragments that were run on agarose. The result was low-precision fragments covering most of the clone. This method was improved by Marra et al. [3] and used extensively, including for the human fingerprint map [4]. Ding et al. [5, 6] proposed high information content fingerprinting (HICF), which uses a two-enzyme method run on a sequencing machine, where each fragment is identified by both a size and one or five labeled bases. As with the method used for *C. elegans*, this produces high-precision fragments covering a subset of the clone. Modifications of this method have been developed by Dupont [7] and by Luo et al. [8], which we will refer to as the Dupont and SNaPshot methods, respectively. This chapter discusses and compares these methods, as well as the techniques for processing HICF fragments and assembling them in FingerPrinted Contigs (FPC) [9, 10].

12.1
Introduction

12.1.1
Review of Agarose Fingerprinting and FPC

In agarose fingerprinting, the clones are digested with a restriction enzyme such as *Hind*III and the resulting fragments are run on an agarose gel. The enzyme typically has a six-base recognition sequence and therefore cuts approximately once every 4^6=4096 bases, leading to an average of 32 fragments for a typical 130 kb BAC clone. From the gel, the fragments are detected as dark-colored bands, and

The Handbook of Plant Genome Mapping.
Genetic and Physical Mapping. Edited by Khalid Meksem, Günter Kahl

ISBN 3-527-31116-5

a migration rate is assigned to each one. This is generally done with Image ([11], www.sanger.ac.uk/software/Image), which is mainly interactive, i. e., a human must determine which are the true bands. The migration rate for each fragment is inversely related to the fragment size. By running a standards lane, it is possible to convert the migration rates back to base pair sizes. Either the rates or the sizes can be used as input into FPC, though the rates are typically used. The real value of the rate is within a tolerance of the measured values. When using migration rates, a fixed tolerance is used, while with sizes, a variable tolerance is used. We will consider only rates in this chapter, and they will frequently be referred to as "bands."

The clone fingerprints are then assembled into contigs by FPC. There are two distinct steps involved in the assembly. First, every clone is compared with every other, and every pair is given an overlap score, which is determined from the numbers of bands in each clone and the number of shared bands. Two bands from two different clones are considered shared if they differ by less than the tolerance, which is a user-supplied parameter to FPC. The score for a pair is intended to be roughly the probability that those two clones would share that many bands by random chance; the lower the score, the more likely it is that the two clones actually overlap in the genome. FPC compares the score against another user-supplied parameter, the cutoff, and if the score is lower, FPC puts those two clones into the same contig.

After FPC joins the clones into contigs, it tries to determine their ordering within the contigs. For this purpose, FPC computes a "consensus band map," or CB map. Ideally, the CB map lists in the correct order all of the bands in the stretch of genomic DNA covered by the contig. In practice, the clone band files contain numerous errors – including extra bands, missing bands, and incorrectly sized bands – which make it impossible to compute an exact CB map. Also, contigs can be wrongly joined by a false-positive overlap, meaning that even with perfect data, a good CB map could not be generated since the contig actually covers two or more disjointed pieces of DNA.

Computing the best solution would take an unacceptable amount of computer time; therefore, the algorithm to build the CB maps is an approximation. FPC makes several attempts to compute a CB map (the exact number is user-specified), each time starting from a different initial clone. Each attempted CB map gets a score, which reflects the fraction of the clone bands that actually align to the map; bands that do not align are referred to as "extra" bands. A clone that contains more than 50% extra bands is labeled as a "Q" (for question) clone. Generally, the presence of more than a few Q clones indicates that a contig has at least one false positive overlap and should be broken into two or more contigs (this is further discussed in Section 12.4.5).

FPC chooses the best-scoring CB map for each contig and records the number of Q clones in each contig. These numbers are shown to the user, who can then choose to run the "DQer" to reduce the number of Q clones. The DQer attempts to break up contigs having more than a given (user-specified) number of Q clones by rerunning the assembly algorithm on those contigs using more stringent cutoff values. Specifically, it reruns the algorithm up to three times, each time with the cutoff reduced by a user-specified step. For example, if the cutoff is set at 1e-12 and the step is set to 1, then all contigs with too many Q clones will be reevaluated at 1e-13. This may split some contigs apart, hopefully eliminating the bad joins and reducing the Q count. Contigs that still have too many Q clones will then be re-evaluated at 1e-14 and, if necessary, again at 1e-15.

This is the procedure followed to build an agarose map. We will now discuss the exact values of the parameters appropriate for agarose projects in order to compare to the HICF projects discussed next.

The tolerance is determined by how accurately the gel image can be scored by band callers; typically, a value of 7 is used (which corresponds very roughly to 15 bp). In these units, the range of possible band sizes is approximately 500–1900, so the "gel length" parameter should be 1400. Gel length is an important parameter because it determines the probability of two random bands overlapping to within the tolerance; for example, two bands picked randomly from a range of 500 to 1900 are much more likely to be within 7 of each other than are two bands picked from a range 10 times larger.

We note that the default gel length on FPC is 3300, which is the value that has been used in most agarose projects. This value is considerably larger than the "correct" value of 1400, which means that the score cannot really be interpreted as a probability (and this interpretation was already somewhat problematic due to factors discussed in Section 12.5.2 below). The success of these projects indicates that the score can function successfully as a coincidence score even if it loses its strict interpretation as a probability; in fact, it is often referred to as a coincidence score.

The best cutoff to use is influenced by two factors. First is the size of the project. The more clones a project has, the more opportunities there are for false overlaps to occur strictly by random chance; therefore, the cutoff needs to be low enough to exclude these false overlaps. Second, since the genome is not random, there are repeated bands that cause clones to share more bands than they would randomly. Because of this, the cutoff has to be lower than it would otherwise be in order to avoid false-positive overlaps. But since the presence of many Q clones provides a way to detect most false positives [10], the cutoff does not have to be low enough to completely eliminate these two factors; instead, the DQer can be run to reduce the false positives. For agarose projects using BAC clones, with gel length 3300, a cutoff of 1e-12 generally gives good results (with subsequent execution of the DQer).

The cutoff determines the approximate percentage overlap that two clones must have in order to be joined, and this in turn determines the expected numbers of contigs and gaps. For example, with tolerance 7 and gel length 3300, a cutoff of 1e-12 means two 30-band clones must share at least 21 bands to be joined (calculated from the Sulston overlap formula [12]). In other words, the clones must overlap by at least 70% in order to be joined.

Letting ϑ denote the required overlap percentage, c the coverage, and N the total number of clones, then the expected number of contigs is approximately $Ne^{-c(1-\theta)}$ [13]. Since this number depends exponentially on ϑ, it is very desirable to reduce ϑ as much as possible, as even a small decrease translates into a significant reduction in the necessary coverage multiple. The 70 % value typical of agarose projects seems quite high, and one of the primary goals of HICF is to lower this number.

Besides the 70 % overlap requirement, the other major disadvantage of the agarose method is the necessity of human band calling. Human band calling is very time-consuming, requiring about 45 minutes per 96-clone plate, and it is also quite prone to error.

12.2 HICF Techniques

12.2.1 Generalities

HICF fingerprinting applies to physical mapping the same technology that has given rise to high-throughput sequencing. In HICF, the clones are digested with restriction enzymes that result in an overhang on one or both ends and then are labeled with one or more fluorescent labels that attach to the overhanging bases. The labels will have different colors depending on the nucleotide they attach to, so that the color or colors of a fragment directly correspond to nucleotides of the overhang (an exception to this rule is the original method of [5]). The fragments are run in a sequencing machine, which determines their sizes and colors, without laborious human band calling. Fragments from different clones are considered to match if their sizes differ by less than the tolerance and if their color labels are the same.

HICF techniques can be divided into two types, according to whether one or multiple bases are labeled on the overhang. The multiple-base labeling methods are really small sequencing reactions in which the overhang of each fragment is sequenced. Clearly this would add greatly to the information content of the fingerprints, if it were practical; however, the analysis of the sequencer output becomes quite difficult in this case, and to our knowledge only single-base labeling has so far been used in practice.

Even with single-base labeling, a fragment can sometimes have more than one label. This cannot be avoided because inevitably some fragments will be cut on both ends by the enzyme that gives an overhang, and the overhangs on each end will be labeled. If the two labels have different colors, the fragment will be detected as two fragments having different colors but of the same size. Whether these should be treated as one fragment or two is discussed further in Section 12.5.2.

In addition to being more automated, HICF is also much more sensitive than agarose for two reasons. First, the fragment sizes are determined to a very high precision, generally giving at least a 0.3-bp precision (the meaning of fractional base pairs will be discussed below). This means that when fragments do match up to within this tolerance, it is less likely to be by chance, so a smaller number of matches are needed to identify a true overlap. Second, there are generally more fragments than in agarose, typically around 100 per clone. More fragments mean more information, which is an advantage as long as there are not so many that they start to randomly match up. With HICF, the high precision plus the use of different color labels to further distinguish fragments means that many more fragments can be used without an unacceptable number of false matches.

Because of this extra sensitivity, HICF is theoretically capable of discriminating much smaller clone overlaps than agarose; for example, a cutoff of 1e-12 requires only about 30 % clone overlaps with typical HICF parameters. This is a tremendous gain in efficiency, although in practice non-randomness of the DNA and the error rate can prevent one from actually using such small overlaps. In the highly repetitive maize and wheat genomes, which have so far been the main subjects of HICF, cutoffs in the range of 1e-25 to 1e-50 have been found to be necessary.

One drawback to HICF, which we will discuss below, is that the HICF bands cannot be accurately reproduced *in silico*, meaning that sequenced clones cannot be mapped back to the physical map. This is unfortunate because placing sequenced clones on the physical map has proved very useful with agarose maps [14]. A second drawback is that the fragments cover only a subset of the clone and thus the fragments cannot be used to approximate the size of the clones, which has been useful in the agarose method.

Several variants of HICF have been proposed in recent years, beginning with Ding et al. in 1999 [5]. This approach required three separate labeling reactions for each clone and was improved upon in 2001 [6] with a method based on a type IIs restriction enzyme. This method was adapted by Dupont [7] and will be described in Section 12.2.2. Another promising variant (see [8]), based on the ABI SNaPshot kit, is described in Section 12.2.3.

12.2.2
Approaches Using Type IIS Enzymes

Ding et al. [6] proposed that the DNA be cleaved by the type IIS 5-cutter enzyme *Hga*I into fragments with 5-bp overhangs. The fragment sizes were further reduced by digestion with the type II 4-cutter enzyme *Rsa*I, which produces blunt ends. This second enzyme is necessary to produce small enough fragments to be run on a sequencing machine. The fragments having an overhanging end were then labeled with either one or five fluorescent labels and run through an ABI 377 sequencer to detect their labels and sizes (Note that most of the resulting fragments are cut on both ends by the 4-cutter and therefore are not labeled at all. These fragments are discarded and do not contribute to the fingerprint). Ding et al. [6] also developed a technique, as described below, that allows HICF data to be used directly within FPC.

Dupont has developed a slight modification of this process [7], which we will refer to as the "Dupont" method. In this method, the type IIS enzyme is the 6-cutter *Ear*I, while the type II is the 4-cutter *Taq*I. In contrast to the Ding method, the 4-cutter cuts with a two-base overhang, where the first base is g. Since the 4-cutter cuts very frequently, the base g is not labeled; hence there are only three color labels in the Dupont method as opposed to four in the original. Also, this method uses only single-base labeling, even though *Ear*I has a three-base overhang.

Dupont has applied this method to construct a complete map of the maize genome using the ABI 377 sequencer. They have reported a significantly smaller number of contigs for the same clone count as compared to the public agarose map; however, their map has not been made public.

Based on this positive report, it was decided that a publicly available maize HICF map should be constructed using the Dupont method. This work is being done at Rutgers and the University of Arizona by a group that includes the present authors (see http://www.genome.arizona.edu/fpc_hicf/maize). We are using the same enzyme set as Dupont, but we chose the ABI 3700 sequencer instead of the 377 in order to increase throughput. We have also found it necessary to build with lower cutoffs (1e-45 to 1e-50) than those reported by Dupont, but it is difficult to resolve this discrepancy since the Dupont data is not available.

For this process, we see approximately 17% error; in other words, approximately 17% of the bands in the final band files appear to be incorrect. This means that two identical clones, prepared separately and fingerprinted on different machines, will have only 66% of their bands overlapping on average, and it does not appear possible to correct this through data processing.

12.2.3
SNaPshot HICF

The SNaPshot process was developed by Applied Biosystems for analyzing SNPs and has been adapted for fingerprinting by Luo et al. [8], who used it to construct a physical map of wheat genome D (the NSF Wheat D Genome Project, http://wheatdb.ucdavis.edu:8080/wheatdb; see also [15]). In SNaPshot, clones are digested with four 6-cutter type II enzymes (*Eco*RI, *Bam*HI, *Xba*I, and *Xho*I) and one 4-cutter (*Hae*III) enzyme whose purpose is again to reduce the fragment sizes. The four enzymes are chosen so that each one has a unique first base on its overhang, meaning that after labeling, the four colors correspond cleanly to ends produced by the four enzymes. Since the type II cut sites lie within their recognition sequences, the sequences surrounding the labeling sites are determined, which may lead to more consistent labeling in this method as compared to type IIs methods. It does appear that noise is reduced in SNaPshot, leading to a higher-quality build. In tests carried out as part of the NSF Wheat D Genome Project, the error rate per fingerprint was measured at just over 13% (Frank You, personal communication), a significant improvement over the 17% rate that we measured for Dupont HICF. However, a conclusive comparison of the two techniques has not been carried out. SNaPshot is currently not supported on 3700 machines, but it is supported on the 3730 and 3100 machines, so high-throughput fingerprinting is possible.

12.3
Processing HICF Data

12.3.1
Peak Scoring

The HICF data come from the ABI sequencing machine in the form of a "trace" file, with the extension ".fsa." This file contains a graph of the intensity recorded for the different colors as a function of time (an "electropherogram"), which is the raw output of the machine. As a fragment goes by the detector, it produces a peak in the graph corresponding to its color label. The goal is to locate these peaks, separate the real ones from spurious ones, and figure out the size (in base pairs) of the fragments that produced them.

If the ABI Genescan Analysis software is set to "auto-analyze," then the software will carry out some of these steps automatically. In particular, it will locate the peaks and assign fragment sizes to them. The peaks, along with their heights, widths, and fragment sizes, are entered into a table known as the "peak table," which is also contained in the trace file (Figure 12.1). With current size standards, the fragments can be sized from 35 bp to 500 bp.

Unfortunately, several hurdles remain. First, the peak table contains many spurious peaks, sometimes hundreds, from which the real peaks must be separated. Generally this is done using a threshold scheme: the real peaks are assumed to be higher than the false ones, so a threshold is chosen and peaks below the thresh-

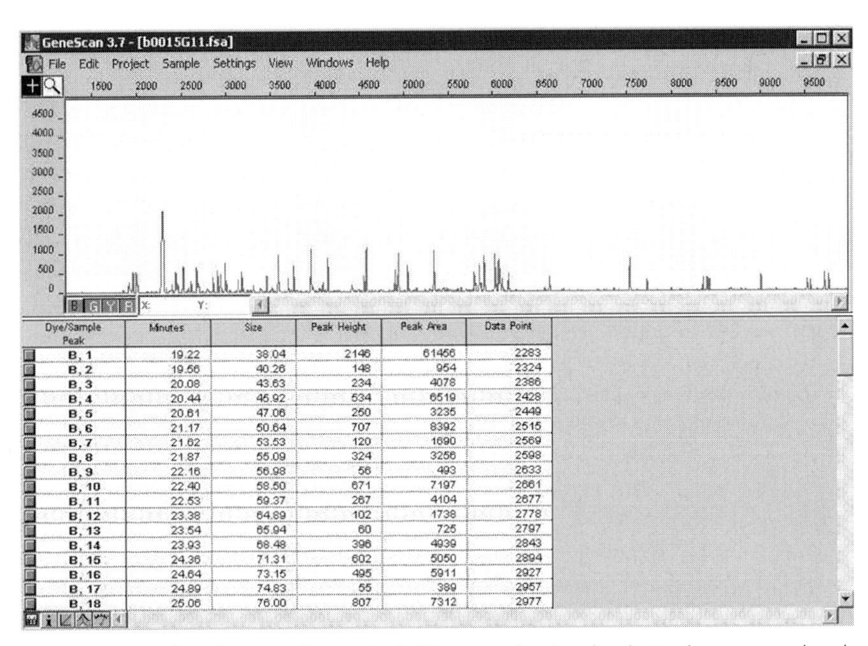

Dye/Sample Peak	Minutes	Size	Peak Height	Peak Area	Data Point
B, 1	19.22	38.04	2146	61456	2283
B, 2	19.56	40.26	148	954	2324
B, 3	20.08	43.63	234	4078	2386
B, 4	20.44	45.92	534	6519	2428
B, 5	20.61	47.06	250	3235	2449
B, 6	21.17	50.64	707	8392	2515
B, 7	21.62	53.53	120	1690	2569
B, 8	21.87	55.09	324	3256	2598
B, 9	22.16	56.98	56	493	2633
B, 10	22.40	58.50	671	7197	2661
B, 11	22.53	59.37	267	4104	2677
B, 12	23.38	64.89	102	1738	2778
B, 13	23.54	65.94	60	725	2797
B, 14	23.93	68.48	396	4939	2843
B, 15	24.36	71.31	602	5050	2894
B, 16	24.64	73.15	495	5911	2927
B, 17	24.89	74.83	55	389	2957
B, 18	25.06	76.00	807	7312	2977

Figure 12.1. Display of an HICF fingerprint in Genescan showing the electropherogram and peak table for the blue channel.

old are discarded. However, choosing the threshold is not so easy. A fixed threshold does not work because the overall strength of the signal can vary dramatically from trace to trace and from color to color. Sometimes the highest peaks will be under 1000, and sometimes they will be 10,000 or more. Because of this variability, usually one or more of the higher peaks are used to normalize the height scale; for example, in our maize project, for each color we take the sixth highest peak of that color and multiply its height by 0.25 to get the threshold for that color. Dupont used a similar scheme, and the GenoProfiler application (discussed below) also implements this idea, with user-specified parameters [16].

Obviously this kind of thresholding scheme has a number of adjustable parameters that need to be optimized on some control data. In our project, we fingerprinted a set of clones repeatedly and chose the parameters that gave the best match of the replicates; however, we saw approximately equal performance for factors from 0.2 to 0.4, and it seems unlikely that there is one "best" parameter set.

A complication present with the 3700, and possibly other capillary-based sequencers, is so-called "front-end loading," which results in spurious peaks at sizes below 75 bp. This problem is most easily resolved by not using fragments smaller than 75 bp.

12.3.2
Sizing of Fragments

The sizing of fragments is carried out by Genescan software, and it is very consistent. For example, we found a maximum standard deviation of 0.15 bp for six identifiable vector bands, measured across five different 3700 machines and approximately 60,000 clones. This consistency allows the use of quite low tolerances for building HICF fingerprints within FPC, as compared to agarose.

Unfortunately, the sizes, although highly consistent, are not very accurate. This has been observed by all HICF groups and is recognized by ABI as well. The error is typically 0–6 bp, up or down; for example, a fragment of an actual size of 100 bp can give rise to a peak anywhere from about 94–106 bp. There is no discernible pattern to the errors, and there is evidence that it depends on the underlying sequence. Luo et al. [15] have observed that a strand and its complement frequently have different mobility; they further tested whether GC content could explain the difference but found an insignificant correlation.

Assuming that the error has some dependence on sequence, the sizes produced by the machines contain extra information beyond just fragment size, and it is this information that allows fractional base pair sizes to have meaning. For example, a fragment sized at 99.7 is almost certainly not the same as one sized at 100.3, but this information would be lost if the machine rounded them both to 100. Thus, sequence-dependent mobility will increase the sensitivity of HICF, but without any deeper understanding of it, we cannot say exactly how much information is gained. Further characterization of this phenomenon would be beneficial.

We emphasize that the sizing error is not at all a problem for comparing fingerprints to each other or for constructing maps; where it becomes a problem is in trying to compare the fingerprints with theoretically expected fragment sizes.

This is one of the obvious approaches to testing and validating the process, but because of the size error it is not as informative as one would like. The error also makes it very difficult to add sequenced clones to the map by *in silico* digestion, as mentioned above.

12.3.3 Quality and Contamination Screening

When processing clone fingerprints, there will be failures, and it is desirable to weed them out through quality checks. As it happens, the cloning process itself provides an excellent check in the form of vector bands. These are the HICF bands produced by digestion of the cloning vector, which will be few in number and the same for every clone from a given library. Requiring the presence of almost all of these bands is a good check on quality (the only reason not to require all of them is that, as mentioned above, the peak scoring is inexact and may sometimes miss a peak that is there but below the threshold).

Also useful as a check are the sizing standards peaks, which are always labeled with a color separate from any of the HICF fragments. If the run was successful, then the standards peaks should be the highest peaks in their color, and they should be sized by Genescan exactly to their predetermined sizes. None of the standards peaks should ever be missing, and if this does happen (which is rare), the clone should be rejected. A more common problem is for a spurious peak to exist that is higher than one or more of the real standards peaks. If more than one or two of these are seen, it indicates a problem and the clone should be rejected.

Other checks can be made on the peak counts, both before and after scoring. Few or no raw peaks is obviously a failure, and this is rather common. Some traces will have very large numbers of raw peaks, but we have not found that this indicates a bad fingerprint. Clones that produce too few scored peaks may as well be rejected, because they will probably just result in singletons anyway, having too few bands to produce a significant overlap with another clone (Clones that do not go into a contig are referred to as singletons).

One problem that can severely degrade the build is chimeric fingerprints, i. e., fingerprints containing fragments from two different clones. These can form if material from one well contaminates another well at any stage of the process. Chimeric fingerprints cause false joins, leading to wrong contigs; in a severe case, all the clones may come together into a single contig. There are two ways to screen this type of contamination. The most thorough screen is to make the assumption that the contamination occurs within plates and then to search for pairs of clones on the same plate that have a suspiciously high overlap. The test can be applied to all pairs or to only neighboring wells. Where to set the threshold is unclear, but lower is probably better since every chimeric clone that slips through is likely to produce a damaging false join. In our project we use the same cutoff we are using in FPC, so effectively we do not allow any two clones from the same plate to join. Genoprofiler, discussed below, also includes this type of plate-by-plate contamination checking.

The second screen takes advantage of the fact that chimeric fingerprints will generally have more bands than normal, since they contain fragments from two clones. Then all fingerprints with more than a given number of bands may be regarded as suspicious and be discarded. To some extent, this is already required by FPC, which will not accept clones having more than 250 bands; however, this number is larger than typical chimerics, which would be expected to have roughly twice the normal band count, or approximately 200 bands for Dupont HICF.

Each of these screens has strengths and weaknesses. The first screen will catch all possible intra-plate contamination, and it also produces useful evidence on whether contamination is actually occurring. However, a chimeric can still slip past this test if its contaminant fails in fingerprinting. Since roughly 15 % of the fingerprints can be expected to fail, this test could miss 15 % of the chimerics, which may not be acceptable. The second screen does not rely on successful fingerprints, but it does require a maximum band count to be chosen, and this is problematic. If the band count limit is too high, chimerics of smaller clones will be passed, while if it is lowered to catch these, many valid clones will also be removed. Probably the best solution is to use a combination of both.

Lastly, we mention that low-quality or noisy fingerprints generally should have little effect on the map even if they are not screened, since they should only result in singletons. Hence, the quality checks need not be too stringent and can be viewed more as a way to track the success rate and to identify problems. Contamination, on the other hand, is very damaging and must be removed if possible.

12.3.4
Vector and Repeat Screening

Screening out vector bands is a good idea in any fingerprint mapping project. The vector bands, which come from the cloning vector, are the same for each clone and therefore lead to false detection of overlaps. The difference in the score is substantial. For example, two 100-band clones matching 60 bands, with tolerance = 7 and gellen = 25,500, will have an overlap score of 7.4e-50; adding four matching vector bands reduces the score to 1.1e-55. This can be compensated by a lower cutoff if all clones use the same cloning vector, but this does not work if the project uses multiple libraries with different vectors. Another group of bands that one may want to screen out are those coming from high-copy repeats, which are especially prevalent in many plant genomes. For an agarose project, the fragments are not much smaller than typical retroelements, and since the retroelements can be fragmented by other retroelement insertions, it seems unlikely that many consistent-sized fragments are produced. However, in HICF the fragments are only 128 bp on average, so they can fit easily in a retroelement, and unless they are badly fragmented or mutated, the bands will appear in the final fingerprint. So there is reason to believe that repeat element bands will play more of a role in HICF than in agarose. In the Wheat D Genome Project, about 30 repeat bands were screened, and we are studying this possibility in our project. Dupont did not do repeat screening for their maize map.

Before screening any repeat (or vector) bands, they must be identified. The easiest way to do this is to fingerprint the vectors by themselves. The repeats can be extracted from whole clones by plotting the occurrence frequency of each band and looking for the spikes. GenoProfiler has a built-in function to carry out this frequency analysis.

Having decided which bands to screen, there are two ways to carry out the screen. FPC can import a file of bands to screen, which it will apply to every newly added clone (cf. Section 12.4.1). FPC carries out the screen only at the time of addition, so if the screening set is changed later, the project must be re-created and all clones re-added. Screening is also available within GenoProfiler, in which case the screening is applied as the band files are created, so a change in screening requires all of the band files to be regenerated.

12.3.5
Packaging HICF Data for FPC

The input to FPC is a file of one or more clone fingerprints, where each fingerprint is represented by a list of positive integers, sorted from smallest to largest. The file can be a "band file," with the extension ".bands," or a "sizes file," with extension ".sizes"; it makes no difference within FPC.

FPC cannot currently accommodate decimal-valued band sizes or multiple colors; hence, the HICF fingerprints must be reprocessed into the correct FPC format. Fortunately, this is easy to do, using a technique proposed by Ding et al. [6]. The idea is to first rescale the sizes by a factor, dropping the decimal, and then add an increment value specific to each color. For example, for three-color HICF, we can first multiply each fragment size by 20, and then we can drop the decimal part while still retaining precision of $1/20 = 0.05$ bp. Since the fragment sizes range from 75–500, after rescaling they will run from 1500–10,000. We then combine all three colors into one list by adding 10,000 to the green bands and 20,000 to the yellow bands; this segregates the different colors into separate ranges so that they do not get matched to each other by FPC. The end result is a list of bands ranging from 0–30,000, with a few gaps corresponding to the excluded 0–75 fragment size range.

For SNaPshot, which has four colors, the procedure is similar. Typically, a scale factor of 10 is used, with color increments of 0, 5000, 10,000, and 15,000. Genoprofiler implements this idea with adjustable parameters.

12.3.6
GenoProfiler

GenoProfiler [16] is a Java software application developed specifically for processing HICF data (http://wheat.pw.usda.gov/PhysicalMapping/tools/genoprofiler; Figure 12.2). GenoProfiler can carry out all of the processing steps described above. Starting from the raw output of ABI sequencers (any model), it will generate FPC-compatible band files with the extension ".sizes." It also includes a number of other

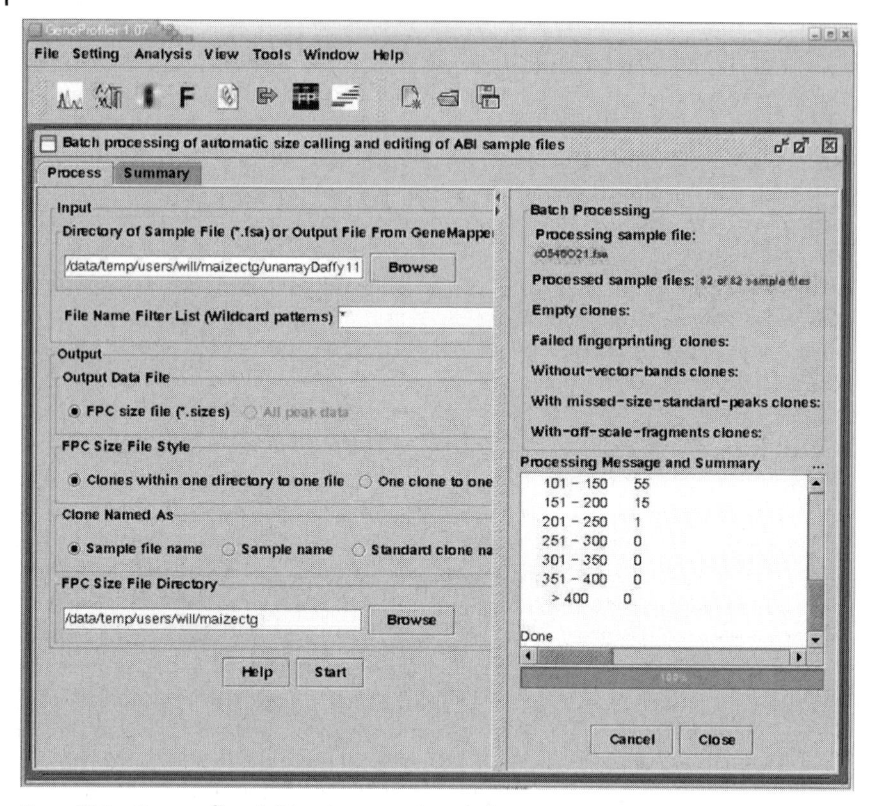

Figure 12.2. Genoprofiler v1.07 main processing window.

features such as fragment frequency analysis and contamination checking. It can process both three-color and four-color HICF data and is currently undergoing active development and improvement.

12.4 Building HICF Maps in FPC

Once the band files are created, they are loaded into FPC and assembled. For HICF projects, it is crucial to use FPC version 7 or later; this can be obtained from http://www.genome.arizona.edu/software/fpc. Version 7 contains several modifications to accommodate HICF, most importantly the use of double-precision numbers to accommodate the very small overlap scores generated by HICF. This version is also vastly superior to previous versions of FPC for reasons unrelated to HICF. It has a new, versatile, and easy-to-use contig graphical display, which allows marker and other data to be viewed in configurable tracks. It also has the Blast Some Sequence (BSS) routine, which simplifies blast searches against sequence located on the FPC map, and an automated minimal tiling path selector (see [14]).

12.4.1
Creating a New Project, Loading the Fingerprints, and Screening Vectors

For the most part, the process of building a map is the same as for agarose projects and is described in the FPC tutorial [17], which is also available from www.genome.arizona.edu/software/fpc, along with a User's Guide and demo. We summarize the procedure here in order to elucidate the differences, but new users should still read the tutorial, as it describes many more important features.

To create a new project, go to the directory where it will be located and make a subdirectory called "Image." Put the band files (or sizes files) to be used into the Image directory. FPC always looks for new fingerprints in this directory. Next, run FPC with no command-line arguments, and the main window will appear (Figure 12.3). Right-click on the File button, choose "Create new project," and enter the project name in the pop-up window.

Now a new, empty project exists. Before adding any fingerprints, the vector band screening should be configured, if desired. To set this up, press the "Configure" button on the Main window, which will bring up the Configure window (Figure 12.4). Enter the file name of vector bands to be screened in the text box labeled "Vector file," and press Close. The file can contain vector bands for several different clone libraries. Details of the file format can be found in the User's Guide (http://www.genome.arizona.edu/software/fpc/faq.html); note that this format is currently being revised for an upcoming release of FPC.

Once the vector screen file is specified, the fingerprints can be loaded. This is done by pressing the "Update .cor" button on the Main window. This causes FPC to load all the fingerprint files from the Image directory. After it loads a file, it transfers it to either the Bands directory or the Sizes directory, depending on the file extension; these directories are created if they do not already exist. These two directories are only for storage – the files in them are not used again by FPC.

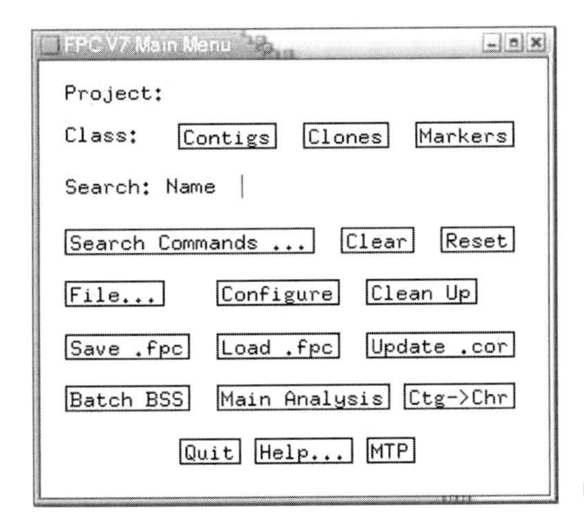

Figure 12.3. FPC Main window.

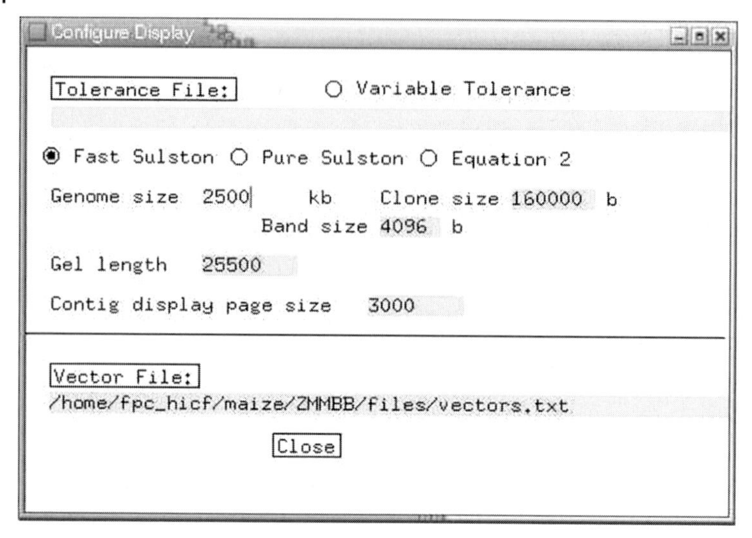

Figure 12.4. FPC configuration window, with typical HICF gel length setting. Also shown is the entry for a vector screening file.

The ".cor" on the "Update .cor" button refers to the file with that extension which FPC creates in order to store the band information. This is a binary format file that cannot be edited. FPC also creates a file with extension ".fpc," which is the primary project file. All the information needed to display the project is in the .fpc file, but since this file does not contain the actual bands, any operation that uses the bands also requires the .cor file. Lastly, a .fpp file is also created, which stores contig configurations. This file is not necessary for running FPC, but if it is missing, the settings will revert to their defaults.

12.4.2
Tolerance and Gel Length

The tolerance setting is found on the FPC Main Analysis window (Figure 12.5), and it tells FPC how close two bands must be to be considered matching. Its value should be set according to the observed size variation of particular bands in the project. For example, as described above, we observed a maximum standard deviation of 0.15 bp for vector bands fingerprinted on 3700 machines; based on this, we chose a tolerance of 0.35 bp, and after rescaling by the factor of 20, we got a tolerance of 7 in the rescaled units. There is no "perfect" value, and anything from 5 to 10 appears to work equally well for us. For SNaPshot projects using a rescale factor of 10, the same calculation gives a tolerance of 3.5, so a tolerance of 3 or 4 would work.

The "gel length" setting is found on the FPC Configuration window, and it should be set to the total number of possible band values. For example, for three-color HICF as described above, with scale factor 20 and increments 10,000

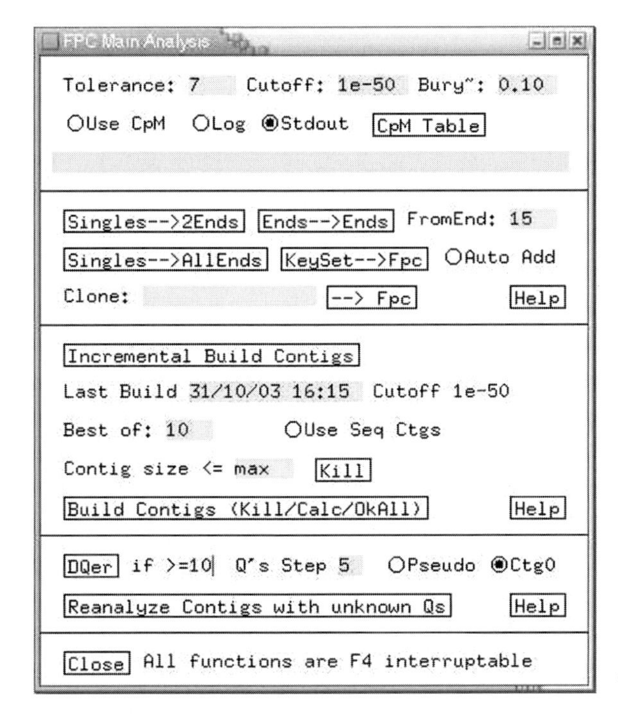

Figure 12.5. FPC Main Analysis window, showing typical HICF cutoff and DQ settings.

and 20,000, the bands lie in three intervals: 1500–10,000, 11,500–20,000, and 21,500–30,000. Each of these has a length of 8500, for a total gel length of 25,500.

12.4.3 Cutoff

The cutoff setting on FPC's Main Analysis window governs how much overlap two clones must have in order to be joined into a contig. As discussed above, there are various theoretical factors that govern its value, but in practice it has to be determined by trial and error. The goal is to find the highest cutoff such that, after the DQer is run, the build has an acceptably low number of bad joins. For agarose projects, cutoffs around 1e-12 are normal, but in HICF, they tend to be much lower, typically 1e-30 to 1e-50. FPC v7 has been altered to use double-precision floating-point numbers in the overlap calculation, which allows these low values to be used without problem.

12.4.4
Building

Once the settings described above have been set, the project is ready to build. To build from scratch, press the button labeled "Build Contigs (Kill/Calc/OkAll)" on the Main Analysis window. This will destroy all the contigs, if any, and rebuild everything.

To build new fingerprints into an existing project without destroying existing contigs, press "Incremental Build Contigs." New fingerprints will be added to existing contigs, some of which may then be merged to form larger contigs.

12.4.5
Build Quality, Q Clones, and the DQer

As discussed in the Introduction, the build may result in wrongly joined contigs containing substantial numbers of Q clones. These Q clones are then reduced by running the DQer, which for v7 has been modified to accept a user-defined step size. For agarose, a step of 1 works well (e.g., 1e-13, 1e-14, 1e-15), but for HICF a step this small has little effect, and the step should be changed to some-

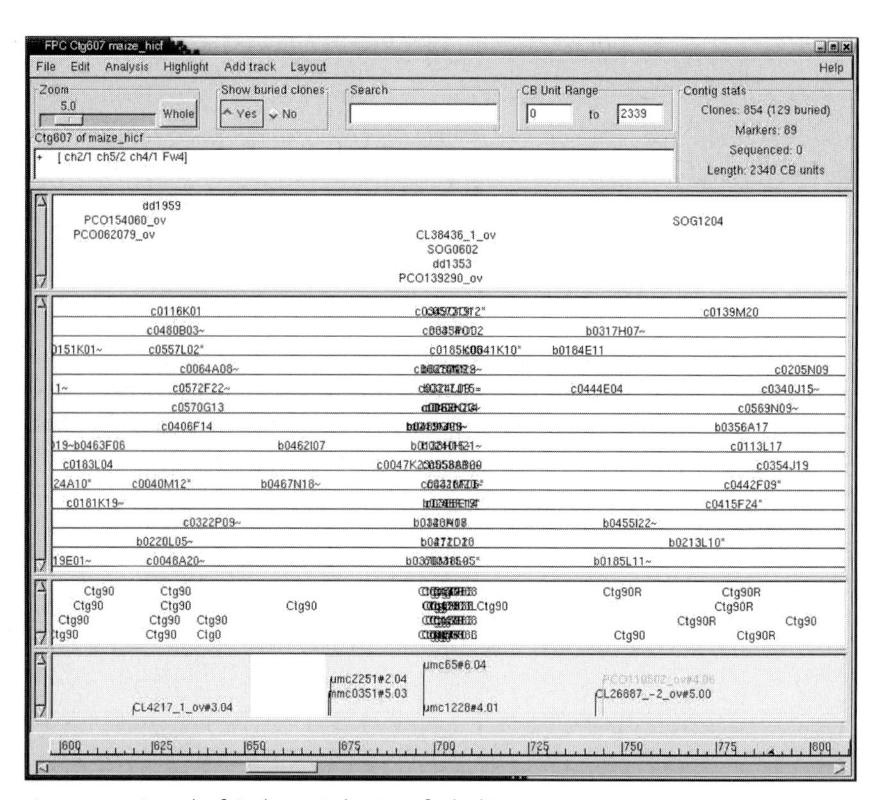

Figure 12.6. A stack of Q clones, indicative of a bad join.

thing like 5 (e. g., 1e-55, 1e-60, 1e-65). If a larger amount of noise is present, as in our experience with the Dupont method, then it also makes sense to raise the number of allowable Q clones from the default of 5. We are currently using 10, and an even higher number may be appropriate.

To elaborate on this, a Q clone is not necessarily wrong. If the fingerprints contain substantial errors, then there will be correctly located clones that still contain more than 50 % "extra" bands and get labeled as Q. The goal of the DQer is not to eliminate all Q clones but rather to break apart badly joined contigs. If a false positive join occurs, then all the clones of one contig are attached incorrectly to another contig; since there is no way to order clones from two different contigs in one linear CB map, the clones of one contig end up in a stack, as illustrated in Figure 12.6. The bands of the stacked clones cannot be ordered, so they all end up as Q clones. Hence, a contig with many Q clones probably has a false join and should be split. Ideally, the DQer settings should be as low as possible while still eliminating most of the bad join Q-stacks. If the settings are more stringent than necessary, valid contigs will be split apart.

It is not unusual for the build, before DQing, to put a substantial fraction of the clones into one contig. If the DQer is not able to break this clump apart into good contigs, then the starting cutoff should be lowered and the build rerun. The benefit of the DQer is that it allows the initial build to be performed with a relatively high cutoff to maximize the number of joins, while still breaking apart most of the wrong joins subsequently using more stringent cutoffs.

The DQer should also be run after each incremental build, because each addition of new clones can cause wrong contig merges, which should be broken apart.

12.4.6 multiFPC

MultiFPC, created by Discovery Biosciences (http://www.discoverybio.com/products/multifpc/multifpc_intro.htm), is an adaptation of FPC v4.6.4 that contains a modified clone overlap formula specifically designed for HICF. The ideas behind these modifications are reasonable and will be discussed further in Section 12.5.2; however, there are some significant drawbacks to multiFPC as it exists now (in version 1.2.01). First, it is adapted from an old version of FPC and therefore lacks many improvements found in FPC v7. Second, it is considerably slower than FPC, taking 13 minutes to build a 930-clone test project versus 1 minute for FPC. Hence, we feel that FPC v7 should be preferred at present for building HICF projects, pending further development of multiFPC.

12.5
Theoretical Aspects

12.5.1
Simulations

We have carried out simulations using rice chromosome 10 to estimate the effectiveness of HICF mapping. Rice has few repeats compared to maize, so it is a good subject for HICF. Our simulations show a great advantage for HICF as compared to agarose mapping when there is no noise. By introducing noise into the simulations, we also get an idea of how the quality degrades in terms of contig-forming power and incidence of Q clones.

For the simulations, we used the rice chromosome 10 pseudomolecule, with all sequence contig gaps removed so that it was just one long sequence. We then created a simulated 15× clone library with 2029 clones having an average insert size of 168 kb and a standard deviation of 12 kb. Lastly, we created one dataset with the clones digested *in silico* with *Hind*III (for the agarose simulation) and a second dataset with the clones digested with the Dupont HICF enzymes.

HICF outperforms agarose greatly when comparing these two simulated projects. Regardless of the cutoff used, there are no Q clones, and at 1e-25 all of the clones come together into one contig. For agarose, by contrast, there are 34 contigs at 1e-12, while at 5e-12 there are still 25 contigs along with 180 Q clones, indicating at least one false positive join.

A similar result is seen with an 8× simulated library of 930 clones. In this case, HICF at 1e-15 forms 7 contigs and 0 Q clones, while agarose at 1e-8 still forms 26 contigs, and at the cost of 200 Qs. From these results it is clear that HICF, without noise and on a genome with few repeats, is vastly more efficient that agarose.

Returning to the 15× library, we also ran simulations with noise. The noise was simulated by simply removing bands at random and replacing them with other, randomly generated bands. For example, 5 % noise means that 5 % of the real bands (on average) were removed and replaced with random bands. Builds were run at 5 %, 10 %, 15 %, and 20 % noise, all at a cutoff of 1e-45. (This stringent cutoff was chosen to ensure that any Q clones generated were due to noise and not to bad joins.)

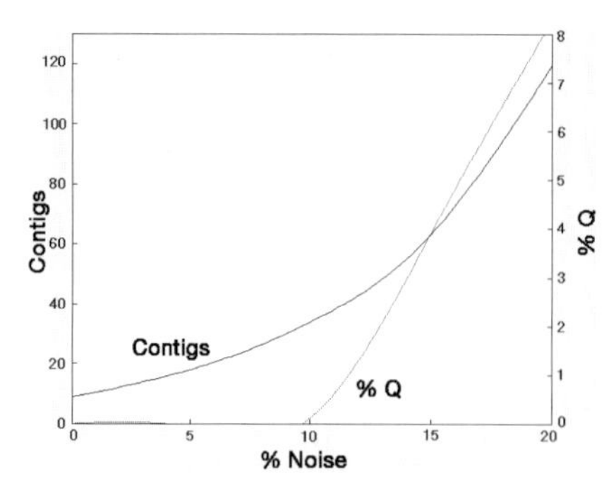

Fig 12.7. Contig count and Q clone percentage in HICF simulations with noise.

As shown in Figure 12.7, the noise degrades the contig-forming power of HICF considerably. Thirty-four contigs are formed at 10 % noise versus nine contigs at 0 %. Indeed, the contig count approximately doubles for every 5 % increase in noise. Qs are minimal until noise reaches 10 %, after which they proliferate rapidly.

In our maize project, after running the DQer to reduce bad joins, we have 6.2 % Q clones, corresponding to about 17 % noise in Figure 12.7. Interestingly, this is exactly the noise estimate we arrived at using clone replicate comparisons, as mentioned above; however, we do not know whether the percentage of Q clones always measures noise so accurately.

12.5.2 Overlap Equations

The methods outlined above, and used by all HICF projects to date, use the Sulston formula [12] without modifications to compute the overlap probability between clones. As discussed, the bands of several colors are simply merged into one numeric range using the scale factors and color increments discussed above. Nevertheless, the novel features of HICF do suggest certain modifications to the formula. We will discuss two types that have been incorporated into the program multiFPC, mentioned above.

One possible modification accounts for the uneven distribution of the HICF fragments. HICF fragments average approximately 128 bp in size, but the range of fragments used is typically around 50–500 bp. At the high end of this range, the density of fragments is much lower than at the low end; therefore, a match of large fragments is considerably less likely to be due to chance. It would make sense to modify the overlap formula to give these matches a higher weight; indeed, this is true for agarose as well and has been studied by J. Hatfield [18]. In this implementation, the idea was found to work well on simulated data, but little improvement was seen with real agarose data.

The authors of multiFPC assert that the effect is stronger in HICF, and this does seem to be the case (Figure 12.8). However, the bias towards smaller sizes is substantial also in agarose, and in both cases the distributions are far from smooth,

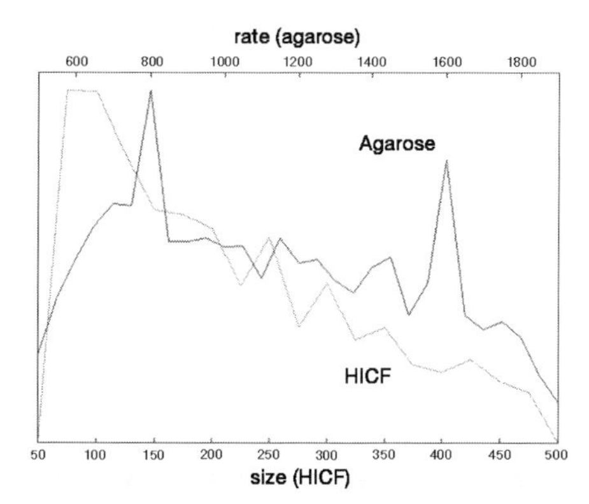

Figure 12.8. Distribution of bands from HICF and agarose maize data. Both distributions fall off to the right, but there are substantial local fluctuations.

which means that any probability calculations assuming smooth distributions can be only approximate.

Another modification, which seems to have been exclusively studied by the multiFPC authors, deals with double-end-labeled fragments. These are the relatively uncommon fragments that are cut on each end by a labeling enzyme instead of having one end cut by the non-labeling 4-cutter. The clone will then have two bands of identical size but possibly of different colors. Clones that share both bands actually share only one fragment, and it is plausible that this should count for less than sharing of two genuinely different fragments. Exactly how much less has been worked out by the authors of multiFPC, and it will be interesting to see the derivation [19].

To conclude we briefly discuss tests carried out by Discovery Biosciences to compare the multiFPC overlap formula with the Sulston formula of FPC (http://www.discoverybio.com/products/multifpc/multifpc_results1.2.01.htm). The results were very favorable to the multiFPC formula, but it seems that much of the poor performance of the Sulston score in these tests was due to the use of single-precision floating-point numbers, which cannot represent the very small overlap scores obtained in HICF. FPC v7, as mentioned above, has been changed to use double-precision numbers, so it is not clear how much performance gap there is at present.

Acknowledgments

This work is supported in part by NSF grants #0213764 and #0211851. The authors express their thanks to Ed Butler, Arvind Bharti, Galina Fuks, Jamie Hatfield, Frank You, Michele Morgante, and Kevin Fengler for many fruitful exchanges.

References

1. Coulson, A., Sulston, J., Brenner, S., and Karn, J. 1986. Towards a physical map of the genome of the nematode *C. elegans*. *Proc. Natl. Acad. Sci. USA*, 83:7821–7825.
2. Olson, M., Dutchik, J., Graham, M., Brodeur, G., Helms, C., Frank, M., MacCollin, M., Scheinman, R., and Frank, T. 1986. Random-clone strategy for genomic restriction mapping in yeast. *Proc. Natl. Acad. Sci. USA* 83:7826–7830.
3. Marra, M., Kucaba, T., Dietrich, N., Green, E., Brownstein, B., Wilson, R., McDonald, K., Hillier, L., McPherson, J., and Waterston, R.. 1997. High throughput fingerprint analysis of large-insert clones. *Genome Research* 7:1072–1084.
4. The International Human Genome Mapping Consortium. 2001. A physical map of the human genome, *Nature* 409:934–941.
5. Ding, Y., Johnson, M., Colayco, R., Chen, Y., Melnyk, J., Schmitt, H., and Shizuya, H.. 1999. Contig assembly of bacterial artificial chromosome clones through multiplexed fluorescent-labeled fingerprinting. *Genomics* 56:237–246.
6. Ding, Y., Johnson, M., Chen, W., Wong, D., Chen, Y-J., Benson, S., Lam, J., Kim, Y-M., and Shizuya, H.. 2001. Five-color-based

high-information-content fingerprinting of bacterial artificial chromosome clones using type IIS restriction endonucleases. *Genomics.* 74:142–154.

7. Faller, M., Fengler, K., Meyers, B., Dolan, M., Tingley, S., and Morgante, M. 2000. Construction of a contig-based physical map of corn using fluorescent fingerprinting techonology. Poster presentation, Plant and Animal Genome VIII Conference.
8. Luo, M-C., Thomas, C., You, F., Hsiao, J., Shu, O., Buell, C., Malandro, M., McGuire, P, Anderson, O., and Dvorak, J. 2003. High-throughput fingerprinting of bacterial artificial chromosomes using the SNaPshot labeling kit and sizing of restriction fragments by capillary electrophoresis. *Genomics.* 82:378–389.
9. Soderlund, C., Longden, I., and Mott, R.. 1997. FPC: a system for building contigs from restriction fingerprinted clones. CABIOS 13:523–535.
10. Soderlund, C., Humphrey, S., Dunhum, A., and French, L. 2000. Contigs built with fingerprints, markers and FPC V4.7. *Genome Research* 10:1772–1787.
11. Sulston, J., Mallett, F., Durbin, R., and Horsnell, T. 1989. Image analysis of restriction enzyme fingerprint autoradiograms. *CABIOS* 5:101–132.
12. Sulston, J., Mallet, F., Staden, R., Durbin, R., Horsnell, T., and Coulson, A. 1988. Software for genome mapping by fingerprinting techniques. *CABIOS* 4:125–132.
13. Lander, E. and Waterman, M.1988. Genomic mapping by fingerprinting random clones: a mathematical analysis. *Genomics* 2:231–239.
14. Engler, F., Hatfield, J., Nelson, W., and Soderlund, C. 2003. Locating sequence on FPC maps and selecting a minimal tiling path. *Genome Research* 13:2152–2163.
15. Luo, M-C., Thomas, C., Deal, K., You, F., Anderson, O., Gu, Y., Li, W., Kuraparthy, V., Gill, B., McGuire, P., Dvorak, J. 2003. Construction of contigs of aegilops tauschii genomic DNA fragments cloned in BAC and BiBAC vectors. Proc. 10th Intern. Wheat Genet. Symp. 1:293–296, Paestum, Italy.
16. You, F., Luo, M-C., Gu, Y., Lazo, G.,Thomas, C., Deal, K., McGuire, P., Dvorak, J., and Anderson, O. 2004. GenoProfiler: automated processing of capillary fingerprinting data. Submitted to Bioinformatics.
17. Engler, F. and Soderlund, C.. 2002. Software for Physical Maps. In Ian Dunham (ed.) *Genomic Mapping and Sequencing*, Horizon Press, Genome Technology series. Norfolk, UK, pp. 201–236.
18. Hatfield, J. 2002. Analyzing restriction fragments for contig assembly. Master's Thesis, Clemson University.
19. Robinson, M., Chen, W., and Hunkapiller, T. In preparation. multiFPC: The use of automated sequencer multicolor data in large-scale clone mapping.

Website References

http://www.genome.arizona.edu/software/fpc/, FPC and WebFPC Download Site
http://www.genome.arizona.edu/fpc_hicf/maize/, AGI/PGIR Maize HICF Map
http://www.discoverybio.com/products/multifpc/multifpc_intro.htm, MultiFPC Site
http://wheat.pw.usda.gov/PhysicalMapping/tools/genoprofiler, GenoProfiler Site
http://wheatdb.ucdavis.edu:8080/wheatdb/, NSF Wheat D Genome Project

13
Reduced Representation Strategies and Their Application to Plant Genomes

Daniel G. Peterson

Overview

Many important crop species have large, highly repetitive genomes that make whole-genome sequencing and assembly technically difficult and/or prohibitively expensive. However, there are a growing number of high-throughput "reduced representation" strategies that allow isolation and study of important and/or interesting sequence subsets from even the largest plant genomes. The following is a review of some of the major reduced representation techniques that are currently being utilized to study plants. In particular, the merits and limitations of strategies for sequencing gene space and/or repetitive elements will be discussed. Additionally, techniques for de novo discovery of DNA polymorphisms will be reviewed. Specific techniques addressed include the following.

1. Expressed sequence tag (EST) sequencing, which is currently the most economical means of elucidating the coding regions of expressed genes (Rudd 2003).
2. Methylation filtration (MF), a gene-enrichment technique based on the observation that, in some plants, genes are hypomethylated compared to repeats (Rabinowicz et al. 1999).
3. Cot-based cloning and sequencing (CBCS), a technique rooted in the principles of DNA renaturation kinetics that allows enrichment for genes, repeats, or any other group of sequences based upon their relative iteration in the genome (Peterson et al. 2002a, 2002b).
4. Reduced representation shotgun (RRS) sequencing, a means of small polymorphism discovery rooted in some of the most basic molecular biology techniques, i. e., restriction enzyme digestion of DNA and agarose gel electrophoresis (Altshuler et al. 2000).
5. Degenerate oligonucleotide primed PCR (DOP-PCR), a small polymorphism discovery tool in which partially degenerate primers are used to amplify a subset of genomic sequences that are then cloned, sequenced, and compared (Jordan et al. 2002).

The Handbook of Plant Genome Mapping.
Genetic and Physical Mapping. Edited by Khalid Meksem, Günter Kahl

ISBN 3-527-31116-5

6. Microsatellite capture techniques that use synthetic oligonucleotides composed of short tandem repeats to discover simple sequence repeats (SSRs) in genomic DNA.

13.1 Introduction

"Everything is simpler than you think and at the same time more complex than you imagine."

Johann Wolfgang von Goethe (1749–1832)

Seed-bearing plants (angiosperms and gymnosperms) exhibit considerable conservation with regard to relative gene order and overall gene repertoire. This conservation is evident in comparisons of even the most distantly related taxa; for instance, loblolly pine and *Arabidopsis* diverged from a common ancestor 300 million years ago, yet 90 % of pine EST contigs have apparent homologues in *Arabidopsis* (Kirst et al. 2003). In contrast, the genome sizes (1C DNA contents) of seed plants exhibit tremendous plasticity, e. g., loblolly pine has a genome 162 times larger than that of *Arabidopsis*, though it is clear that the former does not have or need 162 times as many genes. Of note, species within the same family may exhibit up to 100-fold differences in 1C DNA content, and 10-fold differences have been observed for species in the same genus (Bennett and Leitch 2003). At the extreme, the angiosperms *Fritillaria assyriaca* (a lily) and *Cardamine amara* (a mustard) show a 2123-fold difference in genome size (Bennett and Leitch 2003) although both species possess comparable levels of structural sophistication. The lack of correlation between genome size and structural complexity/gene repertoire in higher eukaryotes is known as the *C-value paradox*, and its evolutionary implications have been a subject of study and debate for decades (see Hartl 2000 and Petrov 2001 for reviews).

The vast majority of genome size variation in seed plants is due to lineage-specific amplification of non-genic "repeat sequences," some of which may be found in thousands or millions of copies per 1C genome. While a few of these repeats have come to serve structural roles (e. g., centromeric and telomeric repeats), most have no known function. Many of the repetitive elements in plant genomes appear to have originated from intergenic proliferation of transposable elements (SanMiguel and Bennetzen 2000), while others have uncertain origins (Lapitan 1992). While polyploidy, gene duplication, and gene loss certainly account for some of the variation in seed plant genome sizes, their contributions to the C-value paradox appear to be rather small (Hartl 2000). For instance, polyploidy accounts for <0.5 % of the >2000-fold variation in plant genome sizes[1], and gene duplications and losses likely account for even less.

1) In the RBG Kew Plant DNA C-Values Database (Bennett and Leitch 2003), seed plant species vary in ploidy from 2× to 20× (10-fold). A 10-fold variation in ploidy divided by the 2123-fold observed variation in DNA content among these same species is 0.5 %.

Whole-genome sequences provide the ultimate dataset by which the DNA of different species can be compared. However, many important crop species have large genomes (Figure 13.1) in which repetitive sequences constitute the bulk of genomic DNA, are highly interspersed with genes, and/or have relatively recent origins (i. e., exhibit little inter-copy divergence), making gene isolation and whole-genome sequence assembly prohibitively difficult and costly (Rudd 2003). In recent years, a variety of "reduced representation" techniques have been developed to isolate and study important and/or interesting sequence subsets from large, repetitive genomes in a cost-effective manner; subsets include (but are not limited to) expressed sequences, low-copy (≈gene-rich) genomic regions, polymorphic DNA markers, and repetitive elements. Use of some reduced representation methods may permit capture and elucidation of a species' "sequence complexity" (SqCx – Figure 13.2) and thus provide most of the benefits of whole-genome sequencing at a fraction of the cost. Other reduced representation techniques allow rapid characterization of DNA polymorphisms without a priori knowledge of genomic sequence, affording inroads into the sequence diversity of under-explored genomes and providing mechanisms for efficient genotyping in those species that enjoy finished sequences.

This chapter focuses on those reduced representation strategies that will help in the sequencing of plant gene space, allow efficient characterization of the repetitive elements of genomes, and permit de novo discovery of DNA polymorphisms. The strengths and limitations of each reduced representation technique are discussed.

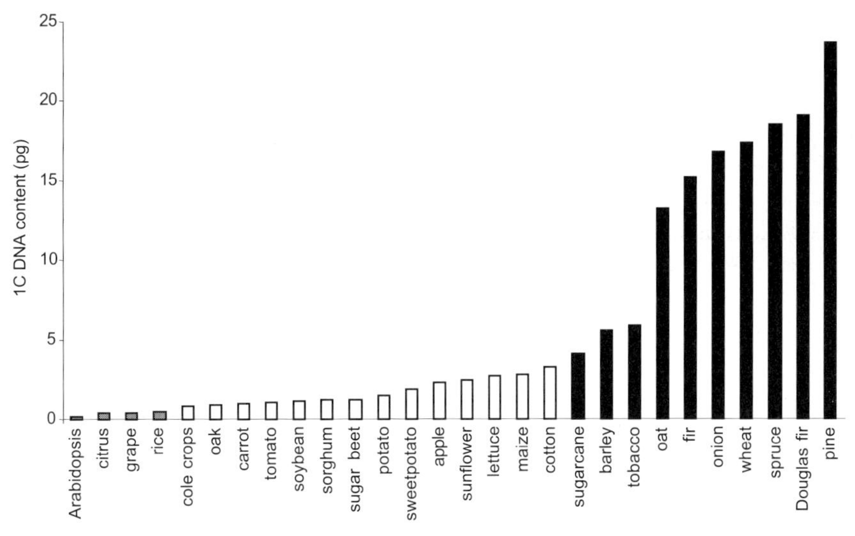

Figure 13.1. Comparative genome sizes of economically important crop species and *Arabidopsis*. All 1C genome size data are from Bennett and Leitch (2003) except the value for *Arabidopsis* (Arabidopsis Genome Initiative 2000). Values for pine, spruce, oak, fir, and Douglas fir are mean genome sizes for their respective genera. Species with genome sizes equal to or smaller than rice are shown in gray, those with genome sizes larger than human (3.26 pg) are shown in black, and those with 1C DNA contents in between rice and human are represented in white.

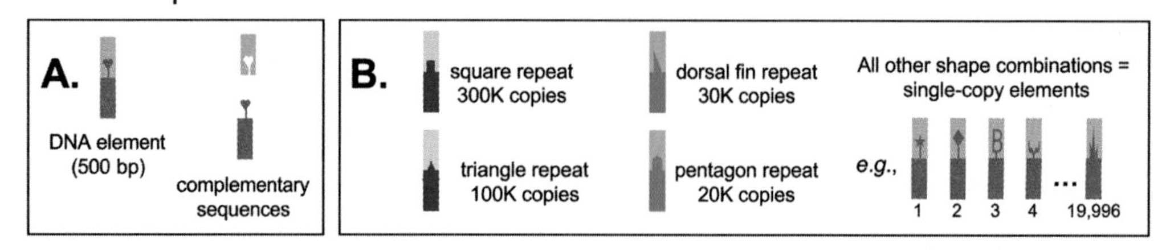

C. Genome size = 500 bp/element * (300,000 + 100,000 + 30,000 + 20,000 + 19,996) = 234,998,000 bp
(square, triangle, dorsal fin, pentagon, single-copy)

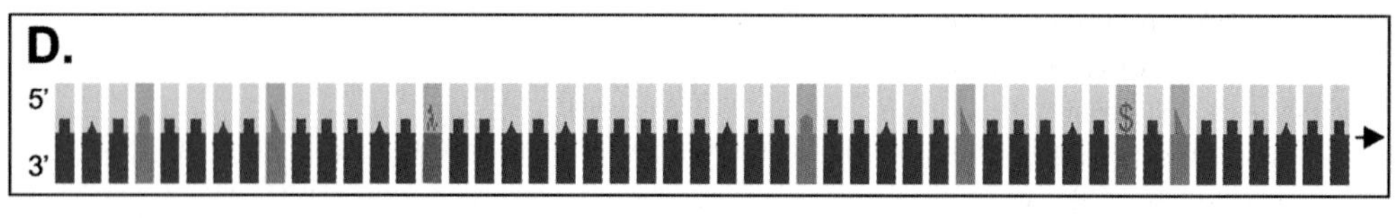

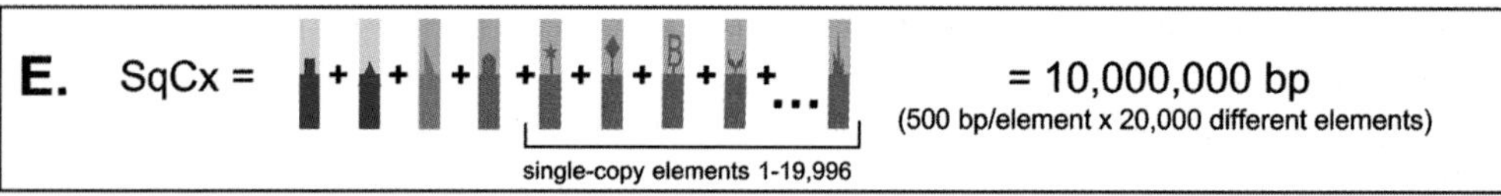

Figure 13.2. Genome size vs. sequence complexity (SqCx). (A) The genome of the hypothetical plant *Planta genericus* is composed of a variety of DNA elements; each element is represented by a rectangle composed of two interlocking (i. e., complementary) pieces/sequences. For simplicity, each element is assumed to be 500 bp in length. (B) The two most highly repetitive (HR; blue) elements in the genome are "square" (300,000 copies/genome) and "triangle" (100,000 copies). Considerably fewer copies are found of the moderately repetitive (MR; green) elements "dorsal fin" (30,000 copies) and "pentagon" (20,000 copies). All other 19,996 elements in the genome are single/low-copy (SL; red) in nature. (C) 63.8 % of the *P. genericus* genome is composed of copies of "square," while "triangle," "dorsal fin," and "pentagon" constitute 21.3 %, 6.4 %, and 4.3 % of the genome, respectively. The remaining 4.2 % of the genome is composed of the 19,996 different SL sequences. (D) A short stretch of *P. genericus* DNA showing the comparative frequency of HR, MR, and SL elements. (E) SqCx is the sum of all the unique sequence information in a genome. While "square" and "triangle" account for the vast majority of *P. genericus* genome size, the contribution of each of these elements to SqCx is negligible [100 × (500 bp 10,000,000 bp) = 0.0005 %]. In contrast, the SL sequences, which constitute only 4.2 % of genome size, account for 99.9 % of SqCx. The goal of gene-enrichment reduced representation techniques is to isolate and sequence those elements that contribute most to SqCx (i. e., SL DNA) with minimum encumbrance from the repetitive elements, which may make up the lion's share of the genome but contribute very little to SqCx.

13.2 Reduced Representation Techniques

13.2.1 EST Sequencing

Reverse transcriptase (Baltimore 1970; Temin and Mizutani 1970) is an RNA-dependent DNA polymerase. In nature, the enzyme is the means by which the genomes of retroviruses (i.e., viruses with RNA genomes) make *complementary DNA* (cDNA) for incorporation into a host's genome. With regard to molecular biology, the use of reverse transcriptase to generate cDNA molecules from isolated mRNA and the subsequent construction of cDNA libraries heralded in the era of gene expression research. Because each cDNA library serves as a "snapshot" of gene expression in the cells from which it was derived, comparison of cDNA libraries from different tissues or developmental stages, or from the same tissues exposed to different environmental stimuli, affords a means of correlating changes in morphology and cellular activity with changes in gene expression.

With the development of PCR and automated sequencing, it became possible to sequence large numbers of end sequences from cDNA clones. The cDNA end sequences, commonly referred to as *expressed sequence tags* (ESTs; Adams et al. 1991), can be utilized as molecular markers, a discovery that has greatly accelerated molecular mapping efforts (e.g., Komulainen et al. 2003). Because cDNA/EST sequencing results in the preferential sequencing of portions of expressed genes, it is a powerful reduced representation technique that allows an economical preliminary means of exploring plant gene space (see Rudd 2003 for review).

13.2.2 Methylation Filtration

DNA methyltransferases are enzymes that add methyl groups onto select bases of DNA. In plants, DNA methyltransferases normally catalyze the transfer of the methyl group from S-adenosyl-L-methionine to the fifth carbon in the pyrimidine ring of cytosine to form 5-methylcytosine (m5C) (Kumar et al. 1994). However, cytosine residues are not methylated indiscriminately. Rather, methylation is highly regulated, and it has been correlated with all kinds of genetic phenomena, including normal control of gene expression (e.g., Finnegan et al. 1998, 2000; Zluvova et al. 2001), imprinting (e.g., Kinoshita et al. 2004; Berger 2004), paramutation (e.g., Lisch et al. 2002), transgene silencing (e.g., Fojtova et al. 2003; Meng et al. 2003), aging (e.g., Fraga et al. 2002), repression of recombination (e.g., Fu et al. 2002), diploidization in allopolyploids (e.g., Liu and Wendel 2003), and rapid epigenetic adaptation in response to major environmental changes (e.g., Fraga et al. 2002; Kovalchuk et al. 2003; Fojtova et al. 2003).

Early attempts at cloning methylated DNA (including m5C-rich DNA from plants) were not particularly successful (see Redaschi and Bickle 1996 for review).

The principal reason for this difficulty was elucidated in the late 1980s when it was shown that some *E. coli* strains possess enzymes that preferentially restrict foreign methylated DNA sequences. Like other restriction endonucleases, the three *E. coli* methylation-specific restriction enzymes (MSREs), known as McrA, McrBC, and Mrr, apparently evolved to protect the bacterium from invading bacteriophages. However, they are quite capable of cleaving methylated plant DNA as well (Redaschi and Bickle 1996). After the discovery of MSREs, *E. coli* strains were engineered with nonfunctional MSRE genes (i. e., with a $mcrA^-$, $mcrBC^-$, and/or mrr^- genotype) to facilitate cloning of methylated DNA.

In some plant species, the presence of a high density of m5C residues (i. e., hypermethylation) is strongly correlated with certain repeat sequences (e. g., retroelements), while low-copy sequences typically contain few or no m5Cs (i. e., they are hypomethylated). A striking example of this type of methylation pattern is seen in maize (Bennetzen et al. 1994).

Recently, Rabinowicz et al. (1999) developed a clever and simple means of preparing genomic libraries enriched in hypomethylated sequences. In short, they cloned mechanically sheared maize DNA into $mcrA^+$, $mcrBC^+$, and/or mrr^+ host strains. Most hypermethylated DNA is cleaved in these strains, and thus the resulting libraries are enriched in hypomethylated (ostensibly gene-rich) DNA. An overview of their technique, methylation filtration (MF), is shown in Figure 13.3. The efficacy of MF in producing gene-enriched genomic libraries has been shown for maize (Rabinowicz et al. 1999; Whitelaw et al. 2003) and claimed for canola and wheat. Methylation filtration is licensed exclusively to Orion Genomics (www.oriongenomics.com) and is marketed under the name GeneThresher.

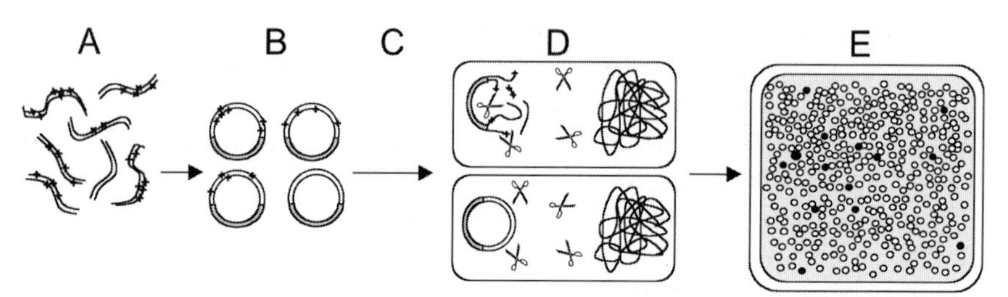

Figure 13.3. Overview of methylation filtration. (A) Plant genomic DNA is mechanically sheared into fragments. For the species shown, repeat sequences are hypermethylated compared to gene sequences. Hypermethylated DNA regions are indicated by small stars along the DNA strands. (B) The genomic DNA fragments are ligated into a vector containing an antibiotic resistance gene and a cloning site that allows alpha-complementation (blue/white selection). (C) The recombinant molecules are used to transform a strain of bacteria with active MSREs (i. e., McrAB, McrC, and/or Mrr gene products). (D) The MSREs, depicted by scissors, cleave plant hypermethylated DNA (upper cell) but do not restrict hypomethylated DNA (lower cell). (E) When challenged with an antibiotic on selective medium, only those bacteria that contain an intact circular plasmid will survive to form colonies. Colonies containing plasmids that lack an insert in their cloning site will appear blue, while those containing an insert in their cloning site will appear white. Archiving white clones results in a library enriched in hypomethylated (ostensibly gene-rich) DNA.

13.2.3
Cot-based Cloning and Sequencing

Much of what is known about eukaryote genome structure stems directly from work done by Roy Britten and colleagues during the 1960s and 1970s.[1] Britten and his collaborators at the Carnegie Institution of Washington empirically studied the re-association of genomic DNA in solution using a technique they called "Cot analysis," and it was through Cot analysis that the repetitive nature of eukaryotic genomes was discovered (Britten and Kohne 1968). In brief, when mechanically sheared DNA in solution is heated to near-boiling temperature, the molecular forces holding complementary base pairs together are disrupted and the two strands of the double helix dissociate or "denature." If the denatured DNA is then slowly returned to a cooler temperature, sequences will begin to "re-associate" (renature) with complementary strands. The temperature at which renaturation occurs can be regulated so that little or no sequence mismatch is tolerated. As predicted by the law of averages, the rate at which a sequence finds a complementary strand with which to hybridize is directly related to that sequence's iteration in the genome (Figure 13.4). In other words, those sequences that are extremely abundant (on average) find complementary strands with which to pair relatively quickly, while single-copy sequences take a much longer time to find complements. In a Cot analysis, a series of DNA samples are allowed to renature to different Cot values; a sample's Cot value is the product of its DNA concentration (C_0), re-association time (t), and, if appropriate, a buffer factor that accounts for the effect of cations on the speed of renaturation (Britten et al. 1974). The amount of re-association at each Cot value is typically determined by using hydroxyapatite (HAP) chromatography to separate double- and single-stranded DNA (dsDNA and ssDNA) and spectrophotometry to quantitate the amount of DNA in ssDNA and dsDNA eluents. A graph showing re-association of genomic DNA as a function of Cot is known as a Cot curve. Through study of Cot curves of different species, Britten et al. discovered that eukaryotic genomes tend to be composed of several distinct kinetic components – specifically, highly repetitive (HR), moderately repetitive (MR), and single/low-copy (SL) DNA. Using a Cot curve as a guide, HAP chromatography can be used to isolate the different kinetic components of a genome (e.g., Britten and Kohne 1968; Goldberg 1978; Kiper and Herzfeld 1978; Peterson et al. 1998) as illustrated in Figure 13.4.

In the late 1990s, I began work as a postdoctoral associate in the lab of Andrew H. Paterson at the University of Georgia. My previous training in Cot analysis and nuclear DNA isolation (Peterson et al. 1997, 1998) coupled with Paterson's expertise in plant genetics/genomics (e.g., Paterson et al. 1995, 2000) eventually led us to develop "Cot-based cloning and sequencing" (CBCS), a synthesis of Cot methods, molecular cloning, and high-throughput DNA sequencing that permits

1) The principles of nucleic acid re-association elucidated by Britten et al. lie at the heart of many molecular techniques utilized today, including PCR, filter hybridization (Southern/Northern blots, colony blots, macroarrays), microarrays, and chip-based re-association experiments (see Goldberg 2001 for review).

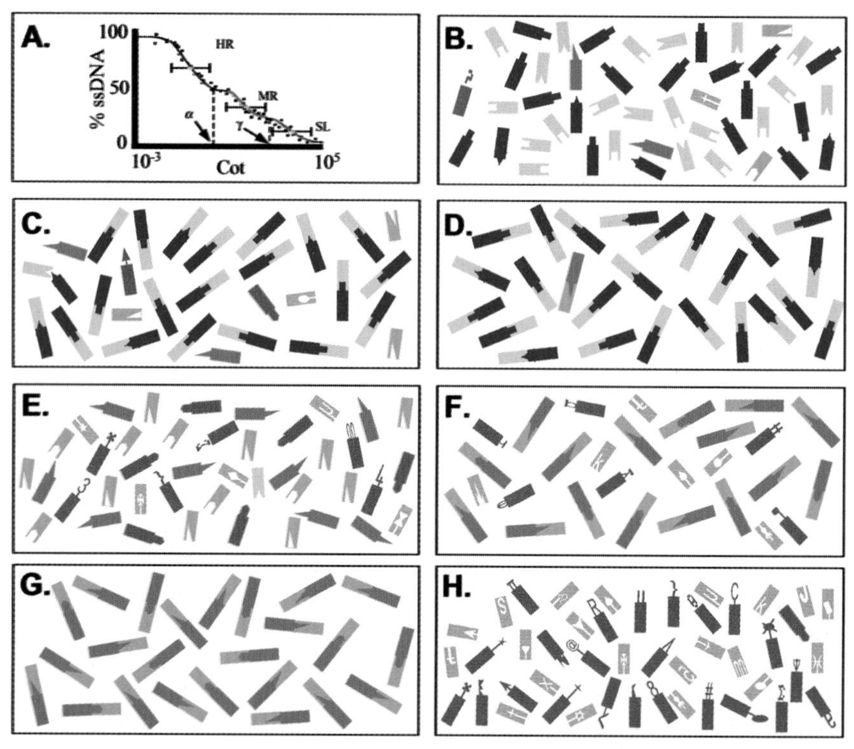

Figure 13.4. Fractionation of the hypothetical *Planta genericus* genome using CBCS. As in Figure 13.1, each element in the *P. genericus* genome is represented by a rectangle composed of two interlocking (i. e., complementary) pieces/sequences. Additionally, all elements are roughly 500 bp in length. (A) A Cot analysis of *P. genericus* reveals that its genome is composed of a fast (blue), a moderately fast (green), and a slow re-associating component. The fraction of the genome in each component and the kinetic complexity (estimated sequence complexity) of each component are determined from mathematical analysis of the Cot curve. Because the elements in a genome re-associate at a rate proportional to their copy number, those sequences in the fast re-associating component represent the most highly repetitive (HR; blue) sequences in the genome, while those in the moderately fast and slow re-associating components represent moderately repetitive (MR; green) and single/low-copy (SL; red) elements, respectively. (B) *P. genericus* DNA is sheared into 450-bp fragments and denatured. This illustration and each subsequent frame show a small, random part of a much larger renaturation reaction. However, each element within a frame is shown in correct proportion to elements from its own and other components. (C) The DNA is allowed to renature to the Cot value "α" (see frame A). As shown in the Cot curve (frame A), at Cot α nearly all HR elements have formed duplexes (based upon "collisions" with complementary strands), but few MR and essentially no SL elements have found complements. Hydroxyapatite (HAP) chromatography then is used to separate (D) double-stranded (renatured) HR DNA from (E) single-stranded DNA (MR and SL DNA). The double-stranded HR DNA is cloned to produce an HRCot library. (F) The single-stranded DNA remaining after the first HAP fractionation is allowed to renature to a Cot value equivalent to γ (see frame A), at which virtually all MR elements have formed duplexes but single-copy elements are still unlikely to have found complements. HAP chromatography is used to separate the (G) double-stranded MR DNA from the (H) single-stranded SL DNA. The MR-enriched DNA fraction (frame G) is cloned to produce an MRCot library. Random primer strand synthesis is used to generate complementary strands for the single-stranded DNA molecules in the SL-enriched fraction (frame H), and the resulting duplexes are cloned to produce an SLCot library.

production and exploration of DNA libraries enriched in genes and/or repeats (Peterson et al. 2002a, 2002b). In CBCS, the results of a Cot curve (or genome parameters determined through other techniques) are used to guide HAP-based fractionation of genomic DNA into its major kinetic components (e. g., HR, MR, and SL DNA). The isolated kinetic components are cloned to produce HRCot, MRCot, and SLCot libraries, respectively, and component libraries are sequenced (Figure 13.4). For those solely interested in sequencing low-copy DNA (i. e., most genes), only an SLCot library need be prepared (e. g., Yuan et al. 2003; Lamoreux et al., submitted).[1] The efficacy of CBCS as a gene space enrichment tool has been demonstrated for sorghum (Peterson et al. 2002a), maize (Yuan et al. 2003; Whitelaw et al. 2003), wheat (Lamoreux et al., in preparation), cotton (Paterson et al., in preparation), and chicken (Wicker et al., in preparation).

Of reduced representation techniques, CBCS is the only one that theoretically permits sequencing of a species' sequence complexity (SqCx; Figure 13.2). Because Cot analysis provides the kinetic complexity (i. e., the estimated SqCx) of each component, the most efficient means of capturing a species' total SqCx is to sequence each Cot library to a depth that provides a high probability that all the elements in that component are sequenced at least once (Peterson et al. 2002a, 2002b). Since almost all of the SqCx of a plant genome will be found in its SLCot library, the vast majority of resources can be devoted to sequencing SLCot clones (Figure 13.4). For many plant genomes, CBCS should allow sequencing of SqCx at a cost of one-quarter to one-twentieth that of traditional shotgun sequencing (see Peterson et al. 2002b).

While in retrospect it seems rather obvious that DNA re-association could be used to create gene-enriched (or repeat-enriched) genomic libraries, there are several factors that likely contributed to the relatively late development of CBCS: (1) while Cot analysis was highly utilized in the 1970s, its popularity waned with the advent of molecular cloning; (2) Cot analysis (especially as it was practiced in the 1970s) was a technically demanding procedure requiring extreme standardization, and thus it was performed only in a handful of labs even in its heyday; (3) Cot analysis is considerably "less forgiving" than molecular cloning, and the tremendous appeal of the latter was such that most leaders in re-association kinetics shifted their research focus; and (4) by the time high-throughput sequencing became possible, many of the original practitioners of Cot analysis had retired.

Since publication of our original CBCS manuscript (Peterson et al. 2002a), we have been continually working to improve all aspects of CBCS. For example, in our initial experiments in sorghum we cloned SL duplexes that were the products of kinetic re-association. Although high stringency was maintained during renaturation to prevent base mismatches, any mismatch making it through to the clon-

1) Working independently, J. L. Bennetzen and colleagues developed "high Cot" sequencing (Yuan et al. 2003), a technique that is essentially identical to enrichment for single/low-copy sequences using CBCS (i. e., sequencing of SLCot clones). Because the publications of Peterson et al. (2002a, 2002b) pre-date Yuan et al. (2003) and because, in our experience, people often confuse the term "high Cot" with "high copy" ("high Cot" DNA is actually composed of low-copy sequences), in this review we will use the terminology established by Peterson et al. (2002a).

ing process presumably would be resolved by the host cell's DNA repair mechanisms, possibly resulting in generation of a sequence that does not exist in the donor genome. While this phenomenon would have minor implications in capturing SqCx, it would limit the usefulness of SLCot clones in the detection of small polymorphisms (e. g., SNPs) within gene and repeat families. By the time our first CBCS paper was published (Peterson et al. 2002a), we had already started using an enzymatic approach to synthesize complementary strands from single-stranded SL DNA templates, thus circumventing the problems associated with cloning re-association products and increasing the utility of CBCS. We are working to successfully adapt a random priming approach to synthesize complementary strands for HR/MR molecules as well.[1)] We are also conducting tests with a recently discovered nuclease that preferentially cleaves dsDNA (Shagin et al. 2002; Zhulidov et al. 2004) and consequently may allow elimination of HAP chromatography, thus speeding up and simplifying the CBCS protocol.

13.2.4
De Novo Polymorphism Discovery

Within a species, the vast majority of gene sequence variation is due to relatively small DNA polymorphisms. There are several types of small polymorphisms that are widely utilized to study plant gene/genome evolution.

1. Single-nucleotide polymorphisms (SNPs): As their name suggests, SNPs are single base pair differences within alleles of a gene. They represent a powerful means of relating the smallest possible changes in DNA sequence to variation in phenotype. SNPs are widely utilized molecular markers in mammals and are becoming more common molecular markers in plants (Schmid et al. 2003; Törjék et al. 2003).
2. Insertion/deletion (indel) polymorphisms: Indels are typically discovered using SNP discovery approaches. Indels are useful genetic markers and are easier to score than SNPs as detection of the latter requires sequencing (or resequencing) while the former can be detected as length differences in PCR products (Bhattramakki et al. 2002).
3. Microsatellites or simple sequence repeats (SSRs): SSRs are tandemly repeated DNA sequences of 1–7 bp. They are abundant in the genomes of most eukaryotes, and because they are frequently polymorphic, codominant, and easily scored, they have been utilized as molecular markers in numerous pursuits including linkage map construction, parentage analysis, population genetics studies, and marker-aided selection (see Fisher et al. 1996 and Dekkers and Hospital 2002 for reviews). SSRs, which technically are a subclass of indels, can be discovered through analysis of large EST sets (e. g., Gupta et al. 2003).

An obvious means of studying SNPs, indels, and SSRs is to use PCR to examine polymorphisms at specific loci within a population. Likewise, genome/EST se-

1) Single-stranded HR/MR DNA has to be immobilized to permit second-strand synthesis.

quence data from one species can be used to study parallel loci in closely related species. However, such research requires synthesis of locus-specific primers and thus a priori knowledge of the nucleotide sequence of the locus being studied. Below is a discussion of several reduced representation techniques that allow SNP, indel, and/or microsatellite discovery without existing sequence data.

13.2.4.1 Reduced Representation Shotgun (RRS) Sequencing

Recently Altschuler et al. (2000) devised a means of studying SNPs and other small polymorphisms at a reasonable number of loci without prior knowledge of genomic sequence. Their technique, which they call reduced representation shotgun (RRS) sequencing, is quite simple and presumably is applicable to most species (see Figure 13.5 for an overview). In brief, genomic DNA from a number of individuals is mixed together, digested with a single restriction enzyme, and size-fractionated by agarose gel electrophoresis. An agarose band containing DNA fragments within a relatively narrow size range (e. g., between 600 and 650 bp) is removed, and DNA fragments from the band are cloned and sequenced. Because most restriction sites will be shared by individuals in the population, the gel slice will likely contain alleles of the same loci. Polymorphisms can be detected by sequence analysis.

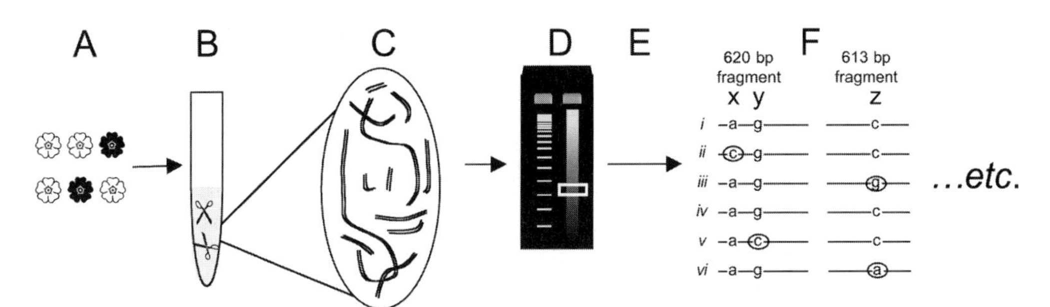

Figure 13.5. Overview of RRS sequencing. (A) DNA is extracted from multiple individuals in a population. (B) All of the DNA samples are placed in the same tube, and a single restriction enzyme (depicted by scissors) is added. (C) After complete digestion, the DNA mixture consists of many fragments of different sizes. As the individuals in the population are likely to share most of the same restriction sites, most homologous DNA sequences are likely to be about the same size. (D) The digested DNA is size-fractionated via electrophoresis, and a small band containing DNA fragments in a relatively narrow size range (e. g., 600–650 bp) is extricated from the gel. The DNA in the gel band is isolated, cloned, and sequenced. (E, F) Sequences are compared using sensitive alignment algorithms. SNPs and very small deletions and insertions can be discovered by comparing those sequences that appear to represent the same locus (i. e., are largely identical). For the 620-bp fragment shown, SNPs are discovered at nucleotides *x* and *y*. For the 613-bp fragment, a SNP is detected at nucleotide *z*.

13.2.4.2 **DOP-PCR**

More than a decade ago, degenerate oligonucleotide PCR (DOP-PCR) was developed as a means of amplifying (more or less) entire genomes. Using human DNA, Telenius et al. (1992) demonstrated that PCR primers with the sequence 5′-CCGACTCGAGNNNNNNATGTGG-3′ (where *N* can represent any one of the four nucleotides) will bind to enough sites throughout the genome to allow amplification of most chromosomal regions. DOP-PCR has proven to be an extremely useful tool in genome studies in which DNA quantity is limited (e.g., Cheung and Nelson 1996; Dietmaier et al. 1999; Buchanan et al. 2000; Kittler et al. 2002).

Recently, Jordan et al. (2002) demonstrated the utility of DOP-PCR as a reduced representation technique for finding SNPs and other small polymorphisms in three different species (human, mouse, and *Arabidopsis*). In short, they produced a series of primers identical to the standard DOP-PCR primer (see above) except that each new primer had one to four additional nucleotides added to its 3′ end. By increasing primer length, Jordan et al. effectively decreased the number of complementary sequences to which the primers were likely to bind, and thus decreased the number of PCR products generated in a particular reaction, i.e., reduced the number of loci amplified. To detect polymorphisms, the sequences of PCR products resulting from amplification using a specific primer or combination of primers were compared across multiple individuals.

13.2.4.3 **SSR Capture**

The first SSR enrichment protocols involved hybridizing "microsatellite-like"[1] oligonucleotides to colony blots of either large- or small-insert genomic clones and identifying those clones that contain a microsatellite. While effective, these approaches are relatively cumbersome and expensive.

As an alternative to library screening, Ostrander et al. (1992) propagated a small-insert phagemid library in a *dut ung E. coli* strain. The *dut ung* genotype results in frequent substitution of dUTP for dTTP during DNA replication. Ostrander et al. then isolated circular single-stranded phagemid DNA from the bacteria using an M13 helper phage. A microsatellite-like primer and *Taq* polymerase then were used to generate second strands for those molecules containing a region complementary to the primer. Introduction of the DNA molecules into wild-type *E. coli* resulted in strong selection against single-stranded, dUTP-rich DNA and, consequently, in enrichment for double-stranded DNA containing microsatellites.

An additional means of isolating SSRs involves cross-linking microsatellite-like oligonucleotides to nylon and hybridizing the membrane with genomic DNA fragments. This effectively captures DNA sequences with regions complementary to the oligonucleotides (Edwards et al. 1996).

Currently, the most popular SSR enrichment techniques are those rooted in PCR-based primer extension (e.g., Fisher et al. 1996; Phan et al. 2000; Waldbieser

1) "Microsatellite-like" sequences are synthetic, single-stranded oligonucleotides that possess the characteristics of microsatellites and may be complementary in whole or part to naturally occurring SSRs.

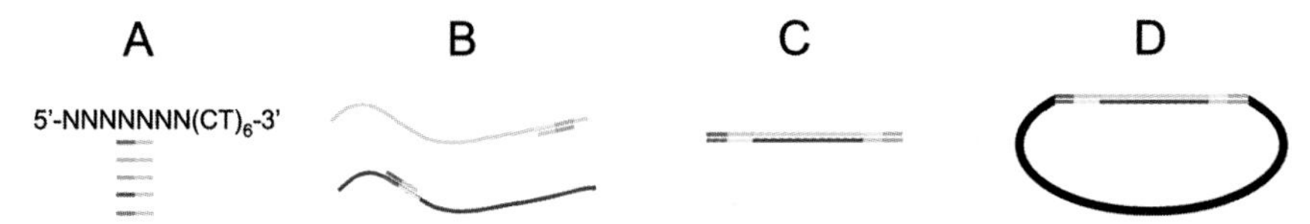

Figure 13.6. A primer extension method for SSR enrichment. The technique shown is similar to that of Fisher et al. (1996). (A) A series of primers are synthesized. Each primer possesses the same SSR-like repeat, i.e., $(CT)_6$, at its 3′ end (gold line) and a partially degenerate seven-base sequence at its 5′ end (green lines) that does not include any additional CT repeats. (B) The primers are used to PCR-amplify those genomic regions flanked by primer binding sites. The conditions of the PCR reaction are kept stringent to make sure that the 5′ ends of the primers actually anneal with complementary sequences (red and pink lines) in the genomic DNA, thus preventing slippage of primers to the 3′ ends of targeted microsatellite loci and subsequent loss of variation in repeat length. (C) Each resulting PCR product has an SSR (gold and yellow paired lines) near its ends. (D) PCR products are cloned and sequenced.

et al. 2003) and those that utilize streptavidin-coated beads to capture re-association hybrids between biotin-labeled, microsatellite-like oligonucleotides and genomic DNA fragments (e.g., Fischer and Bachmann 1998; Hamilton et al. 1999). Primer extension techniques require that (1) two SSRs be in close proximity to one another so that the region between them is amplified with SSR-based primers (e.g., Fisher et al. 1996; Figure 13.6), (2) an SSR be near a repeat sequence so that amplification can be achieved using an SSR-like primer and a repeat-based primer (e.g., Phan

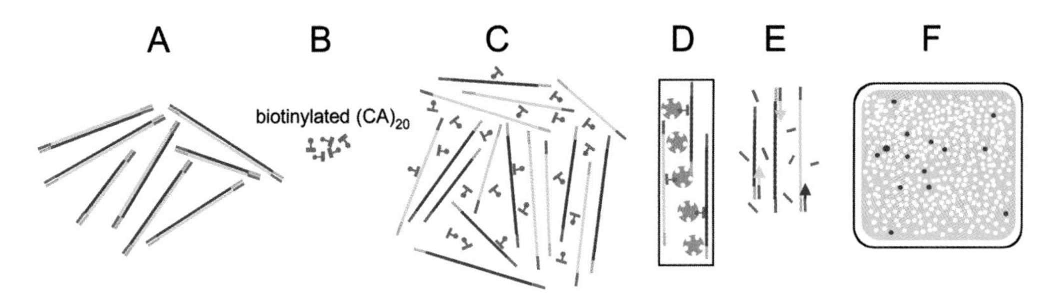

Figure 13.7. Hybrid capture method for SSR enrichment. (A) Genomic DNA is digested into fragments with restriction enzymes that produce blunt-ended cut sites, a linker is ligated to the fragments, and PCR using a primer complementary to one strand of the linker is used to amplify the DNA, thus ensuring that the vast majority of molecules in the reaction have linkers on both ends. (B) One or more single-stranded biotinylated SSR-like sequences are synthesized. In the diagram a biotinylated $(CA)_{20}$ probe is shown. The ball on each $(CA)_{20}$ probe represents biotin. (C) The genomic DNA/linker molecules are denatured and allowed to re-anneal with an excess of the biotin-labeled probe(s). (D) The mixture is loaded onto a column containing streptavidin-coated beads and the column is thoroughly washed. Only those genomic DNA molecules hybridized to a biotin-labeled probe stick to the column. (In one variation, the streptavidin-labeled beads are magnetic and are removed from a slurry using a magnet.) (E) The SSR-containing DNA is eluted from the column (by heating or chemical means) and amplified via PCR with the same primer used in step A. (F) The SSR-containing molecules are cloned and ultimately sequenced.

et al. 2000), or (3) genomic DNA be digested with a restriction enzyme and linkers with a primer-binding site be attached to the ends, allowing PCR amplification using an SSR-like primer and a linker-based primer or two linker-based primers (i. e., Fischer and Bachmann 1998). The biotin-streptavidin "hybrid capture" approaches do not require that any particular sequence be near an SSR (Figure 13.7). Combinations of the primer extension and "hybrid capture" techniques have also been developed (e. g., Paetkau 1999).

13.3 Other Reduced Representation Techniques

In addition to the techniques listed above, there are several other reduced representation techniques that are applicable to a particular species or group of species but are likely to be of limited use to the plant genomics community as a whole. An example of this is the *RescueMu* technique that has been used to isolate genes in maize (see Raizada et al. 2001 and Raizada 2003 for reviews). In brief, *RescueMu* is a plasmid inserted into a maize *Mu1* transposon. The active *RescueMu* element preferentially inserts itself into gene regions in maize (70–90 % of the time). Plasmid rescue then can be used to recover the 5–25 kb of DNA flanking the inserted element. However, *RescueMu* is currently limited to studying certain maize genotypes in which *Mu1* elements are active.

In some plants, certain class II transposons known as "miniature inverted-repeat transposable elements" (MITEs) appear to be preferentially associated with genes (see Feschotte et al. 2002 for review). Based on this observation, a modified AFLP procedure known as transposon display was used to amplify genomic regions containing a MITE known to show an insertion preference for genes (Casa et al. 2000). However, because MITE families can vary widely between species, not all MITEs are preferentially associated with genes, and some genomes have few MITEs (see Casacuberta and Santiago 2003 for review), it is unlikely that transposon display will be useful in sequencing gene space from most plant genomes.

13.4 Discussion

13.4.1 Repeat Sequence Enrichment

While repetitive sequences are often deemed "junk DNA," it is becoming increasingly clear that repeats (especially transposable elements) are one of the principal factors responsible for the evolutionary success of eukaryotes (see Britten 1996 and Wessler 1997). Repetitive DNA influences gene expression and recombination (Assaad et al. 1993; Dorer and Henikoff 1994; Sherman and Stack 1995), and some repeat sequences are involved in maintaining chromosome structure (Lee et al. 1994; Lundblad and Wright 1996). Consequently, studying repeat sequences is necessary to understanding plant genome evolution, gene function, and chromosome organization.

Of the various reduced representation techniques, CBCS is the only technique that allows efficient isolation and characterization of the repetitive sequences of genomes. Its utility as a repeat enrichment tool has been demonstrated for sorghum (Peterson et al. 2002a) and chicken (Wicker et al., submitted). Of note, CBCS can be used to generate a more accurate overview of the repeat content of a genome than sequencing a small number of large-insert clones. For example, at the time the sorghum CBCS research was conducted, there were three sorghum BAC sequences (AF061282, AF114171, and AF124045) in GenBank. Sequencing of sorghum Cot clones revealed that the most abundant repeat sequence in the sorghum genome is a previously unnamed retroelement found only once in the 424,434 bp of sequenced BAC DNA. The prevalence of this element, now known as *Retrosor-6*, was verified by BAC macroarray analysis, and it was shown to comprise roughly 6% of the sorghum genome (Peterson et al. 2002a). Since publication of the sorghum CBCS paper, the amount of sorghum sequence data in GenBank has increased dramatically. As expected, *Retrosor-6* is a common feature of many recent sorghum dbGSS entries, most notably *Sorghum propinquum* genomic clones.

While repetitive DNA is an enormous impediment to gene research in most major crops, knowledge of the sequences and distribution of repeats may circumvent many problems and, indeed, create new research opportunities. For instance, a better understanding of the physical distributions of repetitive DNA families at a resolution compatible with cloning technologies (such as over different BAC clones) may provide the means to identify "gene-rich" genomic domains that are priorities for early sequencing. Additionally, complete physical mapping of large genomes will benefit substantially from, and perhaps even require, a comprehensive knowledge of the sequences and distributions of repetitive DNA families. In this regard, HRCot and MRCot sequences from sorghum have facilitated physical mapping in this species (Peterson et al. 2002a). Identification of repetitive DNA is also valuable for masking repeats out of EST databases, significantly improving the quality of unigene sets.

The principal limitation of CBCS in the study of repeat sequences is that duplexes formed by strand re-association are cloned, a feature that will result in occasional base mismatches and consequently make it difficult to detect small polymorphisms within repeat families. However, as mentioned above, we are working to adapt a second-strand synthesis technique for HRCot and MRCot library construction that may enhance the utility of CBCS.

13.4.2
Sequencing Gene space

Of the reduced representation techniques described above, EST sequencing, methylation filtration, and CBCS represent means of preferentially sequencing all or part of a plant's gene space. Toward this end, each of these techniques has its relative advantages and limitations.

13.4.2.1 EST Sequencing

Compared to methylation filtration and CBCS, EST sequencing is the only reduced representation method that affords insight into both gene expression and gene space. This dual functionality makes EST sequencing a highly cost-effective genome research tool (hence the rapidly growing EST databases for many plant species; Rudd 2003). Additionally, because ESTs correspond to the exonic regions of genes, they are ideal molecular markers: each EST is not simply close to a gene but is part of a gene. However, EST sequencing has definite limitations as a gene space–enrichment technique.

1. The mRNA molecules used in cDNA production have been posttranscriptionally modified; specifically, introns have been removed, a 5′ m^7GpppN cap has been added to each transcript, and 3′ poly-A tails have been appended. It is usually via their poly-A tails that mRNAs are isolated and from oligo(dT) primers that DNA strands are synthesized by reverse transcriptase. However, information corresponding to the 5′ ends of transcripts may be lost due to the limited processivity of reverse transcriptase and/or the inhibition of enzyme movement by mRNA secondary structure (Edery et al. 1995).
2. Since intron removal precedes addition of poly-A tails, little or no information about introns is acquired by sequencing ESTs.
3. Reverse transcriptase is rather prone to mistakes, and consequently cDNA molecules are considerably more likely to contain errors than is cloned genomic DNA (Menendez-Arias 2002). In fact, as many as 3 % of nucleotides in reverse transcriptase-catalyzed strand synthesis reactions are likely to be incorrect (Rudd 2003).
4. Since processed mRNA is the starting material in cDNA library construction, promoter sequences, a crucial portion of gene space, are not found in EST libraries.
5. Plant tissues may be dominated by a few abundant transcripts. For example, a handful of cellular biogenesis genes account for >40 % of transcripts found in *Arabidopsis* pollen (Lee and Lee 2003). While the representation of dominant transcripts in an isolated mRNA population can be reduced using "normalization" strategies (e. g., Ko 1990; Soares et al. 1994; Neto et al. 1997; Zhulidov et al. 2004), gene copy number is not accurately reflected in cDNA libraries even if normalization techniques are employed.

6. Some transcripts are ephemeral, lasting only a few minutes in cells (this is especially true of transcription factor mRNAs; e.g., Gee et al. 1991; Knauss et al. 2003; Yang et al. 2003), and others are simply transcribed at extremely low levels, making their recovery difficult (Ito et al. 2003; Hsu et al. 2004).
7. The representation of genes in a given cDNA library is indicative of gene expression in the source tissue(s) only under specific environmental conditions. In order to have some confidence of acquiring the mRNA-encoding regions of every gene in a genome, one would theoretically have to obtain mRNA from every tissue exposed to every likely environmental condition/stress at every stage of development (Rudd 2003), as well as overcoming the problems mentioned above.
8. Due to the limitations described (specifically points 4, 5, and 6), EST sequencing, even from libraries representing numerous developmental stages and tissues, reaches a point of diminishing returns at 50–70 % transcript coverage. For example, although there are roughly 29,000 *Arabidopsis* genes as determined by genome sequencing (Arabidopsis Genome Initiative 2000), the 178,000 ESTs obtained from 61 different *Arabidopsis* tissues/stages/environments account for only 63 % (i. e., 16,115) of these (Rudd 2003).

Despite its shortcomings, EST sequencing's dual ability to provide sequence information on the coding regions of probable genes and data on differential gene transcription has made it an invaluable tool in genome research. The value of ESTs as molecular markers and the incorporation of EST data into new techniques such as serial analysis of gene expression (SAGE), microarrays, transcriptome chips, etc., indicate that EST sequencing will continue to be an important tool for quite awhile.

13.4.2.2 Methylation Filtration

Methylation filtration is by far the simplest reduced representation technique, a fact that makes it highly appealing to those doing high-throughput genomics. It clearly enriches for gene regions in maize (Rabinowicz et al. 1999; Whitelaw et al. 2003) and several other species (www.oriongenomics.com), and it is possible that it would provide some level of gene enrichment/repeat reduction for many (most?) plant species. In this regard, a recent BLAST analysis (D. Peterson, unpublished results) indicates that of the 50,160 methylation filtered *Sorghum bicolor* sequences in GenBank (as of 24 March 2004), only 1.53 % show significant (E value $<1\times10^{-5}$) homology to *Retrosor-6*, a repetitive element that appears to account for about 6 % of sorghum DNA (see above; Peterson et al. 2002a). Thus MF appears to reduce the level of *Retrosor-6* in sorghum DNA by roughly (768 50,160 =) 3.9-fold.

Potential problems with using MF as a gene enrichment tool are evident by examination of the literature on DNA methylation in plants. First, it is well documented that some plant genes are normally hypermethylated and may become inactivated if hypomethylated. For example, (1) methylation of CGCG sites in several petunia genes is correlated with normal adventitious shoot bud induction, but both gene methylation and adventitious shoot budding are repressed by DNA methylase inhibitors (Prakash et al. 2003); (2) a 261-kb BAC from barley containing the powdery mildew resistance locus *Mla* has been shown to contain an extensive hypermethylated, but transcriptionally active, gene-rich island (Wei et al. 2002); (3) in at least seven ecotypes of *Arabidopsis*, the PAI1 and PAI4 genes involved in tryptophan biosynthesis are hypermethylated when active (Melquist et al. 1999; Bartee and Bender 2001); and (4) in vitro, somatic embryogenesis in carrot is blocked if cells are treated with 5-azacytidine, an agent that causes DNA hypomethylation (LoSchiavo et al. 1989).

Second, there is considerable evidence that changes in methylation are a normal means by which plant genes, especially those involved in development or response to stress, are regulated (Finnegan et al. 2000). For example, (1) in dormant potatoes, large-scale, transient demethylation (50–70 %) of 5′-CCGG-3′ sequences precedes transcription of genes involved in cell division and meristem growth (Law and Suttle 2002); (2) meristematic regions in pine exhibit 35 % DNA methylation in juvenile trees versus >60 % methylation in adult plants, but exposure of adult plants to reinvigoration stimuli causes a decrease in methylation to levels similar to those observed in juveniles (Fraga et al. 2002); (3) a rapid global decrease in DNA methylation occurs during seed germination and shoot apical meristem development in *Silene vulgaris* (Zluvova et al. 2001); (4) methylation appears to be involved in regulation of mRNA genes in numerous plant species (Drozdenyuk et al. 1976; Watson et al. 1987; Follman et al. 1990; Zluvova et al. 2001); (5) vernalization in *Arabidopsis* and tobacco appears to be triggered by demethylation of genes involved in the transition to flowering (Finnegan et al. 2000); (6) in maize, cold stress leads to genome-wide DNA demethylation in root tissues and subsequently to changes in transcription (Steward et al. 2002); (7) endosperm-specific demethylation and activation of specific alpha-tubulin alleles has been reported in maize (Lund et al. 1995); (8) the *P1-Blotched* and *P1-Rhoades* genes of maize show developmentally sensitive changes in methylation (Hoekenga et al. 2000); and (9) tissue-specific differences in DNA methylation have been observed in a number of plants, including tomato (Messeguer et al. 1991) and rice (Xiong et al. 1999).

Finally, the hypermethylation of repeat sequences is by no means a constant or consistent characteristic of plant genomes. For example, (1) the tandem *Hae*III repeat in the grass *Pennisetum glaucum* is hypomethylated (Kamm et al. 1994); (2) when snapdragon is exposed to cold weather, methylation of the transposon Tam3 is reduced and transposition of the element increases (Hashida et al. 2003); (3) in maize, Robertson *Mutator* transposable elements undergo changes in methylation that coincide with changes in their expression (Singer et al. 2001); (4) within a genome, highly repetitive sequences can exhibit differential methylation patterns as demonstrated for the *Zingeria biebersteiniana* centromeric repeat Zbcen1 (Saunders and Houben 2001), the "*Alu*I" satellite repeats of snapdragon (Schmidt and Kudla 1996), and various high-copy tobacco repeats (Kovarik et al. 2000); (5) middle repetitive DNA sequences from maize are found in both hypermethylated and hypomethylated DNA domains (Bennetzen et al. 1994); (6) the maize *suppressor-mutator* transposable element and a *Cucumis melo* satellite repeat exhibit tissue-specific differences in their methylation patterns (Banks and

Fedoroff 1989; Grisvard 1985); and (7) in some instances (e. g., *Arabidopsis* centromeric regions), one strand of the DNA double helix may be hypermethylated compared to its complementary strand (i. e., there can be strand-biased methylation) (Luo and Preuss 2003).

The preceding observations indicate that methylation filtration will result in the loss of methylated genes whether currently active in the source tissue or inactive. Those genes involved in development and stress responses are particularly likely to be eliminated by MF. Additionally, certain repeat sequences will end up in methylation-filtered DNA. In this regard, 33 % of sequences in methylation-filtered *Zea mays* libraries show significant homology to known repeats (Whitelaw et al. 2003).

Undoubtedly, some species are likely to be more amenable to methylation filtration than others. However, it is clear that before investing in MF, one should have a basic knowledge of the DNA methylation patterns found throughout the life cycle of the experimental organism.

13.4.2.3 Cot-based Cloning and Sequencing

In Cot-based cloning and sequencing (CBCS), repetitive and/or low-copy sequences are separated based upon their relative renaturation rates, which are reflective of their relative copy numbers in the genome. Such fractionation is completely independent of gene expression since the DNA used in re-association is mechanically sheared genomic DNA. Likewise, sequence renaturation is independent of methylation patterns (Burtseva et al. 1979). Moreover, the separation of DNA sequences using Cot techniques is a well-established biochemical phenomenon that can be applied to all species regardless of phylogeny (Peterson et al. 2002a, 2002b). Consequently, of the three gene-enrichment techniques, CBCS theoretically provides the most comprehensive and least biased gateway into the low-copy diversity of plant genomes.

CBCS is the most versatile of the reduced representation strategies, as it can be used for enrichment of genes and/or repeats. It is the only reduced representation method that, in and of itself, could theoretically be used to sequence a genome's SqCx (see Peterson et al. 2002a, 2002b for review). In terms of gene enrichment, the parameters used in isolating the SL component can be adjusted to meet investigator goals and can be adapted for specific genomes (see Peterson et al. 2002a and Paterson et al. 2004 for reviews). For example, in those species in which genes are known to be grouped into "islands" (e. g., grasses), increasing the length of sequences used in constructing the SLCot library will decrease the probability that repetitive elements will elute with the SL component and result in longer clones more suitable for bidirectional sequencing. Additionally, allowing re-association to proceed to a higher Cot value will increase the stringency of SL fractionation and decrease potential repeat contamination. However, in making such changes, one runs the risk of "weeding out" short low-copy sequences near or flanked by repetitive elements. Consequently, sequencing clones from multiple SLCot libraries with different insert sizes may provide the greatest coverage of gene space per sequencing dollar (Paterson et al. 2004).

Compared to MF, CBCS appears to be a more stringent method of separating repetitive and low-copy DNA. While MF appears to result in a fourfold reduction in the level of *Retrosor-6* in sorghum DNA (see above), no SLCot clones in a 499-clone library showed homology to *Retrosor-6* (in contrast, 13.4% and 1.7% of HRCot and MRCot clones, respectively, contained *Retrosor-6* sequence). Likewise, a comparison of MF and SLCot libraries from maize indicates that more than twice as many repeat sequences are found in the former as in the latter (Whitelaw et al. 2003).

As with all reduced representation techniques, CBCS has some limitations/drawbacks:

1. The DNA used in renaturation kinetics must be extremely clean, as even low-level contamination from proteins, carbohydrates, and/or secondary compounds can cause serious problems (see Murray and Thompson 1976 for review). Contaminants such as polyphenols can inhibit renaturation by damaging DNA (Peterson et al. 1997). In contrast, carbohydrate contaminants may effectively decrease the area in which re-association takes place and thus artificially speed up the renaturation rate. Some contaminants may absorb light at or near 260 nm and, if undetected, lead to aberrant results. Of note, many molecular biology protocols work well even if "dirty" DNA is used. However, such DNA is not suitable for Cot analysis and CBCS. Additionally, it is necessary that nuclear DNA, not total cellular DNA, be used in Cot analysis/fractionation, as significant organellar DNA contamination will complicate/confuse Cot analysis and decrease library quality. A protocol for isolating highly pure nuclear DNA suitable for Cot analysis and CBCS is available at the Mississippi Genome Exploration Laboratory (MGEL) website (www.msstate.edu/research/mgel/nucl_DNA.htm).
2. CBCS requires a fairly good understanding of re-association kinetics.
3. In a few plant species, there is a high level of non-genic, low-copy DNA. For example, the tomato genome appears to have a considerable proportion of low-copy elements that are not genes (Peterson et al. 1998). For such species, CBCS may provide only marginal benefits compared to whole-genome shotgun sequencing.
4. It has been speculated that SLCot sequencing may result in underrepresentation of members of large gene families (Martienssen et al. 2004), i.e., large gene families may fractionate with MR DNA. While this may be a problem if sequencing is performed using only one stringently prepared SLCot library, it is easily remedied by preparing several SLCot libraries with partially overlapping kinetic ranges.
5. Complementary regions may be found within the same single-stranded, low-copy sequence. Such molecules will "fold back" on themselves and form duplexes at Cot values approaching zero. While fold-back DNA is thought to be a minor component of genomes, low-copy fold-back sequences will be lost during SLCot library preparation. In species where fold-back DNA accounts for

>3 % of a genome, it may be advisable to clone and sequence the fold-back fraction (see Peterson et al. 2002b for further discussion).

Because a Cot analysis can be a relatively difficult, time-consuming process, the natural tendency of many researchers is to forgo actually constructing a Cot curve and skip right to re-association-based fractionation and cloning. If one's goal is solely to enrich for gene space, this may be a justifiable route, especially if it is clear that the experimental genome contains numerous repeats, the organism's genome size is well established, and the DNA used in re-association is known to be free of contamination. For example, Lamoreux et al. (submitted) recently prepared one SLCot library from bread wheat based on data from an existing wheat Cot curve and a second SLCot library based on an estimated $\mathrm{Cot}^1/_2$ value for single-copy DNA as determined from genome size. The experimental $\mathrm{Cot}^1/_2$ value of the SL component and the theoretical $\mathrm{Cot}^1/_2$ value of single-copy DNA were fairly similar and, not surprisingly, the sequence contents of both Cot libraries were indistinguishable. To estimate the $\mathrm{Cot}^1/_2$ of a species' single-copy component, the following formula can be used:

$$\mathrm{Cot}^1/_{2\mathrm{org}} = (\mathrm{Cot}^1/_{2\mathrm{coli}} \times G_{\mathrm{org}}) \quad G_{\mathrm{col}} \qquad (1)$$

where $\mathrm{Cot}^1/_{2\mathrm{org}}$ is the estimated $\mathrm{Cot}^1/_2$ of single-copy DNA for the organism of interest, G_{coli} is the genome size in base pairs of *E. coli*, G_{org} is the 1C DNA content of the organism of interest in base pairs, and $\mathrm{Cot}^1/_{2\mathrm{coli}}$ is the $\mathrm{Cot}^1/_2$ of *E. coli* DNA. Inserting *E. coli*'s known genome size (4,639,221 bp; Blattner et al. 1997) and its $\mathrm{Cot}^1/_2$ value (4.545455 M·sec; Zimmerman and Goldberg 1977) yields the following:

$$\mathrm{Cot}^1/_{2\mathrm{org}} = (4.545455\ \mathrm{M \cdot sec} \times G_{\mathrm{org}}) \quad 4{,}639{,}221\ \mathrm{bp} \qquad (2)$$

Placing a known genome size for an organism into Eq. (2) and solving for $\mathrm{Cot}^1/_{2\mathrm{org}}$ provides the $\mathrm{Cot}^1/_2$ for a theoretical single-copy component. From the predicted $\mathrm{Cot}^1/_2$, one can decide upon a Cot value that will provide a high likelihood that most low-copy elements will be isolated (e.g., Figure 13.4).

13.4.2.4 Integration of Reduced Representation Strategies

Recently, Whitelaw et al. (2003) compared the sequence content of SLCot, methyl-filtered, and unfiltered libraries from maize. Both MF and SLCot libraries showed enrichment of expressed sequences (27 % and 22 % of clones recognizing ESTs, respectively) compared to the unfiltered library (6 % of sequences exhibited homology to ESTs). With regard to repetitive sequences, 14 % of the SLCot sequences possessed significant homology to known repetitive elements, while 33 % of MF sequences recognized repeats. Compared to the MF library, the SLCot library contained a larger proportion of sequences with no significant matches to any database sequences (63 % vs. 39 %), which partly reflects the higher repeat content of the MF library and may also reflect the ability of CBCS to capture elements (e.g., short-lived transcription factors, other low-copy sequences) that may elude EST and MF approaches. Perhaps the most significant finding of the study was that when SLCot and MF sequences were grouped into contigs, 60 % of the MF clones assembled only with other MF clones, while 72 % of SLCot clones assembled only with other SLCot sequences. This suggests that at least in maize, SLCot sequencing and MF may be enriching for largely different low-copy sequence subsets; alternatively, one technique may be enriching for a greater diversity of low-copy sequences than the other. Consequently, Whitelaw et al. suggested that with regard to maize and possibly other large genomes, a combination of MF and SLCot sequencing might provide better results than using one strategy alone.

13.4.3 Polymorphism Discovery

13.4.3.1 RRS Sequencing and DOP-PCR

The vast majority of polymorphism discovery techniques require a priori knowledge of sequences unique to a particular locus (or loci). Two notable exceptions are RRS sequencing and DOP-PCR. Both approaches enrich for a subset of loci more or less at random. If the DNA from many individuals is pooled, polymorphisms can be discovered by sequencing the isolated sequence subset and using computational algorithms to align probable alleles.

RRS sequencing and DOP-PCR have their relative advantages and disadvantages. Both techniques permit SNP and indel discovery. However, neither technique guarantees that polymorphic regions will be found in a sequence subset. Likewise, repetitive sequences are as likely to be sequenced as non-repetitive sequences; repeats are removed from consideration *in silico*. Of the two methods, RRS sequencing is certainly the most straightforward and probably the least expensive, as it does not require primer synthesis and is rooted in techniques common to most molecular biologists. DOP-PCR involves more experimental steps than RRS sequencing (compare methods of Jordan et al. 2002 and Altschuler et al. 2000). However, a typical DOP-PCR experiment will likely provide information about a larger subset of loci than an RRS sequencing experiment. As both techniques are known to be effective, it is likely that an investigator's preference for one over the other may reflect his/her familiarity (or lack thereof) with primer design and PCR.

13.4.3.2 Microsatellite Isolation

There are currently numerous approaches for de novo isolation of microsatellites. However, all the protocols share one thing in common: they rely upon the use of synthetic microsatellite-like oligonucleotides to "find" SSRs in genomic DNA. Of the microsatellite isolation protocols, the "nylon cross-linking" technique (i. e., cross-linking microsatellite-like oligonucleotides to a nylon membrane and incubating the membrane with genomic DNA; e. g., Edwards et al. 1996) has the greatest potential for capturing all potential SSRs, as every possible SSR oligonucleotide can easily fit on a relatively small nylon membrane. However, this approach is not widely used, reportedly because it does not provide satisfactory results for some species and/or some practitioners (Fischer and Bachmann 1998). The elegant approach of Ostrander et al. (1992) is applicable to many organisms, but it requires the use of two bacterial strains and a phage intermediate and thus is less popular than simpler SSR capture techniques. In the "primer extension" protocols, microsatellite-like sequences form part of primer sequences used in PCR (e. g., Figure 13.6). The primer extension protocols can be very fruitful and can be adjusted to meet the needs of most researchers. However, successful application of these protocols requires a fair amount of skill in primer design and some reaction optimization. Likewise, in order to amplify an SSR locus, two primer-binding sites are necessary (see above). In the widely used "hybrid capture" protocols, biotinylated microsatellite-like oligonucleotides are hybridized to genomic DNA, and hybrid molecules are isolated via streptavidin-coated beads (e. g., Figure 13.7). The hybrid capture technique requires only modest PCR skills, and there is no requirement that microsatellites be near each other or near any other element. Thus the hybrid capture techniques will likely supplant primer extension techniques for those interested in straightforward SSR isolation. The primer extension techniques, however, enjoy the advantage of being easily adapted to certain genomes and/or highly specific needs, and thus they will likely continue to be utilized as well.

13.5 Conclusions

Reduced representation strategies provide a means of exploring large, repetitive plant genomes in a cost-efficient manner. Specifically, they allow study of important sequence subsets that otherwise could be obtained only by whole-genome shotgun sequencing. The reduced representation techniques that can be employed in a given situation depend largely on the goals of the scientist and the biology of the experimental organism. EST sequencing and CBCS should be useful in capturing gene space of all seed plants, while MF will likely provide some level of gene enrichment for many plant species and high levels for others. CBCS is currently the only reduced representation technique that allows preferential study of repeat sequences. DOP-PCR, RRS sequencing, and microsatellite enrichment tools afford access to polymorphisms for mapping, molecular breeding, and characterizing genotypic/phenotypic relationships.

As the focus of genomics becomes centered less on small-genome model organisms and more on economically and socially important species, the utilization of existing reduced representation techniques and the demand for new reduced representation strategies will undoubtedly increase. Additionally, the large size of most crop genomes and the smaller pool of funding sources for plant research compared to vertebrate research make it likely that plant biologists will continue to be among the most avid users and developers of reduced representation techniques.

Acknowledgments

Thanks to Andrew H. Paterson for advice and to the U. S. Department of Agriculture (USDA-NRI Award 9901330), the National Science Foundation (NSF-DBI Award 0115903), and the USDA Forest Service (Award SRS-03-CA-11330126-263) for financial support. This chapter has been approved for publication as Journal Article No. BC10511 of the Mississippi Agriculture and Forestry Experiment Station, Mississippi State University.

References

Adams MD, Kelley JM, Gocayne JD, Dubnick M, Polymeropoulos MH, Xiao H, Merril CR, Wu A, Olde B, Moreno RF, Kerlavage AR, McCombie WR, Venter JC (1991) Complementary DNA sequencing: Expressed sequence tags and human genome project. *Science* 252:1651–1656.

Altshuler D, Pollara VJ, Cowles CR, Van Etten WJ, Baldwin J, Linton L, Lander ES (2000) An SNP map of the human genome generated by reduced representation shotgun sequencing. *Nature* 407:513–516.

Arabidopsis Genome Initiative (2000) Analysis of the genome sequence of the flowering plant *Arabidopsis thaliana*. *Nature* 408:796–815.

Assaad FF, Tucker KL, Signer ER (1993) Epigenetic repeat-induced gene silencing (RIGS) in *Arabidopsis*. *Plant Mol. Biol.* 22:1067–1085.

Baltimore D (1970) RNA-dependent DNA polymerase in virions of RNA tumour viruses. *Nature* 226:1209–1211.

Banks JA, Federoff N (1989) Patterns of developmental and heritable change in methylation of the *suppressor-mutator* transposable element. *Dev. Genet.* 10:425–437.

Bartee L, Bender J (2001) Two *Arabidopsis* methylation-deficiency mutations confer only partial effects on a methylated endogenous gene family. *Nucleic Acids Res.* 29:2127–2134.

Bennett MD, Leitch IJ (2003) Plant DNA C-values database (release 2.0, Jan. 2003). http://www.rbgkew.org.uk/cval/homepage.html.

Bennetzen JL, Schrick K, Springer PS, Brown WE, SanMiguel P (1994) Active maize genes are unmodified and flanked by diverse classes of modified, highly repetitive DNA. *Genome* 37:565–576.

Berger F (2004) Imprinting – a green variation. *Science* 303:483–485.

Bhattramakki D, Dolan M, Hanafey M, Wineland R, Vaske D, Register III JC, Tingey SV, Rafalski A (2002) Insertion-deletion polymorphisms in 3′ regions of maize genes occur frequently and can be used as highly informative genetic markers. *Plant Mol. Biol.* 48:539–547.

Blattner FR, Plunkett III G, Bloch CA, Perna NT, Burland V, Riley M, Collado-Vides J, Glasner JD, Rode CK, Mayhew GF, Gregor J, Davis NW, Kirkpatrick HA, Goeden MA, Rose DJ, Mau B, Shao Y (1997) The complete genome sequence of *Escherichia coli* K-12. *Science* 277:1453–1474.

Britten RJ (1996) Cases of ancient mobile element DNA insertions that now affect gene regulation. *Mol. Phylogenet. Evol.* 5: 13–17.

Britten RJ, Kohne DE (1968) Repeated sequences in DNA. *Science* 161:529–540.

Britten RJ, Graham DE, Neufeld BR (1974) Analysis of repeating DNA sequences by reassociation. *Methods Enzymol.* 29:363–405.

Buchanan AV, Risch GM, Robichaux M, Sherry ST, Batzer MA, Weiss KM (2000) Long DOP-PCR of rare archival anthropological samples. *Hum. Biol.* 72:911–925.

Burtseva NN, Romanov GA, Azizov YuM, Banyshin BF (1979) Intragenome distribution of 5-methylcytosine and kinetics of the reassociation of cow blood lymphocyte DNA in the normal state and in chronic lympholeukemia. *Biochemistry-Moscow* 44: 1636–1641.

Casa AM, Brouwer C, Nagel A, Wang L, Zhang Q, Kresovich S, Wessler SR (2000) The MITE family *Heartbreaker* (Hbr): Molecular markers in maize. *Proc. Natl. Acad. Sci. USA* 97: 10083–10089.

Casacuberta JM, Santiago N (2003) Plant LTR-retrotransposons and MITEs: control of transposition and impact on the evolution of plant genes and genomes. *Gene* 311:1–11.

Cheung VG, Nelson SF (1996) Whole-genome amplification using a degenerate oligonucleotide primer allows hundreds of genotypes to be performed on less than one nanogram of genomic DNA. *Proc. Natl. Acad. Sci. USA* 93:14676–14679.

Dekkers JCM, Hospital F (2002) The use of molecular genetics in the improvement of agricultural populations. *Nature Reviews* 3: 22–32.

Dietmaier W, Hartmann A, Wallinger S, Heinmöller E, Kerner T, Endl E, Jauch K-W, Hofstädter F, Rüschoff J (1999) Multiple mutation analyses in single tumor cells with improved whole-genome amplification. *Amer. J. Pathol.* 154:83–95.

Dorer DR, Henikoff S (1994) Expansions of transgene repeats cause heterochromatin formation and gene silencing in *Drosophila*. *Cell* 77:993–1002.

Drozdenyuk AP, Sulimova GE, Vanyushin BI (1976) Changes in base composition and molecular population of wheat DNA on germination. *Mol. Biol. (Moscow)* 10:1378–1386.

Edery I, Chu LL, Sonenberg N, Pelletier J (1995) An efficient strategy to isolate full-length cDNAs based on an mRNA cap retention procedure (CAPture). *Mol. Cell Biol.* 15:3363–3371.

Edwards KJ, Barkder JHA, Daly A, Jones C, Karp A (1996) Microsatellite libraries enriched for several microsatellite sequences in plants. *BioTechniques* 20:758–760.

Feschotte C, Jiang N, Wessler SR (2002) Plant transposable elements: Where genetics meets genomics. *Nat. Rev. Genet.* 3:329–341.

Finnegan EJ, Genger RK, Peacock WJ, Dennis ES (1998) DNA methylation in plants. *Annu. Rev. Plant Physiol. Plant Mol. Biol.* 49:223–247.

Finnegan EJ, Peacock WJ, Dennis ES (2000) DNA methylation, a key regulator of plant development and other processes. *Curr. Opin. Genet. Dev.* 10:217–223.

Fischer D, Bachmann K (1998) Microsatellite enrichment in organisms with large genomes (*Allium cepa* L.). *BioTechniques* 24:796–802.

Fisher PJ, Gardner RC, Richardson TE (1996) Single locus microsatellites isolated using 5′ anchored PCR. *Nucleic Acids Res.* 24:4369–4371.

Fojtova M, Van Houdt H, Depicker A, Kovarik A (2003) Epigenetic switch from posttranscriptional to transcriptional silencing is correlated with promoter hypermethylation. *Plant Physiol.* 133:1240–1250.

Follmann H, Balzer HJ, Schleicher R (1990) Biosynthesis and distribution of methylcytosine in wheat DNA. How different are plant DNA methyltransferases? In: *Nucleic Acid Methylation*. AR Liss Inc., New York: pp. 199–209.

Fraga MF, Rodriguez R, Canal MJ (2002) Genomic DNA methylation-demethylation during aging and reinvigoration of *Pinus radiata*. *Tree Physiol.* 22:813–816.

Fu H, Zheng Z, Dooner HK (2002) Recombination rates between adjacent genic and retrotransposon regions in maize vary by 2 orders of magnitude. *Proc. Natl. Acad. Sci. USA* 99:1082–1087.

Gee MA, Hagen G, Guilfoyle TJ (1991) Tissue-specific and organ-specific expression of soybean auxin-responsive transcripts GH3 and SAURs. *Plant Cell* 3:419–430.

Goldberg RB (1978) DNA sequence organization in the soybean plant. *Biochemical Genetics* 16:45–68.

Goldberg RB (2001) From Cot curves to genomics. How gene cloning established new concepts in plant biology. *Plant Physiol.* 125: 4–8.

Grisvard J (1985) Different methylation pattern of melon satellite DNA sequences in hypocotyl and callus tissues. *Plant Sci.* 39:189–193.

Gupta PK, Rustgi S, Sharma S, Singh R, Kumar N, Balyan HS (2003) Transferable EST-SSR markers for the study of polymorphism and genetic diversity in bread wheat. *Mol. Genet. Genomics* 270:315–323.

Hamilton MB, Pincus EL, Di Fiore A, Fleischer RC (1999) Universal linker and ligation procedures for construction of genomic DNA libraries enriched for microsatellites. *BioTechniques* 27:500–507.

Hartl DL (2000) Molecular melodies in high and low *C. Nat. Rev.* 1:145–149.

Hashida S, Kitamura K, Mikami T, Kishima Y (2003) Temperature shift coordinately changes the activity and the methylation state of transposon Tam3 in *Antirrhinum majus. Plant Physiol.* 132:1207–1216.

Hoekenga OA, Muszynski MG, Cone KC (2000) Developmental patterns of chromatin structure and DNA methylation responsible for epigenetic expression of a maize regulatory gene. *Genetics* 155:1889–1902.

Hsu HC, Tan LY, Au LC, Lee YM, Lieu CH, Tsai WH, You JY, Liu MD, Ho CK (2004) Detection of bcr-abl gene expression at a low level in blood cells of some patients with essential thrombocythemia. *J. Lab. Clin. Med.* 143: 125–129.

Ito T, Sakai H, Meyerowitz EM (2003) Whorl-specific expression of the *SUPERMAN* gene of *Arabidopsis* is mediated by *cis* elements in the transcribed region. *Curr. Biol.* 13:1524–1530.

Jordan B, Charest A, Dowd JF, Blumenstiel JP, Yeh RF, Osman A, Housman DE, Landers JE (2002) Genome complexity reduction for SNP genotyping analysis. *Proc. Natl. Acad. Sci. USA* 99:2942–2947.

Kamm A, Schmidt T, Heslop-Harrison JS (1994) Molecular and physical organization of highly repetitive, undermethylated DNA from *Pennisetum glaucum. Mol. Gen. Genet.* 244:420–425.

Kinoshita T, Muura A, Choi Y, Kinoshita Y, Cao X, Jacobsen SE, Fischer RL, Kakutani T (2004) One-way control of *FWA* imprinting in *Arabidopsis* endosperm by DNA methylation. *Science* 303:521–522.

Kiper M, Herzfeld F (1978) DNA sequence organization in the genome of *Petroselinum sativum* (Umbelliferae). *Chromosoma*: 335–351.

Kirst M, Johnson AF, Baucom C, Ulrich E, Hubbard K, Staggs R, Paule C, Retzel E, Whetten R, Sederoff RR (2003) Apparent homology of expressed genes from wood-forming tissues of loblolly pine (*Pinus taeda* L.) with *Arabidopsis thaliana. Proc. Natl. Acad. Sci. USA* 100:7383–7388.

Kittler R, Stoneking M, Kayser M (2002) A whole-genome amplification method to generate long fragments from low quantities of genomic DNA. *Anal. Biochem.* 300:237–244.

Knauss S, Rohrmeier T, Lehle L (2003) The auxin-induced maize gene *ZmSAUR2* encodes a short-lived nuclear protein expressed in elongating tissues. *J. Biol. Chem.* 278: 23936–23943.

Ko MSH (1990) An "equalized cDNA library" by reassociation of short double-stranded cDNAs. *Nucleic Acids Res.* 18:5705–5711.

Komulainen P, Brown GR, Mikkonen M, Karhu A, García-Gil R, O'Malley D, Lee B, Neale DB, Savolainen O (2003) Comparing EST-based genetic maps between *Pinus sylvestris* and *Pinus taeda. Theor. Appl. Genet.* 107:667–678.

Kovalchuk O, Burke P, Arkhipov A, Kuchma N, James SJ, Kovalchuk I, Pogribny I (2003) Genome hypermethylation in *Pinus silvestris* of Chernobyl- a mechanism for radiation adaptation? *Mutation Res.* 529:13–20.

Kovarik A, Koukalova B, Lim KY, Matyasek R, Lichtenstein CP, Leitch AR, Bezdek M (2000) Comparative analysis of DNA methylation in tobacco heterochromatic sequences. *Chromosome Res.* 8:527–541.

Kumar S, Cheng X, Klimasauskas S, Mi S, Posfai J, Roberts RJ, Wilson GG (1994) The DNA (cytosine-5) methyltransferases. *Nucleic Acids Res.* 22:1–10.

Lapitan NLV (1992) Organization and evolution of higher plant nuclear genomes. *Genome* 35:171–181.

Law DR, Suttle JC (2002) Transient decreases in methylation at 5′-CCGG-3′ sequences in potato (*Solanum tuberosum* L.) meristem DNA during progression of tubers through dormancy precede the resumption of sprout growth. *Plant Mol. Biol.* 51:437–447.

Lee C, Ritchie DBC, Lin CC (1994) A tandemly repetitive, centromeric DNA sequence from the Canadian woodland caribou (*Rangifer tarandus caribou*): its conservation and evolution in several deer species. *Chromosome Res.* 2:293–306.

Lee J-Y, Lee D-H (2003) Use of serial analysis of gene expression technology to reveal changes in gene expression in *Arabidopsis* pollen undergoing cold stress. *Plant Physiol.* 132: 517–529.

Lisch D, Carey CC, Dorweiler JE, Chandler VL (2002) A mutation that prevents paramutation in maize also reverses *Mutator* transposon methylation and silencing. *Proc. Natl. Acad. Sci. USA* 99:6130–6135.

Liu B, Wendel JF (2003) Epigenetic phenomena and the evolution of plant allopolyploids. *Mol. Phylogenet. Evol.* 29:365–379.

LoSchiavo F, Pitto L, Giuliano G, Torti G, Nuti-Ronchi V, Marazziti D, Vergara R, Orselli S, Terzi M (1989) DNA methylation of embryonic carrot cell cultures and its variations as caused by mutation, differentiation, hormones, and hypomethylating drugs. *Theor. Appl. Genet.* 77:325–331.

Lund G, Messing J, Viotti A (1995) Endosperm-specific demethylation and activation of specific alleles of alpha-tubulin genes of *Zea mays* L. *Mol. Gen. Genet.* 246:716–722.

Lundblad V, Wright WE (1996) Telomeres and telomerase: A simple picture becomes complex. *Cell* 87:369–375.

Luo S, Preuss D (2003) Strand-biased DNA methylation associated with centromeric regions in *Arabidopsis*. *Proc. Natl. Acad. Sci. USA* 100:11133–11138.

Martienssen RA, Rabinowicz PD, O'Shaughnessy A, McCombie WR (2004) Sequencing the maize genome. *Curr. Opin. Plant Biol.* 7: 102–107.

Melquist S, Luff B, Bender J (1999) Arabidopsis *PAI* gene arrangements, cytosine methylation and expression. *Genetics* 153:413.

Menendez-Arias L (2002) Molecular basis of fidelity of DNA synthesis and nucleotide specificity of retroviral reverse transcriptases. *Prog. Nucleic Acid Res. Mol. Biol.* 71: 91–147.

Meng L, Bregitzer P, Zhang S, Lemaux PG (2003) Methylation of the exon/intron region in the *Ubi1* promoter complex correlates with transgene silencing in barley. *Plant Mol. Biol.* 53:327–340.

Messeguer R, Ganal M, deVincente MC (1991) High resolution RFLP map around the root knot nematode resistance gene (Mi) in tomato. *Theor. Appl. Genet.* 82:529–536.

Murray MG, Thompson WF (1976) Contaminants affecting plant DNA reassociation. *Carnegie Inst. Wash. Yearbook* 76:255–259.

Neto ED, Harrop R, Correa-Oliveira R, Wilson RA, Pena SDJ, Simpson AJG (1997) Minilibraries constructed from cDNA generated by arbitrarily primed RT-PCR: an alternative to normalized libraries for the generation of ESTs from nanogram quantities of mRNA. *Gene* 186:135–142.

Ostrander EA, Jong PM, Rine J, Duyk G (1992) Construction of small-insert genomic DNA libraries highly enriched for microsatellite repeat sequences. *Proc. Natl. Acad. Sci. USA* 89:3419–3423.

Paetkau D (1999) Microsatellites obtained using strand extension: an enrichment protocol. *BioTechniques* 26:690–697.

Paterson AH, Bowers JE, Burow MD, Draye X, Elsik CG, Jiang C-X, Katsar CS, Lan T-H, Lin Y-R, Ming R, Wright RJ (2000) Comparative genomics of plant chromosomes. *Plant Cell* 12:1523–1539.

Paterson AH, Bowers JE, Chapman BA, Peterson DG, Rong JK, Wicker TM (2004) Comparative genome analysis of monocot and dicots, toward characterization of angiosperm diversity. *Curr. Opin. Biotech.* (in press).

Paterson AH, Schertz KF, Lin Y-R, Liu S-C, Chang Y-L (1995) The weediness of wild plants: Molecular analysis of genes influencing dispersal and persistance of johnsongrass, *Sorghum halepense* (L.) Pers. *Proc. Natl. Acad. Sci. USA* 92:6127–6131.

Peterson DG, Boehm KS, Stack S (1997) Isolation of milligram quantities of DNA from tomato (*Lycopersicon esculentum*), a plant containing high levels of polyphenolic compounds. *Plant Mol. Biol. Reptr.* 15:148–153.

Peterson DG, Pearson WR, Stack SM (1998) Characterization of the tomato (*Lycopersicon esculentum*) genome using *in vitro* and *in situ* DNA reassociation. *Genome* 41:346–356.

Peterson DG, Schulze SR, Sciara EB, Lee SA, Bowers JE, Nagel A, Jiang N, Tibbitts DC, Wessler SR, Paterson AH (2002) Integration of Cot analysis, DNA cloning, and high-throughput sequencing facilitates genome characterization and gene discovery. *Genome Res.* 12:795–807.

Peterson DG, Wessler SR, Paterson AH (2002) Efficient capture of unique sequences from eukaryotic genomes. *Trends Genet.* 18:547–550.

Petrov DA (2001) Evolution of genome size: new approaches to an old problem. *Trends Genet.* 17:23–28.

Phan J, Reue K, Peterfy M (2000) MS-IRS PCR: a simple method for the isolation of microsatellites. *BioTechniques* 28:18–20.

Prakash AP, Kush A, Lakshmanan P, Kumar PP (2003) Cytosine methylation occurs in a CDC48 homologue and a MADS-box gene during adventitious shoot induction in *Petunia* leaf explants. *J. Exp. Bot.* 54:1361–1371.

Rabinowicz PD, Schutz K, Dedhia N, Yordan C, Parnell LD, Stein L, McCombie WR, Martienssen RA (1999) Differential methylation of genes and retrotransposons facilitates shotgun sequencing of the maize genome. *Nature Genet.* 23:305–308.

Raizada MN (2003) *RescueMu* protocols for maize functional genomics. *Methods Mol. Biol.* 236:37–58.

Raizada MN, Nan G-L, Walbot V (2001) Somatic and germinal mobility of the *RescueMu* transposon in transgenic maize. *Plant Cell* 13:1587–1608.

Redaschi N, Bickle TA (1996) DNA restriction and modification systems. In: *Escherichia coli and Salmonella: Cellular and Molecular Biology.* Edited by: Neidhardt FC. *ASM Press, Washington, D.C.*: pp. 773–781.

Rudd S (2003) Expressed sequence tags: alternative or complement to whole-genome sequences? *Trends Plant Sci.* 8:321–329.

SanMiguel P, Bennetzen JL (2000) Evidence that a recent increase in maize genome size was caused by the massive amplification of intergene retrotransposons. *Ann. Bot.* 82: 37–44.

Saunders VA, Houben A (2001) The pericentromeric heterochromatin of the grass *Zingeria biebersteiniana* (2n=4) is composed of Zbcen1-type tandem repeats that are intermingled with accumulated dispersedly organized sequences. *Genome* 44:955–961.

Schmid KJ, Sörensen TR, Stracke R, Törjék O, Altmann T, Mitchell-Olds T, Weisshaar B (2003) Large-scale identification and analysis of genome-wide single-nucleotide polymorphisms for mapping in *Arabidopsis thaliana. Genome Res.* 13:1250–1257.

Schmidt T, Kudla J (1996) The molecular structure, chromosomal organization, and interspecies distribution of a family of tandemly repeated DNA sequences of *Antirrhinum majus* L. *Genome* 39:243–248.

Shagin DA, Rebrikov DV, Kozhemyako VB, Altshuler IM, Shcheglov AS, Zhulidov PA, Bogdanova EA, Staroverov DB, Rasskazov VA, Lukyanov S (2002) A novel method for SNP detection using a new duplex-specific nuclease from crab hepatopancreas. *Genome Res.* 12:1935–1942.

Sherman JD, Stack SM (1995) Two-dimensional spreads of synaptonemal complexes from solanaceous plants. VI. High-resolution recombination nodule map for tomato (*Lycopersicon esculentum*). *Genetics* 141: 683–708.

Singer T, Yordan C, Martienssen RA (2001) Robertson's *Mutator* tranposons in *A. thliana* are regulated by the chromatin-remodeling gene *Decrease in DNA Methylation (DDM1). Genes & Dev.* 15:602.

Soares MB, Bonaldo MDF, Jelene P, Su L, Lawton L, Efstratiadis A (1994) Construction and characterization of the normalized cDNA library. *Proc. Natl. Acad. Sci. USA* 91:9228–9232.

Steward N, Ito M, Yamaguchi Y, Koizumi N, Sano H (2002) Periodic DNA methylation in maize nucleosomes and demethylation by environmental stress. *J. Biol. Chem.* 277: 37741–37746.

Telenius H, Carter NP, Bebb CE, Nordenskjold M, Ponder BA, Tunnacliffe A (1992) Degenerate oligonucleotide-primed PCR: General amplification of target DNA by a single degenerate primer. *Genomics* 13:718–725.

Temin HM, Mizutani S (1970) RNA-dependent DNA polymerase in virions of Rous sarcoma virus. *Nature* 226:1211–1213.

Törjék O, Berger D, Meyer RC, Müssig C, Schmid KJ, Sörensen TR, Weisshaar B, Mitchell-Olds T, Altmann T (2003) Establishment of a high-efficiency SNP-based framework marker set for *Arabidopsis. Plant J.* 36:122–140.

Waldbieser GC, Quiniow SMA, Karsi A (2003) Rapid development of gene-tagged microsatellite markers from bacterial artificial chromosome clones using anchored TAA repeat primers. *BioTechniques* 35:976–979.

Watson JC, Kaufman LS, Thompson WF (1987) Developmental regulation of cytosine methylation in the nuclear ribosomal RNA genes of *Pisum sativum. J. Mol. Biol.* 193: 15–26.

Wei F, Wing RA, Wise RP (2002) Genome dynamics and evolution of the *Mla* (powdery mildew) resistance locus in barley. *Plant Cell* 14:1903–1917.

Wessler SR (1997) Transposable elements and the evolution of gene expression. *Exp. Biol.* 1039:115–122.

Whitelaw CA, Barbazuk WB, Pertea G, Chan AP, Cheung F, Lee Y, van Heeringen S, Karamycheva S, Bennetzen JL, SanMiguel P, Lakey N, Bedford J, Yuan Y, Budiman MA, Resnick A, van Aken S, Utterback T, Riedmuller S, Williams SM, Feldblyum T, Schubert K, Beachy R, Fraser CM, Quackenbush J (2003) Enrichment of gene-coding sequences in maize by genome filtration. *Science* 302: 2118–2120.

Xiong LZ, Xu CG, Saghai Maroof MA, Zhang Q (1999) Patterns of cytosine methylation in an elite rice hybrid and its parental lines, detected by a methylation-sensitive amplification polymorphism technique. *Mol. Gen. Genet.* 261:439–446.

Yang E, van Nimwegen E, Zavolan M, Rajewsky N, Schroeder M, Magnasco M, Darnell Jr JE (2003) Decay rates of human mRNAs: Correlation with functional characteristics and sequence attributes. *Genome Res.* 13:1863–1872.

Yuan Y, SanMiguel PJ, Bennetzen JL (2003) High-Cot sequence analysis of the maize genome. *Plant J.* 34:249–255.

Zhulidov PA, Bogdanova EA, Shcheglov AS, Vagner LL, Khaspekov GL, Kozhemyako VB, Matz MV, Meleshkevitch W, Moroz LL, Lukyanov SA, Shagin DA (2004) Simple cDNA normalization using kamchatka crab duplex-specific nuclease. *Nucleic Acids Res.* 32:e37.

Zimmerman JL, Goldberg RB (1977) DNA sequence organization in the genome of *Nicotiana tabacum. Chromosoma* 59:227–252.

Zluvova J, Janousek B, Vyskot B (2001) Immunohistochemical study of DNA methylation dynamics during plant development. *J. Exp. Botany* 52:2235–2273.

14
Large-scale DNA Sequencing

Christopher D. Town

Abstract

Advances in technology have brought about dramatic advances in DNA sequencing, greatly increasing the throughput and decreasing the cost per base. As a result DNA sequencing has become a readily available commodity that is now regarded as central to understanding the biology of any organism. This chapter provides an overview of the scope, methodology, and expectations of large-scale plant DNA sequencing projects.

14.1
Introduction

DNA sequencing has come a long way in the past decade. Two major advances have been the introduction of fluorescent "Big Dye" chemistry that permit real-time, multi-channel detection of sequenced bases and the introduction of capillary-based machines. As a consequence of these and other advances, the throughput of individual machines and of sequencing centers has increased dramatically and costs have decreased in parallel. Today, DNA sequencing is a readily available commodity to all researchers at a relatively reasonable cost, so that most small- to middle-sized research labs no longer operate their own sequencing equipment. The reduction in cost of both capital equipment and supplies, coupled with a recognition by funding agencies of the importance of DNA sequence as an underpinning for almost all modern research, has led to an increasing number of progressively larger and more ambitious sequencing projects.

The first large-scale plant genome sequencing effort was the *Arabidopsis* Genome Initiative, which was formally launched in 1996 and led to the completion of the genome in December 2000 (AGI 2000). The sequencing project itself was preceded by more than half a decade of physical mapping activities that started with yeast artificial chromosome (YAC) and cosmid libraries before moving on to bacterial artificial chromosomes (BACs). The majority of the sequencing focused on BACs and

The Handbook of Plant Genome Mapping.
Genetic and Physical Mapping. Edited by Khalid Meksem, Günter Kahl

ISBN 3-527-31116-5

was distributed among sequencing centers in the United States, Europe, and Japan. Each center was responsible for specific chromosomes or parts of chromosomes, and each center carried out both sequencing and the initial annotation of the finished sequence. During the four-year period of sequencing, costs decreased by more than a factor of two and have continued to decrease since then.

Another major advance that occurred in the late 1990s was the development of sequence assemblers that could handle large and complex genomes, resulting in the demonstration by the Celera group that whole-genome sequencing strategies, initially pioneered at the Institute for Genomic Research (TIGR) for the *Haemophilus influenzae* genome (Fleischmann et al. 1995), could be applied to larger and more complex genomes: first *Drosophila melanogaster* (Myers et al. 2000) and then the human genome (Venter et al. 2001). Since that time, whole-genome sequencing projects have become progressively more common and BAC-based projects less common. The principles and merits of each will be discussed briefly below.

14.2 Motivation for Sequencing and Choice of Strategy

Motivations for large-scale sequencing include gene discovery, single-nucleotide polymorphism (SNP) discovery and mapping, and comparative genomics. Each of these involves different approaches to the acquisition and management of templates for the sequencing process as well as the downstream analysis.

14.3 EST Sequencing

Sequencing of expressed sequence tags (ESTs) remains an efficient and cost-effective way for gene discovery and the acquisition of species-specific sequence information for use in functional genomics. Unless the aim is to acquire libraries from specific tissues for downstream use, the most efficient way to cover a new species is to make a single library from an RNA pool derived from a range of tissues and treatments. Library normalization can significantly increase the rate of gene discovery by sequencing (e. g., Kuhl et al. 2004), but the cost savings from reduced sequencing must be offset against the cost of normalization, which can be quite substantial.

14.4 Large-scale Sequencing BAC-based Projects

Some examples of BAC-based plant genomics projects are listed in Table 14.1, which is not intended to be exhaustive. In a BAC-based project, one or more BAC libraries are constructed from genomic DNA using distinct restriction enzymes, with the goal of producing 10–20× genome coverage. Whether or not a physical map exists at the outset, the goal is to finish the genome using the minimal number of BACs necessary. This may be based either primarily on a map-as-you-go sequence tag connector (STC) strategy (Venter et al. 1996) or by making more use of preexisting physical maps developed by fingerprinting, etc. (see Chapters X, Y, Z).

BAC-based sequencing has several advantages over the whole-genome approach:

1. It is easily distributed among multiple centers. This encourages multi-institutional and international collaborations and allows each center to identify and feel more invested in its own "territory."
2. It presents fewer assembly problems for two reasons. Software for assembling regions 50–250 kb in size are readily available and can be run on most computers. Repeated sequences that present problems in assembly are restricted to

Table 14.1. Plant genome projects.

Table 14.1a. BAC-based projects.

Species	*Genome size*	*Approx. number of BACs required*	*Status*
Arabidopsis thaliana[a]	~120 Mb	1500	Finished, December 2000
Oryza sativa[b]	~430 Mb	4300	3391 BACs produced 259 Mb of unique draft sequence
Oryza sativa[c]	~430 Mb	4300	High-quality draft, December 2002. Chromosomes 1, 3, 4, and 10 completed; other finishing ongoing
Zea mays[d]	~2,300 Mb	Survey only	500,000 BAC ends; ~100 BACs in progress
Lotus japonicus[e]	~470 Mb	4700	26.2 Mb in 256 BACs
Medicago truncatula[f]	~500 Mb	5000	HTGS phase 3: 24.71 Mb; 211 BACs; HTGS phase 2: 27.08 Mb; 220 BACs; HTGS phase 1: 34.34 Mb; 349 BACs

[a] http://www.tigr.org/tdb/e2k1/ath1/ath1.shtml. (AGI 2000)
[b] http://www.monsanto.com/monsanto/layout/our_pledge/rice_genmo.asp. (Barry 2001).
[c] http://rgp.dna.affrc.go.jp/; see http://www.tigr.org/tdb/e2k1/osa1/ for genome-wide totals. (Feng et al. 2002; Sasaki et al. 2002; The Rice Chromosome 10 Sequencing Consortium 2003)
[d] http://pgir.rutgers.edu/
[e] http://www.kazusa.or.jp/lotus/ (Asamizu et al. 2003)
[f] http://www.genome.ou.edu/medicago_totals.html

Table 14.1b. Whole-genome projects.

Species	*Genome size*	*Sequence completed*	*Fold coverage*	*Total assembled sequence*
Oryza sativa ssp. *japonica*[a]	~430 Mb	2,500 Mb	5.8×	389.8 Mb
Oryza sativa ssp. *indica*[b]	~460 Mb	2,522 Mb	4.2×	361 Mb
Populus trichocarpa[c]	~550 Mb	4,800 Mb	8.7×	Not available
Zea mays[d]	2,300 Mb			
Methyl-filtered	260 Mb[e]	183.2 Mb	0.7×[e]	102.8 Mb
Cot-enriched	284 Mb[e]	179.7 Mb	0.6×[e]	120.7 Mb
MeF+Cot	413 Mb[e]	362.9 Mb	0.9×[e]	202.7 Mb

[a] http://portal.tmri.org/rice/RiceDescription.html. (Goff et al. 2002)
[b] http://www.genomics.org.cn/bgi/english2/index.htm. (Yu et al. 2002)
[c] http://genome.jgi-psf.org/poplar0/poplar0.info.html
[d] http://www.tigr.org/tdb/tgi/maize/. (Whitelaw et al. 2003; Palmer et al. 2003)
[e] Filtered (effective) genome size (Whitelaw et al. 2003)

only those repeats within an individual BAC and are therefore of lesser magnitude.

3. In certain cases, it may present a more cost-effective way for targeting the region of the genome to be sequenced. For example, the *Medicago truncatula* genome has been shown to be organized into two more or less topographically distinct regions – heterochromatic, pericentromeric gene-poor regions and gene-rich euchromatic chromosome arms (Kulikova et al. 2001). By focusing the sequencing on the euchromatic arms on a BAC-by-BAC basis, the Medicago Genome Sequencing Consortium projects that it will be able to capture ~90% of the gene content by sequencing ~50% of the ~500-Mb genome.

14.5 Chromosome-based Sequencing

Several whole-genome projects have used a chromosome-based strategy to try to simplify an assembly process (e. g., *Plasmodium falciparum*) (Gardner et al. 1998). In this technique, the organisms' chromosomes are first fractionated either by pulse-field gel electrophoresis or flow cytometry, and whole chromosome libraries are made, sequenced, and assembled in much the same way as for whole-genome sequencing projects. In principle, this method has the advantage of simplifying the assembly process both by limiting the amount of sequence to be assembled and by restricting the repeated sequences to those present on a particular chromosome. Its limitations are that chromosome fractionation is not very efficient, resulting in the presence of "contaminants" from other chromosomes, and that the amounts of DNA produced by such fractionation techniques may be limiting.

14.6 Whole-genome Shotgun Sequencing

Whole-genome shotgun (WGS) sequencing has several advantages over a BAC-by-BAC approach, the principal of which is its simplicity. Only a few libraries with different insert sizes are required to cover the whole genome, whereas for a BAC-based project, one or two libraries are required for each BAC, and a typical genome project may require several thousand BACs (Table 14.1). In addition, the intrinsic redundancy involved in a BAC-based project, where 10–20% overlaps are sequenced, is avoided. Whole-genome projects can also be distributed over multiple centers, but resources from all centers must be pooled for assembly. Assembly can be carried out by more than one center and provides a valuable check on the robustness of these assemblies, but as larger genomes are approached, the availability of both software and hardware capable of carrying out such assemblies is restricted to fewer and fewer centers.

While whole-genome sequencing and assembly may produce very large assemblies, ordering and orienting these assemblies relies heavily on good underlying physical and genetic maps with embedded sequences such as sequenced genetic markers and BAC ends. The "finishing" of whole-genome sequencing projects, which requires directed closing of gaps and low-quality sequence, is necessarily carried out on a regional basis. Thus, even whole-genome projects where the shotgun phase of sequencing may be shared by several centers in a necessarily random fashion will transition to regional responsibilities (e. g., chromosome-by-chromosome) once the shotgun phase and initial assembly are complete.

14.7 Selective Genome Sequencing by Differential Cloning Strategies

Whole-genome sequencing projects may also use differential cloning strategies to focus the sequencing efforts on specific regions of the genome. Most common are methyl filtration (Rabinowicz et al. 1999) and Cot-based strategies (Peterson et al. 2002). These are described elsewhere in this volume.

14.8 Low-coverage Sequencing

Sequencing genomes to completion, such as was done for *Arabidopsis*, is costly and labor-intensive in the closure (finishing) stage. When the main goal is to discover gene content rather than the precise order and orientation of every gene and the sequences of intergenic regions, low-coverage sequencing can be very effective.

The Lander-Waterman equation (Lander and Waterman 1988) allows the estimation of assembly size and fraction of the genome coverage for different levels of sequencing. Given a genome of length L and sequence reads of length b, the probability that a base remains unsequenced after n reads can be calculated from the Poisson distribution:

$$P(0) = e^{-bn/L} \tag{1}$$

From Eq. (1) we can obtain the total unsequenced bases (the total gap length)

$$\text{total gap length} = Le^{-bn/L} \tag{2}$$

and the total number of connected unsequenced regions (total number of gaps)

$$\text{number of gaps} = ne^{-bn/L} \tag{3}$$

The assembly size for any given level of coverage is given by

$$\text{assembly size} = L[((e^{c\sigma-1})/c) + (1-\sigma)], \tag{4}$$

where c = genome coverage (number of bases sequenced/genome size) and $\sigma = 1-T/L$ (T = number of bases required for a detectable overlap).

Note that the levels of coverage depend on read length and the amount of sequence overlap required between two sequence reads for assembly and also rely on the assumption that the libraries being sequenced are random representations of the genome of origin. Some figures for a typical 120-kb BAC and a hypothetical 1000-Mb genome are shown in Table 14.2. Regardless of genome size, a 3× sequence coverage of either a BAC or whole genome will capture ~95% of the sequence content of whatever is being sequenced. Reconstructions using random sampling of subsets of sequence reads from finished microbial genomes at TIGR show that the Lander-Waterman equation is generally quite accurate.

Table 14.2a. Expectations from sequencing a 120-kb BAC to various levels of sequence coverage.

Sequence reads (n)	*Percentage of BAC unsequenced*	*Base pairs unsequenced*	*No. of double-strand gaps*	*Average gap length (bp)*	*Avg. contig size (bp)*	*Fold coverage*
171	36.79	44,146	67	662	1122	1×
514	4.98	5974	31	194	3599	3×
857	0.67	809	7	112	14,448	5×
1,371	0.03	40	0	0	147,260*	8×

*In theory, the BAC would be closed, in a single contig. In practice, usually a few gaps remain.

Table 14.2b. Expectations from sequencing a 1000-Mb genome to various levels of sequence coverage.

Sequence reads (n)	*Percentage of genome unsequenced*	*Base pairs unsequenced*	*No. of double-strand gaps*	*Average gap length (bp)*	*Avg. contig size*	*Fold coverage*
1,428,571	36.79	367,879,441	564,453	652	1122	1×
4,285,714	4.98	49,787,068	264,363	188	3599	3×
7,142,857	0.67	6,737,947	68,786	98	14,448	5×
11,428,571	0.03	335,463	6788	49	147,260	8×

Low-coverage sequencing has advantages over EST sequencing in that it surveys all potential coding regions, whereas EST projects typically tap into only ~60% of a genome. It will also provide information on promoters, repeat content, etc., but for large genomes and especially for those with high repeat content, it is not cost-effective. When reference genomes are available to provide information on order and orientation of small assemblies, much valuable information can be obtained from low-coverage sequencing (e.g., the 1.5× coverage of the dog genome; Kirkness et al. 2003). However, dog and human are quite closely related, and the extent to which anchor genomes (e.g., *Arabidopsis* as a reference for the Brassicaceae, rice as a reference for the Poales, and *Medicago* and *Lotus* as references for the Fabaceae) are useful for ordering and orienting small assemblies from other plant species remains to be explored.

14.9 BAC End Sequencing

BAC end sequencing, an important component of physical map development, should also be considered as another form of low-coverage sequencing. BAC end sequencing is a little more demanding than normal plasmid-based sequencing because the combination of the low copy vector and large insert leads to low yields of template and therefore additional labor and reagent demands. Thus, BAC end sequencing can be two to five times more costly than sequencing high-copy plasmids or prepared PCR products. However, paired end sequences of BACs form an important part of scaffolding whole-genome shotgun programs as well as in BAC-based genome projects. In addition, the sequence itself provides a more or less random survey of the information content (genes, transposons, repeats) of a new genome.

14.10 The Sequencing Process Itself

14.10.1 Library Production

Whether sequencing BACs, sub-genomic fractions, or whole genomes, shotgun libraries are almost always constructed from size-fractionated, randomly sheared template DNA. Random shearing, accomplished by either nebulization or more sophisticated methods (e.g., Hydroshear, Gene Machines, San Carlos, CA), minimizes bias in the DNA presented for cloning, while knowledge of the insert size range for each library is essential for later steps in the post-assembly process known as scaffolding. Libraries are typically produced in two or more size ranges (e.g., 2–3 kb and 10–12 kb) in either a high or medium copy number vector. Low copy number vectors (F-origin of replication) can also be used if insert stability due to recombination is an issue, but template preparation is then more demanding. Cloning strategies should maximize the number of insert-containing clones. At TIGR this is accomplished by using BstXI linkers that cannot self-ligate. Cloning strategies that require transcription through the insert (e. g., blue-white selection) are more likely to suffer from problems of toxic sequences, the absence of which from a collection of sequences can compromise the subsequent assembly process. Vector systems that minimize insert toxicity by flanking the insert with transcriptional stop signals have been developed (e. g., Lucigen, Middleton, WI) and appear to have some merit, but they have not been widely used to date. To provide long-range linking information, jumping libraries (see Collins and Weissman 1984 for the principle involved) that capture the two ends of large size-selected genomic fragments are more amenable to high-throughput sequencing since their overall insert size does not compromise template production; this type of library is likely to see increased use for whole-genome projects.

14.10.2 Template Production

High-throughput sequencing requires high-throughput and cost-effective template production. Over the last few years, alkaline lysis–based methods carried out robotically have evolved as the most robust and cost-effective. Cell growth and template production are now generally carried out in 384-well format (250-μL cultures), with the resulting template isolation producing enough material for up to about five reactions. For BAC templates for end sequencing, larger culture volumes (1200 μL) and an oxygen atmosphere are often used, and the yield of template (moles) is lower. Other methods for template production, such as Templiphi (Reagin et al. 2003; Amersham Biosciences, Piscataway, NJ), which is an amplification process based on Φ29 DNA polymerase, are also very effective but more costly.

PCR products can be sequenced directly by first treating samples with a combination of exonuclease and shrimp alkaline phosphatase to remove unused nucleotides and primers. Following heat inactivation of these enzymes, the appropriate primer is added and the standard cycle sequencing reaction is performed. For high-throughput analysis of PCR products from diverse regions of the genome (e. g., SNP analysis), it may be desirable to include a common primer sequence at the end of each SNP-specific primer, thereby allowing all samples to be sequenced using the same primer(s) and eliminating possible mix-ups between SNP and sequencing primer.

14.10.3 Sequencing Reactions and Analysis

Because Big Dye reagents for sequencing are quite costly, recent trends have been towards progressively smaller reaction volumes. Currently, large-scale centers employ reaction volumes 1/20 to 1/32 of the nominal volumes proposed by the manufacturer, with corresponding reductions in cost. Final volume of the entire reaction is ~10 µL. For high-throughput centers, all sequencing reactions are analyzed by capillary sequencing equipment, which permits automatic loading, relatively short run times, and accurate sample tracking. The most common tools currently in use are the ABI 3700, ABI 3730 (Applied Biosystems, Foster City, CA), and MegaBACE (Amersham Biosciences, Piscataway NJ) machines. These machines can execute up to 12 runs of 96 samples per day for up to 72 h without operator intervention.

One of the challenges faced by these high-throughput operations is striking the right balance between keeping up with the latest technology, which consistently offers such advantages as higher throughput and longer read lengths, and getting the full value out of capital investments in obsolescing equipment.

14.10.4 Sequencing Capacity and Costs

The National Human Genome Research Institute (NHGRI) expects its designated large-scale sequencing centers to be able to carry out 20 million attempted reads per year at or below a specified cost (at time of writing, $1.50 per read) (see http://grants1.nih.gov/grants/guide/rfa-files/RFA-HG-03-002.html for details). Only a few centers meet these criteria. Following NHGRI's example, other funding agencies in the United States now also impose upper limits on the permitted cost of sequencing reactions. The imposition of these cost limits (at the time of writing, $80,000 per finished megabase) has effectively identified centers that are efficient enough to meet these requirements and thus are able to conduct moderate- to large-scale sequencing projects.

14.10.5 Post-sequencing Data Processing

14.10.5.1 Quality Trimming

Raw sequencing reads are processed by base-calling and trimming software that defines "high-quality" sequence and removes any residual vector. Examples of such software include phred/cross_match (Ewing and Green 1998; http://www.phrap.org/phrap.docs/phred.html) and Lucy (Chou and Holmes 2001). Although details of these processes differ among institutions, the net effect is the same, generating vector-free, high-quality sequence reads that are ready for sequence assembly. The phred quality value (*QV*) is related to the base call error probability by the formula

$$QV = -10 \log_{10}(P_e)$$

where P_e is the probability that the base call is an error. Most sequencing projects use a phred value of 20 as their cutoff for high-quality sequence, which corresponds to a 1 in 100 chance that the base call is incorrect.

14.10.5.2 **Sequence Assembly**

Sequence assembly is carried out by assembly software that assembles sequence reads that fulfill certain criteria of percent identity, minimum overlap, etc. The precise parameters used depend on both the assembler program itself and user-specified values. Typical requirements for assembly would be >97 % identity with at least a 40-nucleotide overlap and not more than 10 bases overhang mismatch at the ends. Each base in an assembly is assigned a composite quality score that reflects the relative accuracy of that base position. Clone mate information (i. e., the necessary relationship between sequence reads that come from the opposite ends of the same clone) is not used in the assembly process, which is strictly sequence-based. Quite a number of sequence assemblers exist both in the public domain and as commercial packages. However, most assemblers (e. g., phrap - www.phrap.org; TIGR assembler – Sutton et al. 1996) are limited in the size of genome they can handle and are best suited to BACs and prokaryotic genomes. To extend the whole-genome shotgun strategy to large eukaryotic genomes that present problems of both size and repeat content, new generation whole-genome assemblers have been developed, first by Celera (Huson et al. 2001) and then at the Whitehead Institute (Batzoglou et al. 2002; Jaffe et al. 2003). Both these programs embody new assembly strategies and algorithms. Several other assembly packages developed and now in use for large whole-genome assembly processes (e. g., Atlas developed by Baylor College of Medicine, Phusion developed by the Sanger Center, and Jazz developed by the Joint Genomes Institute) are hybrids that provide large-scale sequence management but continue to use phrap as the underlying assembly engine. Running any of these assemblers of large genomes requires significant computer power (16–32 Gb of RAM) and highly parallel processing (20 cpu or more), especially for the initial pairwise sequence comparison phase of the assembly process. Thus, as for sequencing, truly large-scale whole-genome projects are limited to a small number of centers with the appropriate computational resources.

14.10.5.3 **Scaffolding**

Scaffolding is the process by which sequence assemblies are linked together based upon clone-mate and insert-size information to develop the most consistent ordering and orientation information for related assemblies; thus, it provides a long-range representation of the genome. Most assembly packages include a scaffolding module, but users have little control over the algorithm used or its output. Bambus, developed at TIGR (Pop et al. 2004), is a general-purpose scaffolder that affords users significant flexibility in controlling the scaffolding parameters. A major advantage of Bambus is that it is able to make use of linking data other than that inferred from clone-mate information. When appropriate to the goals of the project, such scaffolded assemblies provide the starting point for directed closure or finishing reactions designed to close sequencing gaps between assemblies that are known to be physically linked. In addition, sequences embedded within assemblies such as genetic markers and BAC ends can provide links back to physical and genetic maps that may further assist in the orientation, ordering, and association between contigs. With the increasing number of relatively well-finished reference genomes, postulated conservation of synteny across species represents another potential avenue for ordering and orienting assemblies and contigs, at least as a first approximation for directing closure reactions. Bambus is able to use the output from the genome alignment program MUMmer (Delcher et al. 2002) to construct scaffolds of low-coverage genomes (Pop et al. 2004).

14.10.5.4 **Sequence Editing and Gap Closure**

If the goal of the sequencing is to produce high-quality finished sequence over an entire region, some post-assembly work is required. Sequence editing requires inspection of the assembly of chromatogram traces to resolve discrepancies or ambiguities that compromise the assembly or the joining of assemblies. This is generally a manual process and can be time-consuming, but it is particularly important in resolving regions of highly similar repeats where the sequence differences between repeats are of the same order as the intrinsic error rate of the sequencing process (1–2%). To accelerate this process, TIGR has developed an Autoeditor program (Gajer et al. 2004) to accomplish the same goals using a rationale similar to the human process. In test runs using raw sequence reads from finished genomes where sequences had subsequently been manually edited for the final assembly, Autoeditor reduced the number of erroneous base calls by 80%, with an error rate of one mistake per 8828 corrections

After initial assembly, the complete sequence is obtained by generating sequence for the gaps between contigs. Using software such as Autocloser (TIGR) or Autofinish (Gordon et al. 2001), clones are identified where resequencing will extend a short or failed read into a gap, and custom primers are designed to walk off clones into gaps. Larger clones spanning gaps are often sequenced using a transposon-based technology (e.g., GPS, New England Biolabs, Beverly, MA). Difficult-to-sequence regions may be approached with specialized kits with altered nucleotide composition and/or reagents like DMSO, betaine, or their proprietary equivalents, which can aid in sequencing through secondary structure. Microlibraries where a 2–3-kb clone is broken up into 100–500-bp fragments may also facilitate the acquisition of sequence for intransigent regions.

14.11 Conclusions

As illustrated in Table 14.1, the landscape of large-scale sequencing is changing and will undoubtedly continue to change. As high-quality reference genomes are established, low-coverage sequencing of related genomes will become increasingly informative. Incorporation of selection techniques such as methyl filtration and Cot enrichment will allow more cost-effective sequencing of larger genomes. BAC-based sequencing will continue to be useful for focusing on specific regions of the genome.

The truly large-scale centers achieve their goals of high throughput and low cost through a combination of robotic sample preparation, automated capillary sequencing, reductions in reaction volumes, and economies of scale in purchasing, etc., all tightly coupled with a robust laboratory management information system. Sequencing costs have decreased dramatically in the last five years due not only to the above reasons but also to significantly longer read lengths. However, further reductions will be relatively small until there is a paradigm shift in the technology (e. g., either a reduction in scale from microliter to nanoliter reaction volumes or some distinctly novel technology). From an operational point of view, large-scale centers prefer very large projects. It is logistically much easier to manage one project that requires a million reads from a few libraries than to deal with 100 projects that require 10,000 reads each. Given the diversity of plant species being proposed for some level of sequencing, it is likely that the scale of many projects in plant genomics will be in this intermediate range of size and can also be carried out at smaller but comparably efficient centers.

Acknowledgments

I thank all my colleagues at TIGR for their advice, encouragement, and support.

References

(AGI) The *Arabidopsis* Genome Initiative (2000). Analysis of the genome sequence of the flowering plant *Arabidopsis thaliana.* Nature 408, 796–815.

Asamizu, E., Kato, K., Sato, S., Nakamura, Y., Kaneko, T., and Tabata, S. (2003). Structural analysis of a *Lotus japonicus* genome. IV. Sequence features and mapping of seventy-three TAC clones which cover the 7.5 Mb regions of the genome. DNA Research 10, 115–122.

Barry, G. F. (2001). The Use of the Monsanto Draft Rice Genome Sequence in research. Plant Physiol. 125:1164–1165.

Batzoglou S, Jaffe DB, Stanley K, Butler J, Gnerre S, Mauceli E, Berger B, Mesirov JP, Lander ES. (2002). ARACHNE: A whole-genome shotgun assembler". Genome Research, 12:177–189.

Chou, H. H. and Holmes, M. H. (2001). DNA sequence quality trimming and vector removal. Bioinformatics 17:1093–104.

Collins, F. S. and Weissman, S. M. (1984). Directional cloning of DNA fragments at a large distance from an initial probe: a circularization strategy. Proc. Natl Acad Sci USA 81: 6812–6816.

Delcher, A. L., Phillippy, A., Carlton, J. and Salzberg, S. L. (2002). Fast algorithms for large-scale genome alignment and comparison. *Nucleic Acids Res* 30:2478–83.

Ewing, B. and Green, P. (1998). Basecalling of automated sequencer traces using phred. II. Error probabilities. Genome Res. 8, 186–194.

Feng Q, Zhang Y, Hao P, Wang S, Fu G, Huang Y, Li Y, Zhu J, Liu Y, Hu X, Jia P, Zhang Y, Zhao Q, Ying K, Yu S, Tang Y, Weng Q, Zhang L, Lu Y, Mu J, Lu Y, Zhang LS, Yu Z, Fan D, Liu X, Lu T, Li C, Wu Y, Sun T, Lei H, Li T, Hu H, Guan J, Wu M, Zhang R, Zhou B, Chen Z, Chen L, Jin Z, Wang R, Yin H, Cai Z, Ren S, Lv G, Gu W, Zhu G, Tu Y, Jia J, Zhang Y, Chen J, Kang H, Chen X, Shao C, Sun Y, Hu Q, Zhang X, Zhang W, Wang L, Ding C, Sheng H, Gu J, Chen S, Ni

L, Zhu F, Chen W, Lan L, Lai Y, Cheng Z, Gu M, Jiang J, Li J, Hong G, Xue Y, Han B. (2002). Sequence and analysis of rice chromosome 4. Nature. 420:316–20.

Fleischmann, R. D., Adams, M. D., White, O., Clayton, R. A., Kirkness, E. F., Kerlavage, A. R., Bult, C. J., Tomb, J. F., Dougherty, B. A., Merrick, J. M., McKenney, K., Sutton, G., FitzHugh, W., Fields, C., Gocayne, J. D., Scott, J., Shirley, R., Liu, L., Glodek, A., Kelley, J. M., Weidman, J. F., Phillipps, C. A., Spriggs, T., Hedblom, E., Cotton, M. D., Utterback, T. R., Hanna, M. C., Nguyen, D. T., Saudek, D. M., Brandon, R. C., Fine, L. D., Fritchman, J. L., Fuhrmann, J. L., Geoghagen, N. S.M., Gnehm, C. L., McDonald, L. A., Small, K. V., Fraser, C. M., Smith, H. O. and Venter, J. C. (1995). Whole-genome random sequencing and assembly of *Haemophilus influenzae* Rd. Science 269:496–512.

Gajer, P., Schatz, M. and Salzberg, S. L. (2004). Automated correction of genome sequence errors. *Nucleic Acids Research* (in press).

Gardner, M. J., Tettelin, H., Carucci, D. J., Cummings, L. M., Aravind, L., Koonin, E. V., Shallom, S., Mason, T., Yu, K., Fujii, C., Pederson, J., Shen, K., Jing, J., Aston, C., Lai, Z., Schwartz, D. C., Pertea, M., Salzberg, S., Zhou, L., Sutton, G. G., Clayton, R., White, O., Smith, H. O., Fraser, C. M., Adams, M. D., Venter, J. C. and Hoffman, S. L. (1998). Chromosome 2 sequence of the human malaria parasite *Plasmodium falciparum*. Science 282:1126–32.

Goff SA, Ricke D, Lan TH, Presting G, Wang R, Dunn M, Glazebrook J, Sessions A, Oeller P, Varma H, Hadley D, Hutchison D, Martin C, Katagiri F, Lange BM, Moughamer T, Xia Y, Budworth P, Zhong J, Miguel T, Paszkowski U, Zhang S, Colbert M, Sun WL, Chen L, Cooper B, Park S, Wood TC, Mao L, Quail P, Wing R, Dean R, Yu Y, Zharkikh A, Shen R, Sahasrabudhe S, Thomas A, Cannings R, Gutin A, Pruss D, Reid J, Tavtigian S, Mitchell J, Eldredge G, Scholl T, Miller RM, Bhatnagar S, Adey N, Rubano T, Tusneem N, Robinson R, Feldhaus J, Macalma T, Oliphant A, Briggs S. (2002). A draft sequence of the rice genome (*Oryza sativa* L. ssp. *japonica*). Science. 296:92–100.

Gordon D, Desmarais C, Green P. (2001). Automated finishing with autofinish. Genome Res. 11:614-25.

Huson DH, Reinert K, Kravitz SA, Remington KA, Delcher AL, Dew IM, Flanigan M, Halpern AL, Lai Z, Mobarry CM, Sutton GG, Myers EW. (2001). Design of a compartmentalized shotgun assembler for the human genome. Bioinformatics 17 Suppl 1:S132–9.

Jaffe, D. B., Butler, J., Gnerre, S., Mauceli, E., Lindblad-Toh, K., Mesirov, J. P., Zody, M. C., and Lander, E. S. (2003). Whole-genome sequence assembly for mammalian genomes: ARACHNE 2. Genome Research 13:91–96.

Kirkness, E. F., Bafna, V., Halpern, A. L., Levy, S., Remington, K., Rusch, D. B., Delcher, A. L., Pop, M., Wang, W., Fraser, C. M. and Venter, J. C. (2003). The dog genome: survey sequencing and comparative analysis. *Science* 301:1898–903.

Kuhl, J. C. Cheung, F., Yuan, Q., Martin, W., Zewdie, Y., McCallum, J., Catanach, A., Paul Rutherford, P., Sink, K. C., Jenderek, M., Prince, J. D. Town, C.D, and Havey, M. J. (2004). Unique set of 11,008 onion (*Allium cepa*) ESTs reveals expressed sequence and genomic differences between monocot orders Asparagales and Poales. Plant Cell (in press).

Kulikova, O., Gualtieri, G., Geurts, R., Kim, D. J., Cook, D., Huguet, T., de Jong, J. H., Fransz, P. F., Bisseling, T.(2001). Integration of the FISH pachytene and genetic maps of *Medicago truncatula*. Plant J. 27:49–58.

Lander, E. S. and Waterman, M. S. (1988). Genomic mapping by fingerprinting random clones: a mathematical analysis. Genomics 2: 231–239.

Myers EW, Sutton GG, Delcher AL, Dew IM, Fasulo DP, Flanigan MJ, Kravitz SA, Mobarry CM, Reinert KH, Remington KA, Anson EL, Bolanos RA, Chou HH, Jordan CM, Halpern AL, Lonardi S, Beasley EM, Brandon RC, Chen L, Dunn PJ, Lai Z, Liang Y, Nusskern DR, Zhan M, Zhang Q, Zheng X, Rubin GM, Adams MD, Venter JC. (2000). A whole-genome assembly of *Drosophila*. Science 287:2196–204.

Palmer, L. E., Rabinowicz, P. D., O'Shaughnessy, A. L., Balija, V. S., Nascimento, L. U., Dike, S., de la Bastide, M., Martienssen, R. A. and McCombie, W. R.(2003) Maize genome sequencing by methylation filtration. *Science* 302:2115–2117.

Peterson DG, Schulze SR, Sciara EB, Lee SA, Bowers JE, Nagel A, Jiang N, Tibbitts DC, Wessler SR, Paterson AH. (2002). Integration of Cot analysis, DNA cloning, and high-throughput sequencing facilitates genome characterization and gene discovery. Genome Res. 12:795–807.

Pop, M., Kosack, D., and Salzberg, S. (2004) Hierarchical scaffolding with Bambus. Genome Research (in press).

Rabinowicz PD, Schutz K, Dedhia N, Yordan C, Parnell LD, Stein L, McCombie WR, Martienssen RA. (1999). Differential methylation of genes and retrotransposons facilitates shotgun sequencing of the maize genome. Nat Genet. 23:305–8.

Reagin, M. J., Giesler, T. L., Merla, A. L., Resetar-Gerke, J. M., Kapolka, K. M. and Mamone, A. J. (2003). TempliPhi: A sequencing template preparation procedure that eliminates overnight cultures and DNA purification. J Biomol. Tech. 14, 143–148.

The Rice Chromosome 10 Sequencing Consortium (2003). In-depth view of structure, activity, and evolution of rice chromosome 10. *Science* 300:1566–1569.

Sasaki T, Matsumoto T, Yamamoto K, Sakata K, Baba T, Katayose Y, Wu J, Niimura Y, Cheng Z, Nagamura Y, Antonio BA, Kanamori H, Hosokawa S, Masukawa M, Arikawa K, Chiden Y, Hayashi M, Okamoto M, Ando T, Aoki H, Arita K, Hamada M, Harada C, Hijishita S, Honda M, Ichikawa Y, Idonuma A, Iijima M, Ikeda M, Ikeno M, Ito S, Ito T, Ito Y, Ito Y, Iwabuchi A, Kamiya K, Karasawa W, Katagiri S, Kikuta A, Kobayashi N, Kono I, Machita K, Maehara T, Mizuno H, Mizubayashi T, Mukai Y, Nagasaki H, Nakashima M, Nakama Y, Nakamichi Y, Nakamura M, Namiki N, Negishi M, Ohta I, Ono N, Saji S, Sakai K, Shibata M, Shimokawa T, Shomura A, Song J, Takazaki Y, Terasawa K, Tsuji K, Waki K, Yamagata H, Yamane H, Yoshiki S, Yoshihara R, Yukawa K, Zhong H, Iwama H, Endo T, Ito H, Hahn JH, Kim HI, Eun MY, Yano M, Jiang J, Gojobori T. (2002). The genome sequence and structure of rice chromosome 1. Nature 420:312–6.

Sutton, G. G., White, O., Adams, M. D. and Kerlavage, A. R. (1996). TIGR Assembler: A new tool for assembling large shotgun sequencing projects. Genome Science and Technology 1:9–19.

Venter, J. C., Smith, H., and Hood, L. (1996). A new strategy for genome sequencing. Nature 381:364–366.

Venter JC, Adams MD, Myers EW, Li PW, Mural RJ, Sutton GG, Smith HO, Yandell M, Evans CA, Holt RA, Gocayne JD, Amanatides P, Ballew RM, Huson DH, Wortman JR, Zhang Q, Kodira CD, Zheng XH, Chen L, Skupski M, Subramanian G, Thomas PD, Zhang J, Gabor Miklos GL, Nelson C, Broder S, Clark AG, Nadeau J, McKusick VA, Zinder N, Levine AJ, Roberts RJ, Simon M, Slayman C, Hunkapiller M, Bolanos R, Delcher A, Dew I, Fasulo D, Flanigan M, Florea L, Halpern A, Hannenhalli S, Kravitz S, Levy S, Mobarry C, Reinert K, Remington K, Abu-Threideh J, Beasley E, Biddick K, Bonazzi V, Brandon R, Cargill M, Chandramouliswaran I, Charlab R, Chaturvedi K, Deng Z, Di Francesco V, Dunn P, Eilbeck K, Evangelista C, Gabrielian AE, Gan W, Ge W, Gong F, Gu Z, Guan P, Heiman TJ, Higgins ME, Ji RR, Ke Z, Ketchum KA, Lai Z, Lei Y, Li Z, Li J, Liang Y, Lin X, Lu F, Merkulov GV, Milshina N, Moore HM, Naik AK, Narayan VA, Neelam B, Nusskern D, Rusch DB, Salzberg S, Shao W, Shue B, Sun J, Wang Z, Wang A, Wang X, Wang J, Wei M, Wides R, Xiao C, Yan C, Yao A, Ye J, Zhan M, Zhang W, Zhang H, Zhao Q, Zheng L, Zhong F, Zhong W, Zhu S, Zhao S, Gilbert D, Baumhueter S, Spier G, Carter C, Cravchik A, Woodage T, Ali F, An H, Awe A, Baldwin D, Baden H, Barnstead M, Barrow I, Beeson K, Busam D, Carver A, Center A, Cheng ML, Curry L, Danaher S, Davenport L, Desilets R, Dietz S, Dodson K, Doup L, Ferriera S, Garg N, Gluecksmann A, Hart B, Haynes J, Haynes C, Heiner C, Hladun S, Hostin D, Houck J, Howland T, Ibegwam C, Johnson J, Kalush F, Kline L, Koduru S, Love A, Mann F, May D, McCawley S, McIntosh T, McMullen I, Moy M, Moy L, Murphy B, Nelson K, Pfannkoch C, Pratts E, Puri V, Qureshi H, Reardon M, Rodriguez R, Rogers YH, Romblad D, Ruhfel B, Scott R, Sitter C, Smallwood M, Stewart E, Strong R, Suh E, Thomas R, Tint NN, Tse S, Vech C, Wang G, Wetter J, Williams S, Williams M, Windsor S, Winn-Deen E, Wolfe K, Zaveri J, Zaveri K, Abril JF, Guigo R, Campbell MJ, Sjolander KV, Karlak B, Kejariwal A, Mi H, Lazareva B, Hatton T, Narechania A, Diemer K, Muruganujan A, Guo N, Sato S, Bafna V, Istrail S, Lippert R, Schwartz R, Walenz B, Yooseph S, Allen D, Basu A, Baxendale J, Blick L, Caminha M, Carnes-Stine J, Caulk P, Chiang YH, Coyne M, Dahlke C, Mays A, Dombroski M, Donnelly M, Ely D, Esparham S, Fosler C, Gire H, Glanowski S,

Glasser K, Glodek A, Gorokhov M, Graham K, Gropman B, Harris M, Heil J, Henderson S, Hoover J, Jennings D, Jordan C, Jordan J, Kasha J, Kagan L, Kraft C, Levitsky A, Lewis M, Liu X, Lopez J, Ma D, Majoros W, McDaniel J, Murphy S, Newman M, Nguyen T, Nguyen N, Nodell M, Pan S, Peck J, Peterson M, Rowe W, Sanders R, Scott J, Simpson M, Smith T, Sprague A, Stockwell T, Turner R, Venter E, Wang M, Wen M, Wu D, Wu M, Xia A, Zandieh A, Zhu X. (2001). The sequence of the human genome. Science 291:1304–51.

Whitelaw, C. A., Barbazuk, W. B., Pertea, G., A. P. Chan, A. P., Cheung, Y. Lee, Y., Zheng, L., van Heeringen, S., Karamycheva, S., Bennetzen, J. L., SanMiguel, P., Lakey, N., Bedell, J., Yuan, Y., Budiman, M. A., Resnick, A., Van Aken, S., Utterback, T., Riedmuller, S., Williams, M., Feldblyum, T., Schubert, K., Beachy, R., Fraser, C. M. and Quackenbush, J. (2003). Enrichment of gene-coding sequences in maize by genome filtration. Science 302: 2118–2120.

Yu J, Hu S, Wang J, Wong GK, Li S, Liu B, Deng Y, Dai L, Zhou Y, Zhang X, Cao M, Liu J, Sun J, Tang J, Chen Y, Huang X, Lin W, Ye C, Tong W, Cong L, Geng J, Han Y, Li L, Li W, Hu G, Huang X, Li W, Li J, Liu Z, Li L, Liu J, Qi Q, Liu J, Li L, Li T, Wang X, Lu H, Wu T, Zhu M, Ni P, Han H, Dong W, Ren X, Feng X, Cui P, Li X, Wang H, Xu X, Zhai W, Xu Z, Zhang J, He S, Zhang J, Xu J, Zhang K, Zheng X, Dong J, Zeng W, Tao L, Ye J, Tan J, Ren X, Chen X, He J, Liu D, Tian W, Tian C, Xia H, Bao Q, Li G, Gao H, Cao T, Wang J, Zhao W, Li P, Chen W, Wang X, Zhang Y, Hu J, Wang J, Liu S, Yang J, Zhang G, Xiong Y, Li Z, Mao L, Zhou C, Zhu Z, Chen R, Hao B, Zheng W, Chen S, Guo W, Li G, Liu S, Tao M, Wang J, Zhu L, Yuan L, Yang H. (2002). A draft sequence of the rice genome (*Oryza sativa* L. ssp. *indica*). Science 296:79–92.

Glossary

This glossary lists a number of relevant terms for the reader who may not be familiar with the topics of the present volume

A

Accession number: A unique identification code for each sequence deposited in the databanks (e. g., GenBank). This number can be used to search, e. g., GenBank records for a specific sequence.
Anonymous single nucleotide polymorphism (anonymous SNP): Any one of the most frequently occurring single nucleotide polymorphisms that have no known effect on the function of a gene.

B

BAC clone: Any bacterial artificial chromosome vector that carries a genomic DNA insert, typically 100–150 kb and maximally 300 kb in length. Thousands of BAC clones make up a BAC library, which covers the target genome several-fold (5–10×), such that each genomic sequence is represented at least once.
BAC DNA microarray (BAC microarray): The ordered alignment of different bacterial artificial chromosome (BAC) clones, immobilized on supports of minute dimensions (e. g., nylon membranes, silicon, glass, or quartz chips). Each colony harbors DNA fragments of 100–150 kb. Such microarrays are used to isolate genomic DNA that contains a gene or genes of interest, detected by hybridization of radiolabeled or fluorescent gene probes to the microarray.
BAC end pair: The sequence reads from both ends of a bacterial artificial chromosome clone.
BAC end sequencing: The estimation of the sequence of one or both ends of a bacterial artificial chromosome clone. BAC end sequencing can actually be used as a first step to study the sequence of whole genomes, starting from a few completely

The Handbook of Plant Genome Mapping.
Genetic and Physical Mapping. Edited by Khalid Meksem, Günter Kahl

ISBN 3-527-31116-5

sequenced BAC clones ("seeds" or "nucleation clones") that are extended into overlapping BAC clones selected from a set of end-sequenced BACs.

BAC fingerprinting: A variant of the restriction fragment length polymorphism technique, which allows discrimination between different clones of a bacterial artificial chromosome (BAC) library by digestion of the clones with a set of restriction endonucleases.

BAC library: A collection of genomic DNA fragments, typically 100–150 kb and maximally 300 kb in length, that are cloned into a bacterial artificial chromosome vector (e. g., pBelo BAC11).

BAC map (genome fingerprint map): The ordered alignment of bacterial artificial chromosome clones such that a physical map of the genome is constructed.

Bacterial artificial chromosome (BAC): A bacterial cloning vector, based on a single copy F-factor of *E. coli*, that allows cloning of DNA fragments of up to 300 kb (average size: 150 kb). The cloned DNA is structurally stable in the host (does not rearrange) because the low copy number of the F plasmid does not favor recombination of inserts.

Balanced polymorphism: Any stable genetic polymorphism that is maintained in a population by natural selection.

Base pair map: A physical map of a genome, or parts of it, for which the sequence is known base by base.

C

Candidate single nucleotide polymorphism (candidate SNP): Any single nucleotide polymorphism in an exon of a gene that can be expected to have an impact on the function of the encoded protein.

Causative single nucleotide polymorphism (causative SNP): Any single nucleotide polymorphism that is in linkage disequilibrium to a disease phenotype and therefore a responsible candidate for the disease.

Centromere mapping: The localization of the centromere on individual chromosomes by, e. g., fluorescent in situ hybridization.

Chromosome microdissection: The removal of relatively large sections (e. g., half a chromosome arm) from an isolated chromosome using atomic force microscopy or laser microdissection. The microdissected pieces still harbor at least 10^6 base pairs and are used for the establishment of subgenomic libraries in bacterial artificial chromosomes or map-based cloning approaches.

Clone-based map: Any physical map of a genome that is based on the alignment and sequencing of overlapping bacterial artificial chromosome clones.

Clone coverage: The extent to which a genome is represented in clones of, e. g., a BAC library or plasmid library. Clone coverage is a measure of the amount of physical DNA coverage of the target genome.

Clone overlap single nucleotide polymorphism (clone overlap SNP): Any single nucleotide polymorphism detected by sequencing overlaps of two (or more) bacterial artificial chromosomes and comparison of the resulting sequences.

Coding single nucleotide polymorphism (coding SNP, copy SNP, cSNP): A misleading term for a single nucleotide polymorphism that is located within an exon of a gene.
Codominant marker: Any molecular marker that detects both alleles of a particular genomic locus. For example, restriction fragment length polymorphism (RFLP) or sequence-tagged microsatellite site markers belong to this category.
Common haplotype: Any haplotype that is characteristic for a majority of individuals in a distinct population.
Common single nucleotide polymorphism (common SNP): Any single nucleotide polymorphism whose minor allele occurs in more than 10% of the genomes of a population.
Comparative candidate positional cloning: The molecular cloning of candidate genes responsible for a specific phenotype and the detection of single nucleotide polymorphism(s) (SNPs) within these genes in different individuals. First, the linkage of an organism's phenotype to a specific chromosomal position via genetic mapping is established. Then candidate genes in the selected chromosomal region are inferred from homologous region(s) of the map of another related organism using comparative mapping with type I DNA markers (i.e., genes) as landmarks. Finally, type III DNA markers (SNPs) in or around the candidate loci are developed and tested for association with the specific phenotype.
Comparative mapping: The establishment of genetic linkage maps in species of unknown genome composition by using probes from a related species. Since there exists appreciable similarity in the arrangement of genes, groups of genes (microsynteny, or whole chromosomal regions (macrosynteny), RFLP probes from, e.g., rice can be used to map the corresponding sequences on the wheat genome, especially since they are sufficiently homologous to cross-hybridize.
Contig: (a) A set of clones (e.g., bacterial artificial chromosome clones) in a physical map that completely and contiguously cover a genomic segment of interest ("minimal tiling path"). (b) The final product of shotgun sequencing.
Contig map: A library of overlapping clones (contigs) representing a complete stretch of DNA (e.g., a BAC clone, a chromosome). A contig map is the result of contig mapping.

D

***De novo* sequencing:** The sequencing of a complete genome base by base for which neither sequence data nor any fragment libraries (e.g., bacterial artificial chromosome libraries) are yet available.
Di-allelic map (bi-allelic map): A genetic map that is based on molecular markers (e.g., single nucleotide polymorphisms) of which both parental alleles are known.
Domain walking: An *in silico* approach to detecting DNA sequences encoding flanking regions adjacent to a specific domain. The sequence of the latter represents the starting point for a search of adjacent sequences in the databases.

Dominant marker: Any molecular marker that does not discriminate between the alleles of a genetic locus.
Double-hit single nucleotide polymorphism (double-hit SNP): Any single nucleotide polymorphism for which each allele is present in two (or more) samples from a distinct population.

E

Electronic single nucleotide polymorphism (eSNP): Any single nucleotide polymorphism (SNP) derived from expressed sequence tag (EST) databases by data mining. The EST sequences from many different individuals of human or animal or plant species are aligned and the sequence is screened for SNPs. These SNPs are therefore discovered solely *in silico*.
Exonic single nucleotide polymorphism (exon SNP; expressed single nucleotide polymorphism): Any single nucleotide polymorphism that is present in an exon of a gene. Synonymous to expressed single nucleotide polymorphism.
Expression marker: Any expressed sequence, e. g., a cDNA, a tag derived from serial analysis of gene expression (SAGE), or an expressed sequence tag (EST), that has been identified by high-throughput expression profiling (e. g., massively parallel signature sequencing, any of the microarray platforms, or serial analysis of gene expression) and serves as a diagnostic (or even prognostic) marker for a disease.

F

Fiber fluorescent *in situ* hybridization (fiber-FISH): A variant of the conventional fluorescent *in situ* hybridization (FISH) technique for the detection and quantification of target sequences in which different probes labeled with different fluorochromes are hybridized to chromosome fibers. These fibers are generated by molecular combing (see front cover image).
Fingerprinted contig (FPC): A software package that allows assembly of clones into contigs by using either the end-labeled double-digest method or the complete-digest method. To determine whether two clones overlap, the number of shared bands is counted. Two bands are considered "shared" if they own the same value (within a range of tolerance). The probability that n shared bands are coincidental is computed. If the resulting score lies below a user-defined cutoff, the clones are considered to overlap. If two clones have a coincidence score below the threshold but do not overlap, they represent false positives (F^+), and if two clones have a coincidence score above the threshold but do overlap, they represent false negatives (F^-). The cutoff has to be set to minimize the number of F^+ and F^- overlaps.
Fingerprinting (DNA profiling, genetic fingerprinting): The whole repertoire of techniques necessary to create a fingerprint (DNA fingerprint, genetic fingerprint).

For example, a genetic fingerprint represents the highly specific hybridization pattern generated by restriction fragment length polymorphisms (RFLPs) of genomic DNA. Such polymorphisms are the consequence of mutations within a restriction site, leading to either the appearance or loss of restrictable recognition sequences.
Fluorescent *in situ* hybridization (FISH): A technique for the identification of specific sequences of intact chromosomes by hybridization with complementary nucleic acid probes that are covalently bound to fluorochromes. Usually, the biological material (preferably metaphase chromosomes) is squashed on a microscope slide, and the DNA is denatured and then hybridized to the fluorochrome-labeled, single-stranded probe. Fluorescence can then be directly visualized by fluorescence microscopy.
Functional map: (a) A physical map of a genome in which the locations of genes with known (i. e., experimentally proven) functions are depicted. (b) Any graphical depiction of molecules that interact with each other and whose individual functions are experimentally proven. For example, interacting proteins with known functions represent part of a functional map.
Functional polymorphism: Any polymorphism in a gene or a promoter (or non-coding regulatory sequence) that either changes the underlying codon, and hence the amino acid composition of the encoded protein, or alters the sequence of a transcription factor–binding site in a promoter such that binding and subsequent activation of the adjacent gene is prevented.

G

Gene-based single nucleotide polymorphism (gene-based SNP): Any single nucleotide polymorphism that is located in an exon, an intron, or a promoter of a gene.
Gene mapping: The estimation of the linear arrangement of genes and the determination of the relative location of specific genes on specific chromosomes or plasmids and their relative distance from one another. Gene maps may be based on classical genetic recombination analysis or on direct DNA data obtained by DNA sequencing.
Gene surfing: The identification of genic DNA sequences in an anonymous DNA by comparing it against sequences in genome and/or protein databases that already have assigned (and in some cases, proven) functions (e. g., coding sequences). Also, computational programs such as Genescan or GeneWise can predict the occurrence of genes in raw sequence data.
Genetic diversity: The sequence variation in the genomic DNA of two organisms or of two populations of organisms.
Genetic map (linkage map, recombination map): A graph depicting the linear arrangement (i. e., the relative positions) of molecular markers and/or morphological markers along so-called linkage groups, based on their frequency of crossovers or recombinations. The average number of crossover events at two compared loci during meiosis, the genetic distance, is given in centimorgans.

Genetic mapping: A technique for the determination of the linear order of molecular markers or genes (generally, loci) along a stretch of DNA (e. g., a BAC clone, a chromosome). The result of genetic mapping is the construction of a genetic map of a genome.
Genome: The entire genetic material of a virus, a cellular organelle (e. g., a mitochondrion, a chloroplast, or a nucleus), or a cell.
Genome equivalent: A statistical measure for the extent of representation of a genome in a genomic library. For example, the presence of five to six genome equivalents in a library warrants that 99 % of all sequences of the corresponding genome are represented at least once.
Genome map: Any graphical depiction of the linear order of all sequence elements (e. g., genes, promoters, or repetitive DNA) in a genome, as composed by, e. g., genome mapping.
Genomic library: A collection of recombinant DNA molecules derived from genomic (i. e., nuclear) DNA of a single organism, ideally containing all sequences of the genome.
Genomics: A term that describes the whole repertoire of technologies to study the organization of genomes and the functions of their constituents (e. g., genes).

H

Haplotype block (hapblock): Any one of the relatively large genomic regions with which defined sequences (e. g., specific genes, but also single nucleotide polymorphisms) are associated.
Haplotype: The linear arrangement of alleles along a region in DNA (e. g., a BAC clone, a restriction fragment, or a chromosome). In laboratory slang, a haplotype can also be an individual with a specific arrangement of alleles in a given piece of its DNA (e. g., a gene; also called "block" or "haplotype block"). Such haplotypes can, e. g., be defined on the basis of specific single nucleotide polymorphisms on a chromosomal segment (in diploids: on the corresponding segments of homologous chromosomes) that requires repeated sequencing of the target region. If the target sequence from different individuals is then compared, the haplotypic organization becomes apparent. Haplotype analysis is used for the establishment of genetic risk profiles and the prediction of the clinical reaction of an individual towards pharmaceutical compounds (e. g., drugs).
Haplotype block: A specific arrangement of adjacent alleles in a given region of genomic DNA (usually in the range of 10,000–100,000 bp) that is inherited as a "block," probably because its recombination frequency is lower than in other parts of the genome. In practice, a haplotype block is characterized by a series of single nucleotide polymorphisms in linkage disequilibrium.
Haplotype map: A variant of a genome map in which the haplotype blocks of the genome of an organism are depicted.
Haplotype mapping: The process of establishing a haplotype map.

Haplotype signature: Any characteristic configuration of specific alleles in an organism. Haplotype signatures can be established for a collection of organisms (e. g., human patients suffering from the same disease) and a disease allele can be identified.

Haplotype single nucleotide polymorphism (htSNP; haplotype tag SNP): Any single nucleotide polymorphism contained within a haplotype block. Not to be confused with high-throughput SNP (htSNP).

Haplotyping: The determination of the haplotype of an individual.

HAPPY mapping (*hap*loid equivalents of DNA and *p*olymerase chain reaction): An *in vitro* technique for the determination of the order and spacing of DNA markers directly on genomic DNA and the establishment of physical maps of any density, without need for parental lines, polymorphic markers, and segregating populations. In short, high-molecular-weight genomic DNA (>2 Mb) of a target organism is first isolated and randomly sheared or γ-irradiated, the resulting random fragments are size-fractionated (450–550 kb) by pulsed-field gel electrophoresis, and the sized fractions are diluted and aliquoted into 96-well microtiter plates such that each aliquot contains less than one haploid genome equivalent ("mapping panel"). Panels may differ in the sizes of the contained genomic fragments, which in turn determine both the resolution of the resulting map and the maximum distance over which linkage between consecutive markers can be detected. For example, a so-called short-range panel with fragments less than about 100 kb will detect linkage between markers up to 50 kb apart ("high-resolution"), whereas a panel with fragments of about 150 kb and more allows monitoring of linkage over distances up to about 100 kb ("low-resolution"). This mapping panel, whatever its type, is then amplified by conventional polymerase chain reaction techniques using random 15mer primers. The resulting amplicons are then amplified by PCR with a forward internal primer and a corresponding reverse primer for each selected DNA marker (e. g., sequence-tagged sites or genic markers). Since many markers are to be localized, this second amplification works with a multitude of primer combinations ("whole-genome PCR"). If two (or more) markers are physically linked in the genome, they will be localized on one single fragment, i. e., will remain linked after random DNA breakage in one aliquot. The products are then analyzed by agarose gel electrophoresis. From the resulting data, associations of markers can be calculated (LOD scores), from which a linear physical map of the target genome can be constructed by, e. g., using *Dgmap* or similar software. Happy mapping then replaces the *in vivo* processes of crossing-over and segregation by *in vitro* breakage of DNA and dilution into aliquots containing less than a haploid genome equivalent.

Haplotype-tagged single nucleotide polymorphism: Any single nucleotide polymorphism that is identified (or "tagged") from larger SNP databases, located in a specific genomic region, and used to define haplotypes and haplotype structure in that region.

High-density bacterial artificial chromosome grid (high-density BAC grid): A laboratory term for any nitrocellulose or nylon membrane onto which different bacterial artificial chromosome clones are spotted at high density. Such membranes may contain a complete genome (depending on its size) and are used to fish out sequences (e. g., genes) via hybridization with radioactively labeled probes.

High-density map (high-resolution map, high-density genetic map, high-resolution genetic map, high-resolution physical map): Any genetic or physical map of a genome that contains a large number of mapped markers (e. g., molecular markers) such that markers are spaced at recombination frequencies of 1–5 cM or about 100–500 kb physical distances. The term is relatively vague, since some laboratories consider a map a high-density map if marker density is 0.5 cM or less (genetic map) and 5–10 kb or less (physical map).

High-density mapping (high-resolution genetic mapping, high-resolution physical mapping): The process of establishing a high-density map.

High information content BAC fingerprinting (HICF): A technique for the detection of minimum overlaps between neighboring BAC clones that is based on the 1–5 base terminal sequences of the clones. The clones are first restricted with the type IIS restriction endonuclease *Hga* I in combination with the type II restriction enzyme *Rsa* I. The 5-base overhangs generated by *Hga* I are sequenced using modified fluorescent dideoxy terminator sequencing reagents. Using an in-lane size standard labeled with a fifth dye, fragments are characterized by both size and sequence of the terminal bases. HICF increases the power to detect clone overlaps and helps to accurately assemble contigs, to close gaps, to verify existing contigs, and to establish minimal tiling paths with fewer clones.

Homologous recombination (legitimate recombination): The exchange of sequences between two DNA molecules (e. g., two homologous chromosomes) that involves loci with complete or far-reaching base sequence homology.

Homology map: Any genetic or physical map that compares gene sequences in homologous segments of two (or more) genomes (or organisms).

Homosequential linkage map: A linkage map of a genome A that shares a similar or even identical marker order with a linkage map (or parts of it) from genome B.

Hybridization: In gene technology, the formation of duplex molecules from complementary single-strands (DNA-DNA, DNA-RNA, RNA-RNA). Hybridization is used to detect sequence homologies between two different nucleic acid molecules. Usually a single-stranded DNA is labeled either radioactively or non-radioactively and is used as a probe that may anneal to homologous sequences in other single-stranded nucleic acid molecules. The resulting hybrid molecules can be detected by autoradiography or various other techniques, depending on the kind of label used.

I

***In silico* map:** A physical map generated by *in silico* mapping.

***In silico* mapping:** A variant of comparative mapping that capitalizes on the use of genome sequence databases for the establishment of physical maps of an unsequenced genome (or part of it). For example, the rice (*Oryza sativa*) genome is completely sequenced and the sequence is deposited in the databases. These databanks can be screened for an interesting gene, the gene located on a bacterial artificial chromosome clone, and this clone (or its ortholog, i. e., a homologous clone from another species) can be used to establish a physical map of the region around the same gene in the *Hordeum vulgare* (rye) genome. *In silico* mapping exploits the macrosynteny and/or microsynteny between related (and also less related) genomes.

***In silico* single nucleotide polymorphism (*in silico* SNP, isSNP):** Any single nucleotide polymorphism that is identified *in silico* by mining overlapping sequences in expressed sequence tags or genomic databases. Since isSNPs represent "virtual" polymorphisms, they have to be validated by resequencing the region in which they occur.

***In situ* hybridization (ISH):** A technique for the identification of specific DNA sequences on intact chromosomes (or RNA sequences in a cell) by hybridization of radioactively labeled or fluorescent complementary nucleic acid probes (frequently, synthetic oligonucleotides) to denatured metaphase or interphase chromosomes. The hybridizing loci are then detected by either autoradiography or laser excitation of the fluorochrome and emission light capture in, e. g., charge-coupled device cameras.

Integrated map: The combination of a genetic and a physical map, i. e., an integration of genetic linkage data and the physical distance between markers or genes. The term is also used to describe a map in which the linkage data from two different segregating populations of the same species are combined or in which maps established with different molecular marker techniques are merged.

Interphase nucleus mapping: A variant of the conventional fluorescent *in situ* hybridization technique for the visualization of specific genes in interphase nuclei (in which the chromosomes are invisible). The gene is detected by a fluorochrome-labeled gene probe that hybridizes to the target gene. The location of the hybridized probe (i. e., the locus of the gene) can then be visualized by laser-induced excitation of the fluorochrome.

Inter-retrotransposon amplified polymorphism (IRAP): Any difference in DNA sequence between two genomes detected by PCR-mediated amplification between two neighboring retrotransposons using a left and right outward-facing primer complementary to the conserved regions within the long terminal repeats. Any observed polymorphism is the consequence of mutations primarily in the region between the two retrotransposons as well as in one or both long terminal repeats. Since retrotransposons are ubiquitous elements of eukaryotic genomes, the IRAP technique produces multiple bands, some of which are polymorphic.

Intronic single nucleotide polymorphism (intronic SNP, intron SNP): Any single nucleotide polymorphism that occurs in introns of eukaryotic genes. Intron SNPs are more frequent than SNPs in coding regions.

L

Linkage: The close physical association of two or more genes on the same chromosome such that they are inherited together.
Linkage analysis: The estimation of the frequency of crossovers or recombinations between DNA sequences (recombination frequency). It is used to establish the position of a particular sequence (e. g., a gene) on a genetic map.
Linkage drag: The association between a desirable (positive) and an undesirable (negative) trait or gene.
Linkage group (LG): Any group of genes or DNA markers that are contiguous on a linear chromosome map (i. e., are located on the same chromosome) and show a high degree of linkage. Usually, a linkage group is equivalent to all DNA markers or genes on the same chromosome. Therefore, the number of linkage groups is identical to the number of chromosomes in a haploid set.
Linkage group homology: The presence of identical markers on linkage groups from two different crosses, so that a homology can be inferred.
Linked marker: Any molecular marker(s) located near a target gene on the same chromosome such that the recombination frequency between them approaches zero. Linked markers are exploited for the isolation of the linked gene(s) via positional cloning.
Locus (plural: loci): The position of a gene (generally DNA sequence) on a chromosome or a genetic map.

M

Macrosynteny: The conserved order of large genomic blocks (in the megabase range) in the genomes of related (but also unrelated) species, as detected by, e. g., chromosomal in situ suppression hybridization or fluorescent *in situ* hybridization.
Map: A graphical description of genetically or physically defined positions on a circular (e. g., plasmid) or linear DNA molecule (e. g., chromosome) and their relative locations and distances. A map may show the distribution of specific restriction sites (restriction map), genes (gene map), markers (marker map), chromosome markers (chromosome map), or the distance between two loci (e. g., a marker and a gene) in base pairs (physical map) or centimorgans (genetic map). The term is now also used for the illustration of peptide-peptide, peptide-protein, and protein-protein interaction networks in a cell or an organelle and for the intracellular distribution of low-molecular-weight cellular compounds (metabolites).
Mapping: A general term for the plotting of specific positions (e. g., genes) along a strand of DNA.
Marker-assisted breeding (MAB): The use of molecular markers for the development of new plant varieties, e. g., by marker-assisted selection.
Marker-assisted introgression: A technique to facilitate introgression of desirable genes into target organisms. First, molecular markers are identified, that are

closely linked to the gene encoding the trait of interest. There the fate of the linked markers in the progeny of sexual crosses are monitored. The presence of the markers in a progenial organism is then taken as indicator for the presence of the linked gene. Marker-assisted introgression therefore avoids lengthy evaluation processes (e. g., the continuous monitoring of the phenotype of plants in the field over several years).

Marker-assisted selection (MAS, marker-based selection, marker-mediated selection): The selection of individual organisms (e. g., plants or animals) carrying a desirable gene with the assistance of molecular marker(s) linked to the gene. Such markers are identified by linkage analysis.

Megabase (Mb): One million nucleotides or nucleotide pairs, equivalent to 1000 kb.

Megabase cloning: A technique for the cloning of extremely large fragments of DNA (in the range from one to several megabases) into suitable vectors.

Megabase mapping: The construction of a linear genetic or physical map using markers that are within one million bases (megabases) from each other.

Microsatellite (short tandem repeat, STR; simple sequence repeat, SSR; repetitive simple sequence, RSS; simple repetitive sequence, SRS): Any one of a series of short (2–10 bp) middle-repetitive, tandemly arranged, and highly variable (hypervariable) DNA sequences dispersed throughout fungal, plant, and animal genomes. For example, $(A)_n$, $(AT)_n$, $(ATT)_n$, and $(AATTA)_n$ are such microsatellites. Hypervariability of certain microsatellites leads to human diseases (e. g., microsatellite expansion diseases such as Huntington chorea) but can also be exploited for the generation of so-called sequence-tagged microsatellite markers.

Microsatellite polymorphism (short tandem repeat polymorphism, STRP): Any difference in the number of microsatellite repeats at corresponding genomic loci in two (or more) different genomes.

Minimal tiling path: Any map or table showing the placement and order of a set of clones (e. g., bacterial artificial chromosome clones) that completely and contiguously cover a specific segment of DNA.

Missense single nucleotide polymorphism (missense SNP): Any single nucleotide polymorphism that occurs in the coding region of a gene and changes the amino acid sequence of the encoded protein. Such missense SNPs, if responsible for a functional change in, e. g., a protein domain, may cause diseases.

Molecular marker (DNA marker): Any specific DNA segment whose base sequence is different (polymorphic) in different organisms and is therefore diagnostic for each of them. Molecular markers can be visualized by either hybridization (e. g., in DNA fingerprinting or restriction fragment length polymorphism) or polymerase chain reaction techniques. Ideal molecular markers are highly polymorphic between two organisms, inherited codominantly, distributed evenly throughout the genome, and visualized easily.

N

Next generation screening (NGS): A variant of the classical screening for single nucleotide polymorphisms (or other mutations) that uses high-density glass microarrays onto which thousands of PCR-amplified single loci or gene fragments of patients are immobilized. Each spot on such microarrays corresponds to a single locus of a particular patient and contains the specific allele of this patient. These NGS microarrays are hybridized with synthetic, fluorescently labeled, allele-specific oligonucleotides complementary to the disease alleles, and the hybridization event is detected by laser excitation of the corresponding fluorochrome. Three signal intensities identify healthy (weak signals), carrier (intermediate signals), and disease (strong signals) genotypes. The use of multiple fluorochromes (e. g., cyanin 5 and cyanin 3) allows the screening of samples from up to 10,000 different patients on a single NGS array and the screening for 12–20 disease loci, whereas the classical microarray format permits screening of only one patient per chip. NGS therefore determines the genotypes of multiple patients in a single test and is used for blood typing, HLA analysis, forensic medicine, and research into hereditary hearing loss and infectious diseases.
Non-coding single nucleotide polymorphism (ncSNP): A misleading term for any single nucleotide polymorphism that occurs in a non-coding region of the genome (e. g., an intron). NcSNPs are the most frequent types of SNPs in eukaryotic organisms.
Non-synonymous single nucleotide polymorphism (nsSNP): Any single nucleotide polymorphism that occurs in a coding region of a eukaryotic gene and changes the encoded amino acid. NsSNPs may cause the synthesis of a nonfunctional protein and therefore may be involved in diseases.

P

Physical map: The linear arrangement of genes or other sequences on a chromosome (or part of it) as determined by techniques other than genetic recombination. Map distances are expressed as numbers of nucleotides (or nucleotide pairs) between identifiable genomic sites (e. g., markers, genes).
Physical mapping: The technologies to produce a physical map.
Population-specific single nucleotide polymorphism (population-specific SNP): Any single nucleotide polymorphism that is present in one population and absent in another. For example, the colonization of Polynesia or the Americas led to the development of single base pair exchanges that did not occur in ancestral groups of hominids in Asia. These SNPs can therefore be considered to be specific to Polynesian or American Indian populations, respectively.
Primer extension: A technique to detect the so-called single nucleotide polymorphisms (SNPs) in target DNA. In short, the target (e. g., a gene) is first amplified with specific primers in a conventional polymerase chain reaction and subsequently denatured. Then, a single-stranded primer is annealed to the single-stranded target

DNA such that the primer ends exactly at the SNP site. After annealing, the duplex exposes a 3′ OH-group for an extension catalyzed by DNA polymerase in the presence of all four dideoxynucleoside triphosphates (ddNTPs) rather than deoxyribonucleoside triphosphates (dNTPs), each labeled with a specific fluorochrome. The matching ddNTP will then be incorporated and stops extension. The incorporated ddNTP is then identified by the specific fluorescence emission of its fluorochrome. A comparison with the wild-type sequence at the SNP site allows for identification of the type of SNP.

Promoter single nucleotide polymorphism (promoter SNP, pSNP): Any single nucleotide polymorphism that occurs in the promoter sequence of a gene. If a pSNP prevents the binding of a transcription factor to its recognition sequence in the promoter, the promoter becomes partly dysfunctional.

Pulsed-field gel electrophoresis (PFGE): A technique for the electrophoretic separation of DNA molecules from the size of ordinary restriction fragments (<10 kb) to intact chromosomes of up to 20 million base pairs (20 Mb). The DNA is alternately subjected to two electrical fields at different angles for specific time intervals called pulse time (τ). Under appropriate conditions, reorientation of DNA segments with different size and topology is different, which leads to the separation of the different segments.

R

Random amplified polymorphic DNA (RAPD): Any DNA segment that is PCR amplified using short oligodeoxynucleotide primers of arbitrary nucleotide sequence and that is polymorphic between two genomes. The amplified products are separated by agarose gel electrophoresis, and the polymorphism is detected as a band on the ethidium bromide-stained agarose gels. The term RAPD is also deliberately used for the technique to detect the random amplified polymorphic DNA.

Rare single nucleotide polymorphism (rare SNP): Any single nucleotide polymorphism (SNP) that occurs at a frequency of less than 1 % in a population.

Recombination (genetic recombination): The combination of genes in a progeny molecule, cell, or organism in a pattern different from that of the parent molecules, cells, or organisms. Recombination is based on the exchange of DNA sequences between two chromosomes or on the novel association of genes in a recombinant arising from independent assortment of unlinked genes, from crossover between linked genes, or from intra-cistronic crossover.

Reference single nucleotide polymorphism (refSNP, rsSNP, rsID, SNP ID): Any single nucleotide polymorphism at a specific site of a genome (or part of a genome, e. g., a BAC clone) that serves as reference point for the definition of other SNPs in its neighborhood. The rsID number ("tag") is assigned to each refSNP at the time of its submission to the databanks.

Regulatory single nucleotide polymorphism (regulatory SNP, rSNP): A relatively rare single nucleotide polymorphism that affects the expression of a gene (or several genes). Usually this SNP is located in the promoter of the gene.

Resequencing by hybridization (RBH): A variant of the sequencing by hybridization technique that allows the detection of a single nucleotide polymorphism (or another type of mutation) in a target region of a genome. In short, a SNP is first detected in a genome-wide SNP mapping project, its linkage disequilibrium with a target region (e. g., conferring a disease phenotype) is determined, and the region is amplified in a conventional polymerase chain reaction with bracketing primers. The amplification product or an oligonucleotide with the complementary sequence is then spotted onto a microarray together with thousands of other targets and hybridized to test samples of genomic DNA (also cDNA) from patients. Hybridizing probes are then resequenced and the SNP is verified.
Restriction: The exclusion of foreign DNA from bacterial cells by restriction endonuclease-catalyzed recognition and degradation.
Restriction endonuclease (restriction enzyme): Any bacterial enzyme that recognizes specific target nucleotide sequences ("recognition site") in double-stranded DNA and catalyzes the breakage of internal bonds between specific nucleotides within these targets or within a specific distance from there. Restriction generates double-stranded breaks with either cohesive or blunt ends. Restriction endonucleases are part of bacterial restriction-modification systems that exclude foreign DNA. The cell's own DNA is protected by methylation of cytosyl residues in the recognition sites.
Restriction fragment length polymorphism (RFLP): The variation in the length of DNA fragments produced by a specific restriction endonuclease from genomic DNAs of two (or more) individuals of a species. RFLPs are generated by genomic rearrangements or mutations that either create or delete recognition sites for the specific endonuclease. Since RFLP technology is labor-intensive and involves radioactivity, it is superseded by PCR-based techniques.

S

Saturation mapping: The enrichment of specific regions of an already-established genetic map with molecular markers such that marker density is extremely high. A saturated map is a prerequisite for map-based cloning of a gene of interest. Saturation mapping produces a high-resolution map and usually incorporates markers generated with different marker systems.
Sequence-tagged microsatellite site (STMS): A locus-specific, codominant molecular marker, used for, e. g., DNA fingerprinting or genetic and physical mapping, that is generated by sequencing microsatellite-flanking sequences, designing specific PCR primers towards these flanks, and using them to amplify genomic DNA. In most, though not all, cases, only one amplification product can be detected, which represents an STMS.
Single base extension (SBE): A technique for the detection of single nucleotide polymorphisms that uses primers ending directly adjacent to the SNP mismatch and is used to incorporate the complementary, fluorescently labeled ddNTP in a conventional polymerase chain reaction with *Thermus aquaticus* DNA polymerase.

After incorporation of the ddNTP, the reaction is terminated and the incorporated nucleotide can be detected by laser-induced fluorescence. In one single reaction, several SNPs can be discovered if primers with different 5′ tails and different fluorochromes are used (e. g., ddATP: green fluorescence; ddCTP: blue; ddGTP: yellow; ddTTP: orange).

Single-hit single nucleotide polymorphism (single-hit SNP): Any single nucleotide polymorphism for which each allele is present in only one sample from a distinct population.

Single nucleotide polymorphism (SNP): Any polymorphism between two genomes that is based on a single nucleotide exchange, a small deletion, or an insertion. SNPs represent the most frequent mutations in genomes, and their relatively good genome coverage and genome-wide distribution in both coding and non-coding regions recommend SNPs as highly informative markers for mapping procedures.

Single nucleotide polymorphism database (dbSNP): A database for single nucleotide polymorphisms (i. e., single-base exchanges), short deletions, and insertion polymorphisms. Website: http://www.ncbi.nlm.nih.gov/SNP. HGVbase maintains an extensive list of other SNP databases at http://hgvbase.cgb.ki.se/databases.htm

Single nucleotide polymorphism density (SNP density): The frequency of single nucleotide polymorphisms per unit length of genomic DNA (usually SNPs per kilobase). SNP density is different for different regions of genomes. For example, regional heterogeneity of SNP density is characteristic for different chromosomes of *Anopheles gambiae*, possibly caused by the introgression of divergent Mopti and Savanna cytotypes (chromosomal forms). SNPs are therefore distributed along the chromosomes in a bimodal way: one mode contains about one SNP per 10 kb, the other about one SNP per 200 bp. SNP density is high in intergenic and intronic regions, as compared with genic SNPs.

SNP cluster: The accumulation of single nucleotide polymorphisms in relatively small genomic regions. Such clusters can extend over 1 kb (or more), are relatively rare but very old, and originate from ancestral chromosome fragments inherited in the extant species.

SNP frequency: The frequency with which single nucleotide polymorphisms occur along a defined stretch of DNA or a chromosome (usually expressed as SNP per kilobase). SNP frequency varies between genomic regions in the same individual and between related individuals. For example, in some regions of the maize genome, SNP frequency is about 1/65 bp, in other regions 1/85 bp. Generally, SNP frequency is much higher than the frequency of insertions or deletions (1/250 bp).

SNP map: The linear arrangement of single nucleotide polymorphisms along a specific region in two homologous genomes.

SNP-rich segment: Any one of a series of genomic regions in which single nucleotide polymerphisms (SNPs) occur at a considerably higher frequency than in the rest of the genome. For example, in the mouse genome, such SNP-rich segments contain about 40 SNPs per 10 kb, whereas other regions contain much fewer SNPs (e. g., the intermitting sequences harbor about 0.5 SNPs per 10 kb).

SNP scanning: The *in silico* search for single nucleotide polymorphisms in a sequenced stretch of DNA (e.g., a BAC clone, a genomic segment, or, in extreme cases, whole genomes).
SNP scoring: The search for single nucleotide polymorphisms in two (or more) DNA sequences (in extreme cases, genomes) and their characterization and use for genome analysis (e.g., establishment of a single nucleotide polymorphism map).
Strong overlap: The number of bases matched between the two clones (e.g., bacterial artificial chromosome clones) determined by using the strictest matching criteria.
Synonymous single nucleotide polymorphism (synonymous SNP, synSNP): Any single nucleotide polymorphism that occurs in an exon but does not change the amino acid composition of the encoded protein.
Synteny-based positional cloning: Any positional cloning of a gene (or genes) in the genome of organism A for which mapping information from the genome of organism B is used. For example, if molecular markers bracket the gene of interest in the genetic map, can be located on a physical map (e.g., constructed with bacterial artificial chromosomes), and allow isolation of the target gene of organism B, then the same molecular markers can be exploited to tag the orthologous gene in organism A such that it can be isolated by the same techniques. This approach is based on synteny.
Synteny mapping: The localization of DNA sequences (e.g., whole chromosome arms, large genomic DNA regions, genes) in a target genome that are syntenic to regions in a reference genome.

T

Tag single nucleotide polymorphism (tag SNP): Any one of two (or more) strongly associated (i.e., commonly inherited) single nucleotide polymorphism loci that are characteristic for a specific haplotype. For example, in a specific genomic region, usually many SNPs are present at a frequency specific for this region (e.g., on average about one SNP per 300 bases in the human genome; however, there is much variation in SNP frequency across the genome). Determination of all SNPs in the selected region usually produces a SNP distribution profile, which requires extensive sequencing in many different genotypes. The selection of only a few distinct SNPs from this region (the so-called tag SNPs) for genotyping predicts the remainder of the common SNPs in this region, and only the tag SNPs need to be known to identify each of the common haplotypes in a population.
Telomere mapping: The localization of telomeres on the ends of chromosomes by, e.g., fluorescent *in situ* hybridization and the identification of other sequence elements within the telomeric (and also sub-telomeric) region.
Tetra-allelic single nucleotide polymorphism (tetra-allelic SNP): Any single nucleotide polymorphism of which four different alleles are present in a population.

Tiling path: The ordered arrangement of BAC clones using sequence overlaps of neighboring clones such that they completely cover the corresponding region of the chromosome.

Transcriptome mapping: The procedure to establish a genetic map with cDNA fragments that differ in size in the respective parents and segregate in the progeny. Usually the resulting map reveals maternal or paternal inheritance of the fragments. Since the cDNA fragments are displayed on Northern-type gels, interesting fragments can be directly isolated and sequenced, and the sequence can be annotated such that the underlying gene can be identified.

Tri-allelic single nucleotide polymorphism (tri-allelic SNP): Any single nucleotide polymorphism of which three different alleles exist in a population.

Type I DNA marker: Any molecular marker derived from gene sequences. Such genic markers are usually highly conserved, and therefore exhibit little sequence polymorphism, and are less suitable for the analysis of pedigrees and population diversity.

Type III DNA marker: Any one of a series of bi-allelic single nucleotide polymorphism markers from sequences within or between coding regions that are highly informative for pedigree, family, or population screens within species, but usually are not informative for comparative ortholog identification between orders.

Type II DNA marker: Any molecular marker derived from hypervariable microsatellite sequences. Such markers are highly informative in forensic analysis, pedigree, and population studies. Type II markers are almost evenly dispersed throughout eukaryotic genomes, and each locus exists as many alleles in a population.

Index

The Handbook of Plant Genome Mapping.
Genetic and Physical Mapping. Edited by Khalid Meksem, Günter Kahl

ISBN 3-527-31116-5

G

H

N

O

P

Q

R